Lecture Notes in Computer Science

Lecture Notes in Artificial Intelligence   13996

Founding Editor

Jörg Siekmann

Series Editors

Randy Goebel, *University of Alberta, Edmonton, Canada*
Wolfgang Wahlster, *DFKI, Berlin, Germany*
Zhi-Hua Zhou, *Nanjing University, Nanjing, China*

The series Lecture Notes in Artificial Intelligence (LNAI) was established in 1988 as a topical subseries of LNCS devoted to artificial intelligence.

The series publishes state-of-the-art research results at a high level. As with the LNCS mother series, the mission of the series is to serve the international R & D community by providing an invaluable service, mainly focused on the publication of conference and workshop proceedings and postproceedings.

Ngoc Thanh Nguyen · Siridech Boonsang ·
Hamido Fujita · Bogumiła Hnatkowska ·
Tzung-Pei Hong · Kitsuchart Pasupa ·
Ali Selamat
Editors

Intelligent Information and Database Systems

15th Asian Conference, ACIIDS 2023
Phuket, Thailand, July 24–26, 2023
Proceedings, Part II

Springer

Editors
Ngoc Thanh Nguyen 🆔
Wrocław University of Science
and Technology
Wrocław, Poland

Hamido Fujita 🆔
Iwate Prefectural University Iwate
Iwate, Japan

Tzung-Pei Hong 🆔
National University of Kaohsiung
Kaohsiung, Taiwan

Ali Selamat 🆔
Malaysia Japan International Institute
of Technology
Kuala Lumpur, Malaysia

Siridech Boonsang 🆔
King Mongkut's Institute of Technology
Ladkrabang
Bangkok, Thailand

Bogumiła Hnatkowska 🆔
Wrocław University of Science
and Technology
Wrocław, Poland

Kitsuchart Pasupa 🆔
King Mongkut's Institute of Technology
Ladkrabang
Bangkok, Thailand

ISSN 0302-9743   ISSN 1611-3349 (electronic)
Lecture Notes in Artificial Intelligence
ISBN 978-981-99-5836-8   ISBN 978-981-99-5837-5 (eBook)
https://doi.org/10.1007/978-981-99-5837-5

LNCS Sublibrary: SL7 – Artificial Intelligence

This Springer imprint is published by the registered company Springer Nature Singapore Pte Ltd.
The registered company address is: 152 Beach Road, #21-01/04 Gateway East, Singapore 189721, Singapore

Paper in this product is recyclable.

Preface

ACIIDS 2023 was the 15th event in a series of international scientific conferences on research and applications in the field of intelligent information and database systems. The aim of ACIIDS 2023 was to provide an international forum for research workers with scientific backgrounds in the technology of intelligent information and database systems and their various applications. The conference was hosted by King Mongkut's Institute of Technology Ladkrabang, Thailand, and jointly organized by Wrocław University of Science and Technology, Poland, in cooperation with IEEE SMC Technical Committee on Computational Collective Intelligence, European Research Center for Information Systems (ERCIS), University of Newcastle (Australia), Yeungnam University (South Korea), International University - Vietnam National University HCMC (Vietnam), Leiden University (The Netherlands), Universiti Teknologi Malaysia (Malaysia), Nguyen Tat Thanh University (Vietnam), BINUS University (Indonesia), and Vietnam National University, Hanoi (Vietnam). ACIIDS 2023 occurred in Phuket, Thailand, on July 24–26, 2023.

The ACIIDS conference series is already well established. The first two events, ACIIDS 2009 and ACIIDS 2010, took place in Dong Hoi City and Hue City in Vietnam, respectively. The third event, ACIIDS 2011, occurred in Daegu (South Korea), followed by the fourth, ACIIDS 2012, in Kaohsiung (Taiwan). The fifth event, ACIIDS 2013, was held in Kuala Lumpur (Malaysia), while the sixth event, ACIIDS 2014, was held in Bangkok (Thailand). The seventh event, ACIIDS 2015, occurred in Bali (Indonesia), followed by the eighth, ACIIDS 2016, in Da Nang (Vietnam). The ninth event, ACIIDS 2017, was organized in Kanazawa (Japan). The 10th jubilee conference, ACIIDS 2018, was held in Dong Hoi City (Vietnam), followed by the 11th event, ACIIDS 2019, in Yogyakarta (Indonesia). The 12th and 13th events were planned to be on-site in Phuket (Thailand). However, the global pandemic relating to COVID-19 resulted in both editions of the conference being held online in virtual space. ACIIDS 2022 was held in Ho Chi Minh City as a hybrid conference, and it restarted in-person meetings at conferences.

These two volumes contain 65 peer-reviewed papers selected for presentation from 224 submissions. Papers included in this volume cover the following topics: data mining and machine learning methods, advanced data mining techniques and applications, intelligent and contextual systems, natural language processing, network systems and applications, computational imaging and vision, decision support, control systems, and data modeling and processing for industry 4.0.

The accepted and presented papers focus on new trends and challenges facing the intelligent information and database systems community. The presenters showed how research work could stimulate novel and innovative applications. We hope you find these results valuable and inspiring for future research work. We would like to express our sincere thanks to the honorary chairs for their support: Arkadiusz Wójs (Rector of Wrocław University of Science and Technology, Poland), Moonis Ali (Texas State

University, President of International Society of Applied Intelligence, USA), Komsan Maleesee (President of King Mongkut's Institute of Technology Ladkrabang, Thailand).

We thank the keynote speakers for their world-class plenary speeches: Saman K. Halgamuge from The University of Melbourne (Australia), Jerzy Stefanowski from Poznań University of Technology (Poland), Siridech Boonsang from King Mongkut's Institute of Technology Ladkrabang (Thailand), and Masaru Kitsuregawa from The University of Tokyo (Japan).

We cordially thank our main sponsors, King Mongkut's Institute of Technology Ladkrabang (Thailand), Wrocław University of Science and Technology (Poland), IEEE SMC Technical Committee on Computational Collective Intelligence, European Research Center for Information Systems (ERCIS), University of Newcastle (Australia), Yeungnam University (South Korea), Leiden University (The Netherlands), Universiti Teknologi Malaysia (Malaysia), BINUS University (Indonesia), Vietnam National University (Vietnam), and Nguyen Tat Thanh University (Vietnam). Our special thanks go to Springer for publishing the proceedings and to all the other sponsors for their kind support.

Our special thanks go to the program chairs, the special session chairs, the organizing chairs, the publicity chairs, the liaison chairs, and the Local Organizing Committee for their work towards the conference. We sincerely thank all the members of the International Program Committee for their valuable efforts in the review process, which helped us to guarantee the highest quality of the selected papers for the conference. We cordially thank all the authors and other conference participants for their valuable contributions. The conference would not have been possible without their support. Thanks are also due to the many experts who contributed to the event being a success.

July 2023

Ngoc Thanh Nguyen
Siridech Boonsang
Hamido Fujita
Bogumiła Hnatkowska
Tzung-Pei Hong
Kitsuchart Pasupa
Ali Selamat

Organization

Honorary Chairs

Arkadiusz Wójs   Rector of Wrocław University of Science and Technology, Poland

Moonis Ali   Texas State University, President of International Society of Applied Intelligence, USA

Komsan Maleesee   President of King Mongkut's Institute of Technology Ladkrabang, Thailand

General Chairs

Ngoc Thanh Nguyen   Wrocław University of Science and Technology, Poland

Suphamit Chittayasothorn   King Mongkut's Institute of Technology Ladkrabang, Thailand

Program Chairs

Hamido Fujita   Iwate Prefectural University, Japan

Tzung-Pei Hong   National University of Kaohsiung, Taiwan

Ali Selamat   Universiti Teknologi Malaysia, Malaysia

Siridech Boonsang   King Mongkut's Institute of Technology Ladkrabang, Thailand

Kitsuchart Pasupa   King Mongkut's Institute of Technology Ladkrabang, Thailand

Steering Committee

Ngoc Thanh Nguyen (Chair)   Wrocław University of Science and Technology, Poland

Longbing Cao   University of Science and Technology Sydney, Australia

Suphamit Chittayasothorn   King Mongkut's Institute of Technology Ladkrabang, Thailand

Ford Lumban Gaol   Bina Nusantara University, Indonesia

Tzung-Pei Hong   National University of Kaohsiung, Taiwan
Dosam Hwang   Yeungnam University, South Korea
Bela Stantic   Griffith University, Australia
Geun-Sik Jo   Inha University, South Korea
Hoai An Le Thi   University of Lorraine, France
Toyoaki Nishida   Kyoto University, Japan
Leszek Rutkowski   Częstochowa University of Technology, Poland
Ali Selamat   Universiti Teknologi Malaysia, Malaysia
Edward Szczerbicki   University of Newcastle, Australia

Special Session Chairs

Bogumiła Hnatkowska   Wrocław University of Science and Technology,
                                       Poland
Arit Thammano   King Mongkut's Institute of Technology
                                       Ladkrabang, Thailand
Krystian Wojtkiewicz   Wrocław University of Science and Technology,
                                       Poland

Doctoral Track Chairs

Marek Krótkiewicz   Wrocław University of Science and Technology,
                                       Poland
Nont Kanungsukkasem   King Mongkut's Institute of Technology
                                       Ladkrabang, Thailand

Liaison Chairs

Sirasit Lochanachit   King Mongkut's Institute of Technology
                                       Ladkrabang, Thailand
Ford Lumban Gaol   Bina Nusantara University, Indonesia
Quang-Thuy Ha   VNU-University of Engineering and Technology,
                                       Vietnam
Mong-Fong Horng   National Kaohsiung University of Applied
                                       Sciences, Taiwan
Dosam Hwang   Yeungnam University, South Korea
Le Minh Nguyen   Japan Advanced Institute of Science and
                                       Technology, Japan
Ali Selamat   Universiti Teknologi Malaysia, Malaysia

Organizing Chairs

Kamol Wasapinyokul   King Mongkut's Institute of Technology
                            Ladkrabang, Thailand
Krystian Wojtkiewicz   Wrocław University of Science and Technology,
                            Poland

Publicity Chairs

Marcin Jodłowiec   Wrocław University of Science and Technology,
                            Poland
Rafał Palak   Wrocław University of Science and Technology,
                            Poland
Nat Dilokthanakul   King Mongkut's Institute of Technology
                            Ladkrabang, Thailand

Finance Chair

Pattanapong Chantamit-O-Pas   King Mongkut's Institute of Technology
                               Ladkrabang, Thailand

Webmaster

Marek Kopel   Wrocław University of Science and Technology,
                            Poland

Local Organizing Committee

Taravichet Titijaroonroj   King Mongkut's Institute of Technology
                               Ladkrabang, Thailand
Praphan Pavarangkoon   King Mongkut's Institute of Technology
                               Ladkrabang, Thailand
Natthapong Jungteerapanich   King Mongkut's Institute of Technology
                               Ladkrabang, Thailand
Putsadee Pornphol   Phuket Rajabhat University, Thailand
Patient Zihisire Muke   Wrocław University of Science and Technology,
                               Poland

Thanh-Ngo Nguyen	Wrocław University of Science and Technology, Poland
Katarzyna Zombroń	Wrocław University of Science and Technology, Poland
Kulwadee Somboonviwat	Kasetsart University Sriracha, Thailand

Keynote Speakers

Saman K. Halgamuge	University of Melbourne, Australia
Jerzy Stefanowski	Poznań University of Technology, Poland
Siridech Boonsang	King Mongkut's Institute of Technology Ladkrabang, Thailand
Masaru Kitsuregawa	University of Tokyo, Japan

Special Sessions Organizers

ADMTA 2023: Special Session on Advanced Data Mining Techniques and Applications

Chun-Hao Chen	National Kaohsiung University of Science and Technology, Taiwan
Bay Vo	Ho Chi Minh City University of Technology, Vietnam
Tzung-Pei Hong	National University of Kaohsiung, Taiwan

AINBC 2023: Special Session on Advanced Data Mining Techniques and Applications

Andrzej W. Przybyszewski	University of Massachusetts Medical School, USA
Jerzy P. Nowacki	Polish-Japanese Academy of Information Technology, Poland

CDF 2023: Special Session on Computational Document Forensics

Jean-Marc Ogier	La Rochelle Université, France
Mickaël Coustaty	La Rochelle Université, France
Surapong Uttama	Mae Fah Luang University, Thailand

CSDT 2023: Special Session on Cyber Science in Digital Transformation

Dariusz Szostek	University of Silesia in Katowice, Poland
Jan Kozak	University of Economics in Katowice, Poland
Paweł Kasprowski	Silesian University of Technology, Poland

CVIS 2023: Special Session on Computer Vision and Intelligent Systems

Van-Dung Hoang	Ho Chi Minh City University of Technology and Education, Vietnam
Dinh-Hien Nguyen	University of Information Technology, VNU-HCM, Vietnam
Chi-Mai Luong	Vietnam Academy of Science and Technology, Vietnam

DMPCPA 2023: Special Session on Data Modelling and Processing in City Pollution Assessment

Hoai Phuong Ha	UiT The Arctic University of Norway, Norway
Manuel Nuñez	Universidad Complutense de Madrid, Spain
Rafał Palak	Wrocław University of Science and Technology, Poland
Krystian Wojtkiewicz	Wrocław University of Science and Technology, Poland

HPC-ComCon 2023: Special Session on HPC and Computing Continuum

Pascal Bouvry	University of Luxembourg, Luxembourg
Johnatan E. Pecero	University of Luxembourg, Luxembourg
Arijit Roy	Indian Institute of Information Technology, Sri City, India

LRLSTP 2023: Special Session on Low Resource Languages Speech and Text Processing

Ualsher Tukeyev	Al-Farabi Kazakh National University, Kazakhstan
Orken Mamyrbayev	Institute of Information and Computational Technologies, Kazakhstan

Senior Program Committee

Ajith Abraham	Machine Intelligence Research Labs, USA
Jesús Alcalá Fernández	University of Granada, Spain
Lionel Amodeo	University of Technology of Troyes, France
Ahmad Taher Azar	Prince Sultan University, Saudi Arabia
Thomas Bäck	Leiden University, The Netherlands
Costin Badica	University of Craiova, Romania
Ramazan Bayindir	Gazi University, Turkey
Abdelhamid Bouchachia	Bournemouth University, UK
David Camacho	Universidad Autónoma de Madrid, Spain
Leopoldo Eduardo Cardenas-Barron	Tecnológico de Monterrey, Mexico
Oscar Castillo	Tijuana Institute of Technology, Mexico
Nitesh Chawla	University of Notre Dame, USA
Rung-Ching Chen	Chaoyang University of Technology, Taiwan
Shyi-Ming Chen	National Taiwan University of Science and Technology, Taiwan
Simon Fong	University of Macau, China
Hamido Fujita	Iwate Prefectural University, Japan
Mohamed Gaber	Birmingham City University, UK
Marina L. Gavrilova	University of Calgary, Canada
Daniela Godoy	ISISTAN Research Institute, Argentina
Fernando Gomide	University of Campinas, Brazil
Manuel Grana	University of the Basque Country, Spain
Claudio Gutierrez	Universidad de Chile, Chile
Francisco Herrera	University of Granada, Spain
Tzung-Pei Hong	National University of Kaohsiung, Taiwan
Dosam Hwang	Yeungnam University, South Korea
Mirjana Ivanovic	University of Novi Sad, Serbia
Janusz Jeżewski	Institute of Medical Technology and Equipment ITAM, Poland
Piotr Jędrzejowicz	Gdynia Maritime University, Poland
Kang-Hyun Jo	University of Ulsan, South Korea
Janusz Kacprzyk	Systems Research Institute, Polish Academy of Sciences, Poland
Nikola Kasabov	Auckland University of Technology, New Zealand
Muhammad Khurram Khan	King Saud University, Saudi Arabia
Frank Klawonn	Ostfalia University of Applied Sciences, Germany
Joanna Kolodziej	Cracow University of Technology, Poland
Józef Korbicz	University of Zielona Gora, Poland
Ryszard Kowalczyk	Swinburne University of Technology, Australia

Bartosz Krawczyk	Virginia Commonwealth University, USA
Ondrej Krejcar	University of Hradec Králové, Czech Republic
Adam Krzyzak	Concordia University, Canada
Mark Last	Ben-Gurion University of the Negev, Israel
Hoai An Le Thi	University of Lorraine, France
Kun Chang Lee	Sungkyunkwan University, South Korea
Edwin Lughofer	Johannes Kepler University Linz, Austria
Nezam Mahdavi-Amiri	Sharif University of Technology, Iran
Yannis Manolopoulos	Open University of Cyprus, Cyprus
Klaus-Robert Müller	Technical University of Berlin, Germany
Saeid Nahavandi	Deakin University, Australia
Grzegorz J Nalepa	AGH University of Science and Technology, Poland
Ngoc-Thanh Nguyen	Wrocław University of Science and Technology, Poland
Dusit Niyato	Nanyang Technological University, Singapore
Manuel Núñez	Universidad Complutense de Madrid, Spain
Jeng-Shyang Pan	Fujian University of Technology, China
Marcin Paprzycki	Systems Research Institute, Polish Academy of Sciences, Poland
Hoang Pham	Rutgers University, USA
Tao Pham Dinh	INSA Rouen, France
Radu-Emil Precup	Politehnica University of Timisoara, Romania
Leszek Rutkowski	Częstochowa University of Technology, Poland
Jürgen Schmidhuber	Swiss AI Lab IDSIA, Switzerland
Björn Schuller	University of Passau, Germany
Ali Selamat	Universiti Teknologi Malaysia, Malaysia
Andrzej Skowron	Warsaw University, Poland
Jerzy Stefanowski	Poznań University of Technology, Poland
Edward Szczerbicki	University of Newcastle, Australia
Ryszard Tadeusiewicz	AGH University of Science and Technology, Poland
Muhammad Atif Tahir	National University of Computing & Emerging Sciences, Pakistan
Bay Vo	Ho Chi Minh City University of Technology, Vietnam
Dinh Duc Anh Vu	Vietnam National University HCMC, Vietnam
Lipo Wang	Nanyang Technological University, Singapore
Junzo Watada	Waseda University, Japan
Michał Woźniak	Wrocław University of Science and Technology, Poland
Farouk Yalaoui	University of Technology of Troyes, France

| Sławomir Zadrożny | Systems Research Institute, Polish Academy of Sciences, Poland |
| Zhi-Hua Zhou | Nanjing University, China |

Program Committee

Muhammad Abulaish	South Asian University, India
Bashar Al-Shboul	University of Jordan, Jordan
Toni Anwar	Universiti Teknologi PETRONAS, Malaysia
Taha Arbaoui	University of Technology of Troyes, France
Mehmet Emin Aydin	University of the West of England, UK
Amelia Badica	University of Craiova, Romania
Kambiz Badie	ICT Research Institute, Iran
Hassan Badir	École Nationale des Sciences Appliquées de Tanger, Morocco
Zbigniew Banaszak	Warsaw University of Technology, Poland
Dariusz Barbucha	Gdynia Maritime University, Poland
Maumita Bhattacharya	Charles Sturt University, Australia
Leon Bobrowski	Białystok University of Technology, Poland
Bülent Bolat	Yildiz Technical University, Turkey
Mariusz Boryczka	University of Silesia in Katowice, Poland
Urszula Boryczka	University of Silesia in Katowice, Poland
Zouhaier Brahmia	University of Sfax, Tunisia
Stéphane Bressan	National University of Singapore, Singapore
Peter Brida	University of Žilina, Slovakia
Piotr Bródka	Wrocław University of Science and Technology, Poland
Grażyna Brzykcy	Poznań University of Technology, Poland
Robert Burduk	Wrocław University of Science and Technology, Poland
Aleksander Byrski	AGH University of Science and Technology, Poland
Dariusz Ceglarek	WSB University in Poznań, Poland
Somchai Chatvichienchai	University of Nagasaki, Japan
Chun-Hao Chen	Tamkang University, Taiwan
Leszek J. Chmielewski	Warsaw University of Life Sciences, Poland
Kazimierz Choroś	Wrocław University of Science and Technology, Poland
Kun-Ta Chuang	National Cheng Kung University, Taiwan
Dorian Cojocaru	University of Craiova, Romania
Jose Alfredo Ferreira Costa	Federal University of Rio Grande do Norte (UFRN), Brazil

Ireneusz Czarnowski	Gdynia Maritime University, Poland
Piotr Czekalski	Silesian University of Technology, Poland
Theophile Dagba	University of Abomey-Calavi, Benin
Tien V. Do	Budapest University of Technology and Economics, Hungary
Rafał Doroz	University of Silesia in Katowice, Poland
El-Sayed M. El-Alfy	King Fahd University of Petroleum and Minerals, Saudi Arabia
Keiichi Endo	Ehime University, Japan
Sebastian Ernst	AGH University of Science and Technology, Poland
Nadia Essoussi	University of Carthage, Tunisia
Usef Faghihi	Université du Québec à Trois-Rivières, Canada
Dariusz Frejlichowski	West Pomeranian University of Technology, Szczecin, Poland
Blanka Frydrychova Klimova	University of Hradec Králové, Czech Republic
Janusz Getta	University of Wollongong, Australia
Daniela Gifu	University "Alexandru Ioan Cuza" of Iaşi, Romania
Gergo Gombos	Eötvös Loránd University, Hungary
Manuel Grana	University of the Basque Country, Spain
Janis Grundspenkis	Riga Technical University, Latvia
Dawit Haile	Addis Ababa University, Ethiopia
Marcin Hernes	Wrocław University of Business and Economics, Poland
Koichi Hirata	Kyushu Institute of Technology, Japan
Bogumiła Hnatkowska	Wrocław University of Science and Technology, Poland
Bao An Mai Hoang	Vietnam National University HCMC, Vietnam
Huu Hanh Hoang	Posts and Telecommunications Institute of Technology, Vietnam
Van-Dung Hoang	Quang Binh University, Vietnam
Jeongkyu Hong	Yeungnam University, South Korea
Yung-Fa Huang	Chaoyang University of Technology, Taiwan
Maciej Huk	Wrocław University of Science and Technology, Poland
Kha Tu Huynh	Vietnam National University HCMC, Vietnam
Sanjay Jain	National University of Singapore, Singapore
Khalid Jebari	LCS Rabat, Morocco
Joanna Jędrzejowicz	University of Gdańsk, Poland
Przemysław Juszczuk	University of Economics in Katowice, Poland
Krzysztof Juszczyszyn	Wrocław University of Science and Technology, Poland

Mehmet Karaata	Kuwait University, Kuwait
Rafał Kern	Wrocław University of Science and Technology, Poland
Zaheer Khan	University of the West of England, UK
Marek Kisiel-Dorohinicki	AGH University of Science and Technology, Poland
Attila Kiss	Eötvös Loránd University, Hungary
Shinya Kobayashi	Ehime University, Japan
Grzegorz Kołaczek	Wrocław University of Science and Technology, Poland
Marek Kopel	Wrocław University of Science and Technology, Poland
Jan Kozak	University of Economics in Katowice, Poland
Adrianna Kozierkiewicz	Wrocław University of Science and Technology, Poland
Dalia Kriksciuniene	Vilnius University, Lithuania
Dariusz Król	Wrocław University of Science and Technology, Poland
Marek Krótkiewicz	Wrocław University of Science and Technology, Poland
Marzena Kryszkiewicz	Warsaw University of Technology, Poland
Jan Kubicek	VSB -Technical University of Ostrava, Czech Republic
Tetsuji Kuboyama	Gakushuin University, Japan
Elżbieta Kukla	Wrocław University of Science and Technology, Poland
Marek Kulbacki	Polish-Japanese Academy of Information Technology, Poland
Kazuhiro Kuwabara	Ritsumeikan University, Japan
Annabel Latham	Manchester Metropolitan University, UK
Tu Nga Le	Vietnam National University HCMC, Vietnam
Yue-Shi Lee	Ming Chuan University, Taiwan
Florin Leon	Gheorghe Asachi Technical University of Iasi, Romania
Chunshien Li	National Central University, Taiwan
Horst Lichter	RWTH Aachen University, Germany
Igor Litvinchev	Nuevo Leon State University, Mexico
Doina Logofatu	Frankfurt University of Applied Sciences, Germany
Lech Madeyski	Wrocław University of Science and Technology, Poland
Bernadetta Maleszka	Wrocław University of Science and Technology, Poland

Marcin Maleszka	Wrocław University of Science and Technology, Poland
Tamás Matuszka	Eötvös Loránd University, Hungary
Michael Mayo	University of Waikato, New Zealand
Héctor Menéndez	University College London, UK
Jacek Mercik	WSB University in Wrocław, Poland
Radosław Michalski	Wrocław University of Science and Technology, Poland
Peter Mikulecky	University of Hradec Králové, Czech Republic
Miroslava Mikusova	University of Žilina, Slovakia
Marek Milosz	Lublin University of Technology, Poland
Jolanta Mizera-Pietraszko	Opole University, Poland
Dariusz Mrozek	Silesian University of Technology, Poland
Leo Mrsic	IN2data Ltd Data Science Company, Croatia
Agnieszka Mykowiecka	Institute of Computer Science, Polish Academy of Sciences, Poland
Pawel Myszkowski	Wrocław University of Science and Technology, Poland
Huu-Tuan Nguyen	Vietnam Maritime University, Vietnam
Le Minh Nguyen	Japan Advanced Institute of Science and Technology, Japan
Loan T. T. Nguyen	Vietnam National University HCMC, Vietnam
Quang-Vu Nguyen	Korea-Vietnam Friendship Information Technology College, Vietnam
Thai-Nghe Nguyen	Cantho University, Vietnam
Thi Thanh Sang Nguyen	Vietnam National University HCMC, Vietnam
Van Sinh Nguyen	Vietnam National University HCMC, Vietnam
Agnieszka Nowak-Brzezińska	University of Silesia in Katowice, Poland
Alberto Núñez	Universidad Complutense de Madrid, Spain
Mieczysław Owoc	Wrocław University of Business and Economics, Poland
Panos Patros	University of Waikato, New Zealand
Maciej Piasecki	Wrocław University of Science and Technology, Poland
Bartłomiej Pierański	Poznań University of Economics and Business, Poland
Dariusz Pierzchała	Military University of Technology, Poland
Marcin Pietranik	Wrocław University of Science and Technology, Poland
Elias Pimenidis	University of the West of England, UK
Jaroslav Pokorný	Charles University in Prague, Czech Republic
Nikolaos Polatidis	University of Brighton, UK
Elvira Popescu	University of Craiova, Romania

Piotr Porwik	University of Silesia in Katowice, Poland
Petra Poulova	University of Hradec Králové, Czech Republic
Małgorzata Przybyła-Kasperek	University of Silesia in Katowice, Poland
Paulo Quaresma	Universidade de Évora, Portugal
David Ramsey	Wrocław University of Science and Technology, Poland
Mohammad Rashedur Rahman	North South University, Bangladesh
Ewa Ratajczak-Ropel	Gdynia Maritime University, Poland
Sebastian A. Rios	University of Chile, Chile
Keun Ho Ryu	Chungbuk National University, South Korea
Daniel Sanchez	University of Granada, Spain
Rafał Scherer	Częstochowa University of Technology, Poland
Yeong-Seok Seo	Yeungnam University, South Korea
Donghwa Shin	Yeungnam University, South Korea
Andrzej Siemiński	Wrocław University of Science and Technology, Poland
Dragan Simic	University of Novi Sad, Serbia
Bharat Singh	Universiti Teknologi PETRONAS, Malaysia
Paweł Sitek	Kielce University of Technology, Poland
Adam Słowik	Koszalin University of Technology, Poland
Vladimir Sobeslav	University of Hradec Králové, Czech Republic
Kamran Soomro	University of the West of England, UK
Zenon A. Sosnowski	Białystok University of Technology, Poland
Bela Stantic	Griffith University, Australia
Stanimir Stoyanov	University of Plovdiv "Paisii Hilendarski", Bulgaria
Ja-Hwung Su	Cheng Shiu University, Taiwan
Libuse Svobodova	University of Hradec Králové, Czech Republic
Jerzy Swiątek	Wrocław University of Science and Technology, Poland
Andrzej Swierniak	Silesian University of Technology, Poland
Julian Szymański	Gdańsk University of Technology, Poland
Yasufumi Takama	Tokyo Metropolitan University, Japan
Zbigniew Telec	Wrocław University of Science and Technology, Poland
Dilhan Thilakarathne	Vrije Universiteit Amsterdam, The Netherlands
Diana Trandabat	University "Alexandru Ioan Cuza" of Iași, Romania
Maria Trocan	Institut Superieur d'Electronique de Paris, France
Krzysztof Trojanowski	Cardinal Stefan Wyszyński University in Warsaw, Poland
Ualsher Tukeyev	al-Farabi Kazakh National University, Kazakhstan

Contents – Part II

Speech and Text Processing

Resource Management and Optimization

Contents – Part I

Data Mining and Machine Learning

Knowledge Integration and Analysis

A New Data Transformation and Resampling Approach for Prediction of Yield Strength of High-Entropy Alloys

Nguyen Hai Chau[1](✉) ⓘ, Genki Sato[2], Kazuki Utsugi[2],
and Tomoyuki Yamamoto[3]

[1] Faculty of Information Technology, VNU University of Engineering and
Technology, 144 Xuan Thuy, Cau Giay, Hanoi, Vietnam
`chaunh@vnu.edu.vn`
[2] Graduate School of Fundamental Science and Engineering,
Waseda University, Tokyo 169-8555, Japan
`{fuwakmykgy,kazuki.36}@fuji.waseda.jp`
[3] Kagami Memorial Research Institute for Materials Science and Technology,
Waseda University, Tokyo 169-0051, Japan
`tymmt@waseda.jp`

Abstract. This paper presents a new approach of data transformation and resampling for prediction of yield strength of high-entropy alloys (HEAs) at room temperature. Instead of directly predicting yield strength of HEAs using common predictors, such as valence electron concentration, electronegativity difference etc., the approach predicts change of yield strength of HEAs based on changes of the predictors after transforming and resampling the original dataset. Experimental results show that on the transformed and resampled dataset, a multiple linear model has the coefficient of determination $R^2 = 0.55$, and mean R^2 of best automatic machine learning-built models is 0.85. The corresponding numbers on the original dataset are 0.44 and 0.75, respectively.

1 Introduction

High-entropy alloys (HEAs) are compounds that are usually made by mixing four or more principal elements with near-equal atomic concentration, often from 5% to 35% [16,21]. This is the most important difference of HEAs from conventional alloys that have one principal element and additional elements at much lower atomic concentration. HEAs typically exhibit high strength, excellent corrosion and wear resistance, and good thermal stability.

Yield strength is an important mechanical property of HEAs. It represents the amount of stress that a HEA can withstand before it starts to deform permanently. Recent research has shown that HEAs possess superior yield strength compared to conventional alloys, making them a potential class for many applications. Because of the large number of compositions of HEAs, designing of HEAs with desired properties including yield strength is hard for materials scientists. Miracle et al. showed that there are 219 million base alloys that have

N. T. Nguyen et al. (Eds.): ACIIDS 2023, LNAI 13996, pp. 3–13, 2023.
https://doi.org/10.1007/978-981-99-5837-5_1

3–6 elements [15]. If atomic concentration of the element varies, the number of alloys will become more than 592 billion [15]. At present, only a small fraction of HEAs is investigated. Thus, building prediction models to narrow down search space of possible HEAs with desired properties receives much research attention. The prediction models can be classified into three categories: first principles calculation [11,14,18,20], machine learning [1,4] and a combination of the two [9,10,13]. First principles calculation is the conventional method in materials science, while machine learning is a new approach that is increasingly used in materials science in recent years. There are representative research results of yield strength prediction. Yin et al. used DFT calculation for prediction of yield strength of RhIrPdPtNiCu alloy and achieved predicted value 583 MPa over 527 MPa measure valued at room temperature [22]. Bhandari et al. used Random Forests to predict the yield strengths of MoNbTaTiW and HfMoNbTaTiZr at 800 °C, 1200 °C and 1500 °C with high accuracy of 95% [1]. Klimenko et al. combined machine learning, phenomenological rules and CALPHAD (Calculation of Phase Diagram) approaches (Calculation of Phase Diagram) [8] to predict the yield strength of the Al-Cr-Nb-Ti-V-Zr HEA system and obtained prediction error of 11% to 13.5% [10]. Giles et al. developed an intelligent framework based on machine learning techniques to predict yield strength of refractory HEAs, a subclass of HEAs, at different temperatures, and achieved cross-validation $R^2 = 0.895$ [4]. Liu et al. propose an integration approach of mechanistic models with machine learning ones to predict the temperature dependent yield strength of HEAs that form in the single-phase face-centered cubic structure [13].

These researches share common points: focusing on certain kinds of HEAs, and the space formed by predictors, such as valence electron concentration, electronegativity difference etc., and the yield strength is used as it is. No transformation or resampling techniques are applied. In this paper, we propose a different approach to predict yield strength of HEAs. While still using the common physical quantities as the predictors, we propose an approach to transform and resample a HEA dataset. On the transformed and resampled dataset, we built a multiple linear regression model for explanation purpose and machine learning models for prediction. The models have better performance than they are built on the original dataset.

The next sections of the paper are organized as follows. Datasets of yield strength of HEAs are described in Sect. 2. The transformation and resampling approach is presented in Sect. 3. Experimental results and conclusion are in Sects. 4 and 5, respectively.

2   Data Description

Two datasets of yield strength of HEAs are collected from a paper Gorsse et al. [5] and its corrigendum [6]. The first one provides yield strength data of HEAs measured at room temperature while the second one provides yield strength of HEAs measured from 25 °C to 1600 °C. This paper uses the first dataset only.

The first dataset has the tabular format with 377 observations of yield strength measured at room temperature and/or phase of 328 HEAs. The unit of measurement of yield strength is MPa (Megapascal). There are 199 observations of 177 HEAs having not-null yield strength and not-null phase values. Other observations are not considered to conduct data analysis in the paper. Denote the set of 199 observations of 177 HEAs as YS1 dataset. Columns (or variables) of the YS1 contain the following information: HEA composition (or chemical formula), publication reference, phase, density, hardness value, type of tests, yield strength, maximum yield strength, elongation and maximum elongation. Thus, the YS1 is not ready for prediction of yield strength.

To use the YS1 as the dataset for training analysis and prediction models of yield strength, additional variables are added. The variables are: valence electron concentration (VEC), electronegativity difference ($\Delta\chi$), atomic size difference (δ), mixing entropy (ΔS_{mix}), and mixing enthalpy (ΔH_{mix}). They are defined as

$$\mathrm{VEC} = \sum_{i=1}^{n} c_i \mathrm{VEC}_i, \tag{2.1}$$

$$\Delta\chi = \sqrt{\sum_{i=1}^{n} c_i (\chi_i - \bar{\chi})^2}, \tag{2.2}$$

$$\delta = 100 \times \sqrt{\sum_{i=1}^{n} c_i \left(1 - \frac{r_i}{\bar{r}}\right)^2}, \tag{2.3}$$

$$\Delta H_{\mathrm{mix}} = \sum_{i=1, i<j}^{n} 4 H_{ij} c_i c_j, \tag{2.4}$$

and

$$\Delta S_{\mathrm{mix}} = -R \sum_{i=1}^{n} c_i \ln c_i, \tag{2.5}$$

where c_i ($0 < c_i < 1$), VEC_i, and r_i are the atomic concentration, VEC and atomic radius of each element in an HEA, respectively. H_{ij} is the enthalpy of atomic pairs calculated by Miedema's model [17] and R is the gas constant. $\bar{\chi}$ and $\bar{r}$ are weighted Pauling electronegativity and atomic radius written as

$$\bar{\chi} = \sum_{i=1}^{n} c_i \chi_i \tag{2.6}$$

and

$$\bar{r} = \sum_{i=1}^{n} c_i r_i, \tag{2.7}$$

respectively. The five variables will be used to predict or analyze yield strength (the outcome), denoted as **ys**. Table 1 shows samples of the YS1, of which AlCoCrFeNi has two observations while the others have one. Variables of the YS1 are composition, five predictors, phase and the **ys** outcome. In the next section, statistical analysis of the YS1 dataset will be presented.

3   Statistical Analysis of the YS1 Dataset

Table 1. Samples of the YS1 dataset.

composition	vec	delta	deltachi	deltahmix	deltasmix	phase	ys
AlCoCrFeNi	7.20	5.78	0.12	−12.32	13.38	BCC	1051
AlCoCrFeNi	7.20	5.78	0.12	−12.32	13.38	BCC	1373
AlC0.1CoCrFeNi	7.14	8.01	0.16	−15.35	13.92	BCC+Im	957
AlC0.2CoCrFeNi	7.08	9.67	0.19	−18.14	14.22	BCC+Im	906
Al0.7Co0.3CrFeNi	7.20	5.54	0.12	−11.40	12.80	FCC+BCC+B2	2033
AlCoCrFeNiSi	6.67	6.61	0.12	−27.33	14.90	BCC+Im	2411
CoCrFeMnNi	8.00	3.27	0.14	−4.16	13.38	FCC	208

There are two types of HEAs composition: equiatomic and non-equiatomic. A composition that has equal atomic concentration for all elements is equiatomic, otherwise non-equiatomic. In Table 1, AlCoCrFeNi, AlCoCrFeNiSi and CoCr-FeMnNi are equiatomic, the others are not. Table 1 also shows that there is one change of from AlCoCrFeNi to AlCoCrFeNiSi: the second composition has Si added, and there are two changes from AlCoCrFeNi to Al0.7Co0.3CrFeNi: atomic concentration of Al and Co changes from 1 to 0.7 and 0.3, respectively. In such situations, it is said that Al0.7Co0.3CrFeNi and AlCoCrFeNiSi are derived from AlCoCrFeNi [19], or in more detailed, AlCoCrFeNiSi and Al0.7Co0.3CrFeNi are 1-derived and 2-derived from AlCoCrFeNi, respectively. AlCoCrFeNi is called the base HEA.

In the YS1 dataset, there are 52 equiatomic HEAs and 125 non-equiatomic HEAs. Of 125 non-equiatomic HEAs, 109 are derived from equiatomic ones. By investigating several equiatomic HEAs and their 1-derived ones, we found that the differences of yield strength of 1-derived HEAs and that of their base HEAs are proportional with change of atomic concentration in percentage. An example of the changes is shown in Fig. 1.

Any change in atomic concentration of elements in a HEA composition causes changes of five predictors given in Eq. (2.1–2.5). Thus, we propose an approach to transform and resample the YS1 dataset as follows. Firstly, HEAs groups are generated from the YS1, where each one contains a base equiatomic HEA and its 1-, 2-, and 3-derived HEAs. Then for each group, a linear data transformation is applied. Since a non-equiatomic HEA can be derived from multiple equiatomic HEAs, generating the groups is also resampling the YS1 dataset. The generation of the groups, i.e. resampling, and data transformation are described in Pseudocode 1.

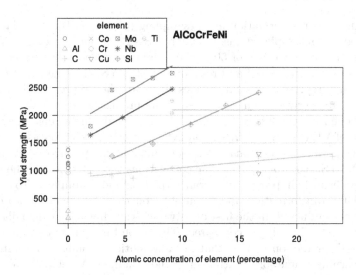

Fig. 1. Yield strength of AlCoCrFeNi and its 1-derived HEAs versus percentage of changed/added/removed Al, C, Co, Cr, Cu, Mo, Nb, Si, and Ti elements.

Pseudocode 1: Transformation and resampling of a HEA dataset of yield strength.

    Input: A HEA dataset YS with structure described in Table 1.
    Output: A transformed and resampled HEA dataset YS-TS with the same structure
            as YS except the **phase** variable.

1   YS-TS = $\emptyset$;
2   **for do**
3 | each distinct equiatomic HEA
4   **end**
5   H in YS D={All observations of 1-, 2-, and 3-derived HEAs of H};
6   E={All observations of H in YS};
7   Calculate mean values $\overline{VEC}, \overline{\delta}, \overline{\Delta\chi}, \overline{\Delta H_{\mathrm{mix}}}, \overline{\Delta S_{\mathrm{mix}}}$, and $\overline{ys}$ of $\forall e \in E$;
8   $G = D \cup E$;
9   **for** *each $g \in G$* **do**
10 | $VEC_g = VEC_g - \overline{VEC}$;
11 | $\delta_g = \delta_g - \overline{\delta}$;
12 | $\Delta\chi_g = \Delta\chi_g - \overline{\Delta\chi}$;
13 | $\Delta H_{\mathrm{mix}_g} = \Delta H_{\mathrm{mix}_g} - \overline{\Delta H_{\mathrm{mix}}}$;
14 | $ys_g = ys_g - \overline{ys}$;
15   **end**
16   YS-TS = YS-TS $\cup$ G;
17   **if** *YS-TS contains duplicated tuples ($VEC = 0$, $\delta = 0$, $\Delta\chi = 0$, $\Delta H_{mix} = 0$,*
       $\Delta S_{mix}, ys = 0$) **then**
18 | Keep only one of the tuples in YS-TS;
19   **end**
20   return YS-TS;

Applying the transformation and resampling process on the YS1 dataset, a transformed and resampled dataset, namely YS1-TS, is generated. The YS1-TS consists of 2160 observations of 50 equiatomic and 109 non-equiatomic HEAs. After the YS1 is transformed and resampled to the YS1-TS, the five predictors and **ys** do not represent absolute values. Instead, they represent the changes when HEAs are derived from their bases. The prediction and analyzing problems mentioned in Sect. 2 becomes prediction and analyzing change of **ys** based on changes of VEC, δ, $\Delta\chi$, ΔH_{mix}, and ΔS_{mix} in the YS1-TS dataset. Figure 2 shows differences of correlations of the outcome and the predictors on the YS1 and the YS1-TS. It is obvious that the correlations the outcome and the predictors are significantly improved after transformation and resampling. To compare effect of data transformation and resampling on the YS1 dataset, two experiments will be conducted. In the first one, two multiple linear models built on the YS1 and the YS1-TS will be compared. The second experiment compares two sets of "best" prediction models that are automatically built by an automatic machine learning framework written in Python, namely AutoGluon [3], on the two datasets. Details of the experiments are presented in Sect. 4.

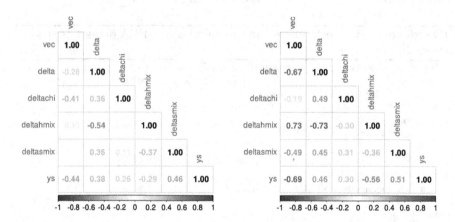

Fig. 2. Correlation matrices of the predictors and yield strength of two HEA datasets. Left: the YS1 dataset, right: the YS1-TS dataset.

4   Experimental Results

This section presents experimental results and models performance comparison. On each of the YS1 and the YS1-TS datasets, a multiple linear regression model and a set of "best" prediction models, generated by AutoGluon automatic machine learning framework [3], are built. Denote the multiple linear model and the set of prediction models on the YS1 as **lm1** and **AG1**, respectively. On the YS1-TS, they are **lm2** and **AG2**. All models use the same predictors VEC, δ, $\Delta\chi$, ΔH_{mix}, and ΔS_{mix}.

4.1   Comparison of Multiple Linear Models

The linear models lm1 and lm2 are built using **brm** function in **brms** package [2]. The **brm** uses Bayesian approach to build models. lm1 and lm2 are described in Tables 2 and 3, respectively. They are expressed as

$$ys \approx 275.3 - 160.7\text{VEC} - 0.67\delta + 360.1\Delta\chi - 14.4\Delta H_{\text{mix}} + 130.6\Delta S_{\text{mix}} \quad (4.1)$$

and

$$ys \approx -1.254 - 595.7\text{VEC} - 50\delta + 1643\Delta\chi - 13.6\Delta H_{\text{mix}} + 71.25\Delta S_{\text{mix}}. \quad (4.2)$$

Coefficients of determination R^2 of lm1 and lm2 are 0.44 and 0.55, respectively. The lm2 is better than lm1 because of its higher R^2. Furthermore, all corresponding credible intervals 95% (CI95s) of regression coefficients for predictors of lm2 do not contain 0. These are indicated in the **CI95 low** and **CI95 high** columns of Table 3. Thus, using lm2 model, one can explain that change of yield strength of a HEA with reference to its base HEA will be positive when changes of $\Delta\chi$ and ΔS_{mix} are positive, and when changes of VEC, δ, and ΔH_{mix} are negative. The amounts of changes are defined by regression coefficients in Eq. (4.2), equivalently in the **Estimate** column of Table 3.

This explanation is not fully possible for lm1 because corresponding CI95s of regression coefficients for δ and $\Delta\chi$ contain 0 as shown in the **CI95 low** and **CI95 high** columns of Table 2.

Table 2. The lm1 model ($R^2 = 0.44$).

	Estimate	Est.Error	CI95 low	CI95 high
Intercept	275.30	315.56	−337.46	893.49
vec	−160.73	22.79	−205.42	−116.02
delta	−0.67	18.39	−37.15	35.09
deltachi	360.13	462.72	−546.49	1265.08
deltahmix	−14.41	5.22	−24.63	−4.33
deltasmix	130.61	21.14	89.13	171.87

4.2   Comparison of AutoGluon-generated Models

Two sets of AutoGluon-generated models on the YS1 and YS1-TS, denoted as AG1 and AG2, are created using AutoGluon framework in the second experiment [3]. The creation procedure is described in the Pseudocode 2.

Table 3. The lm2 model ($R^2 = 0.55$).

	Estimate	Est.Error	CI95 low	CI95 high
Intercept	−1.25	11.01	−22.42	20.12
vec	−595.67	24.05	−643.23	−549.58
delta	−49.69	4.96	−59.50	−40.18
deltachi	1642.88	145.34	1363.62	1928.53
deltahmix	−13.56	1.74	−17.13	−10.27
deltasmix	71.25	5.75	59.85	82.43

Pseudocode 2: Prediction models building using AutoGluon.

Input: A HEA dataset YS with structure described in Table 1, except the phase variable.

Output: A set S of best AutoGluon-generated prediction models and their test performance: R^2, root mean squared error (RMSE).

1 $S = \emptyset$;
2 Create a set I of 30 distinct natural numbers;
3 **for** *each* $i \in I$ **do**
4 $\quad$ Set i as the initial value for the random number generator;
5 $\quad$ Randomly split YS into a training TR_i and a test TE_i dataset;
6 $\quad$ // Ratio of sizes of TR_i and TE_i is 8:2
7 $\quad$ Use AutoGluon to create a set of candidate models M on the TR_i;
8 $\quad$ // Predictors: VEC, δ, $\Delta\chi$, ΔH_{mix}, and ΔS_{mix}; outcome: ys
9 $\quad$ // 10-fold cross-validation is used to create M
10 $\quad$ // Criterion to select a candidate model is R^2
11 $\quad$ **for** *each candidate model* $m \in M$ **do**
12 $\quad\quad$ Use m to predict yield strength on the TE_i dataset;
13 $\quad\quad$ Calculate R^2 and RMSE of the prediction;
14 $\quad$ **end**
15 $\quad$ Select $m_{\mathrm{best}} \in M$ that has the highest R^2;
16 $\quad$ $S = S \cup \{(m_{\mathrm{best}}, \text{its corresponding } R^2 \text{ and RMSE})\}$;
17 **end**
18 **return** S;

The I set of initial values of the random number generator is used for both YS1 and YS1-TS. Applying Pseudocode 2 on YS1 and YS1-TS, two sets of models AG1 and AG2 are created, respectively. It is interesting that all models in AG1 are WeightedEnsemble_L2 and all models of AG2 are WeightedEnsemble_L3. Models in AG1 and AG2 will be evaluated and compared using Bayesian estimation. Tool for this task is BESTmcmc, a core function of **BEST** package [12]. The function works as a replacement for the Student's t-test [7].

Table 4, 5 evaluates and compares performance of models in AG1 and AG2. The evaluation and comparison criteria are R^2 and RMSE. As shown in columns 1, 4, 5, and 6 in the first row (**mu1**) of Table 4, AG1 models' R^2 has 0.75 mean value and $[0.71, 0.78]$ high density interval 95% (HDI95) [12]. Those of AG2 are 0.85 and $[0.84, 0.86]$, respectively.

Table 4. Comparison of R^2 of models in AG1 and AG2.

	mean	median	mode	HDI	HDIlo	HDIup
mu1	0.75	0.75	0.75	95.00	0.71	0.78
mu2	0.85	0.85	0.85	95.00	0.84	0.86
muDiff	−0.10	−0.10	−0.10	95.00	−0.14	−0.07

Comparison details are in the third row (**muDiff**) of Table 4: performance of models in AG1 is 0.1 lower than that of models in AG2 by average, and the difference has [−0.14, −0.07] HDI95. Because the HDI95 does not contain 0, one can conclude that AutoGluon-generated models, built on the transformed and resampled YS1-TS dataset, have better performance than those built on the YS1. The performance improvement in R^2 is 0.1.

Table 5 is interpreted in the same manner. The first two rows of the table show similar performance of models in AG1 and AG2. Comparison results in the third row of Table 5 show that **muDiff** has [−25.95, 21.11] HDI95 consisting of 0. Thus, there is no performance improvement in RMSE.

Table 5. Comparison of RMSE of models in AG1 and AG2.

	mean	median	mode	HDI	HDIlo	HDIup
mu1	294.96	294.97	294.69	95.00	272.14	317.04
mu2	297.57	297.57	297.67	95.00	290.34	304.68
muDiff	−2.61	−2.63	−3.09	95.00	−25.95	21.11

5   Conclusions

We have proposed a new approach to transform and resample a dataset of yield strength of high-entropy alloys (HEAs) at room temperature, then built an explainable multiple linear model and a set of "best" prediction models.

The linear model has a modest coefficient of determination ($R^2 = 0.55$), but it is better than it is built on the original dataset ($R^2 = 0.44$). Furthermore, the HDI95s of regression coefficients of the five predictors do not contain 0, therefore the linear model can be used to explain change of yield strength of HEAs with reference to their base using changes of the predictors.

On the transformed and resampled dataset, best AutoGluon-generated models have mean $R^2 = 0.85$ that is 0.1 better than those on the original dataset.

Using this approach, the prediction of yield strength of HEAs starts from a dataset of known equiatomic HEAs and their derived ones. After transforming and resampling the dataset, change of yield strength of new derived HEAs with reference to their base HEAs can be roughly explained and predicted using the multiple linear model. More accurate yield strength prediction can be achieved using more complex models, for example, AutoGluon-generated ones. Since a new HEA can be derived from multiple distinct base HEAs, it can have multiple predicted values of yield strength, forming a predicted distribution.

Acknowledgement. This work was partly carried out at the Joint Research Center for Environmentally Conscious Technologies in Materials Science (project No. 02210, Grant No. JPMXP0621467974) at ZAIKEN, Waseda University.

References

1. Bhandari, U., Rafi, M.R., Zhang, C., Yang, S.: Yield strength prediction of high-entropy alloys using machine learning. Mater. Today Commun. **26** (2021). https://doi.org/10.1016/j.mtcomm.2020.101871
2. Bürkner, P.C.: brms: An r package for bayesian multilevel models using stan. J. Stat. Softw. **80** (2017). https://doi.org/10.18637/jss.v080.i01
3. Erickson, N., et al.: Autogluon-tabular: Robust and accurate automl for structured data (2020). https://arxiv.org/abs/2003.06505
4. Giles, S.A., Sengupta, D., Broderick, S.R., Rajan, K.: Machine-learning-based intelligent framework for discovering refractory high-entropy alloys with improved high-temperature yield strength. NPJ Comput. Mater. **8**, 235 (2022). https://doi.org/10.1038/s41524-022-00926-0
5. Gorsse, S., Nguyen, M.H., Senkov, O.N., Miracle, D.B.: Database on the mechanical properties of high entropy alloys and complex concentrated alloys. Data Brief **21**, 2664–2678 (2018). https://doi.org/10.1016/j.dib.2018.11.111
6. Gorsse, S., Nguyen, M.H., Senkov, O.N., Miracle, D.B.: Corrigendum to database on the mechanical properties of high entropy alloys and complex concentrated alloys, Data Brief **21**, 2664–2678 (2018). (data in brief (2018) 21 (2664–2678), (s235234091831504x), (https://doi.org/10.1016/j.dib.2018.11.111)) (2020). https://doi.org/10.1016/j.dib.2020.106216
7. Gosset, S.W.S.: The probable error of a mean. Biometrika **6** (1908). https://doi.org/10.1093/biomet/6.1.1
8. Kaufman, L., Bernstein, H.: Computer calculation of phase diagrams. Academic Press Inc, With special reference to refractory metals (1970)
9. Kim, G., et al.: First-principles and machine learning predictions of elasticity in severely lattice-distorted high-entropy alloys with experimental validation. Acta Materialia **181**, 124–138 (2019). https://doi.org/10.1016/j.actamat.2019.09.026
10. Klimenko, D., Stepanov, N., Li, J., Fang, Q., Zherebtsov, S.: Machine learning-based strength prediction for refractory high-entropy alloys of the al-cr-nb-ti-v-zr system. Materials **14** (2021). https://doi.org/10.3390/ma14237213
11. Koval, N.E., Juaristi, J.I., Díez Muiño, R., Alducin, M.: Elastic properties of the TiZrNbTaMo multi-principal element alloy studied from first principles. Intermetallics **106**, 130–140 (2019). https://doi.org/10.1016/j.intermet.2018.12.014
12. Kruschke, J.K.: Bayesian estimation supersedes the t test. J. Exp. Psychol. Gener. **142** (2013). https://doi.org/10.1037/a0029177
13. Liu, S., Lee, K., Balachandran, P.V.: Integrating machine learning with mechanistic models for predicting the yield strength of high entropy alloys. Journal of Applied Physics **132** (2022). https://doi.org/10.1063/5.0106124

14. Ma, D., Grabowski, B., Körmann, F., Neugebauer, J., Raabe, D.: Ab initio thermo-dynamics of the cocrfemnni high entropy alloy: importance of entropy contributions beyond the configurational one. Acta Materialia **100**, 90–97 (2015). https://doi.org/10.1016/j.actamat.2015.08.050

15. Miracle, D.B.: High entropy alloys as a bold step forward in alloy development (2019). https://doi.org/10.1038/s41467-019-09700-1

16. Miracle, D.B., Senkov, O.N.: A critical review of high entropy alloys and related concepts. Acta Materialia **122**, 448–511 (2017). https://doi.org/10.1016/j.actamat.2016.08.081

17. Takeuchi, A., Inoue, A.: Classification of bulk metallic glasses by atomic size difference, heat of mixing and period of constituent elements and its application to characterization of the main alloying element. Mater. Trans. **46**(12), 2817–2829 (2005). https://doi.org/10.2320/matertrans.46.2817

18. Tian, F., Delczeg, L., Chen, N., Varga, L.K., Shen, J., Vitos, L.: Structural stability of NiCoFeCrAl$_x$ high-entropy alloy from ab initio theory. Phys. Rev. B **88**(8), 085128 (2013). https://doi.org/10.1103/PhysRevB.88.085128

19. Tsai, M.H., Yeh, J.W.: High-entropy alloys: a critical review. Mater. Res. Lett. **2**(3), 107–123 (2014). https://doi.org/10.1080/21663831.2014.912690

20. Yao, H.W., Qiao, J.W., Hawk, J.A., Zhou, H.F., Chen, M.W., Gao, M.C.: Mechanical properties of refractory high-entropy alloys: Experiments and modeling. J. Alloys Compd. **696**, 1139–1150 (2017). https://doi.org/10.1016/j.jallcom.2016.11.188

21. Ye, Y.F., Wang, Q., Lu, J., Liu, C.T., Yang, Y.: High-entropy alloy: challenges and prospects. Mater. Today **19**(6), 349–362 (2016). https://doi.org/10.1016/j.mattod.2015.11.026

22. Yin, B., Curtin, W.A.: First-principles-based prediction of yield strength in the rhirpdptnicu high-entropy alloy. NPJ Comput. Mater. **5** (2019). https://doi.org/10.1038/s41524-019-0151-x, https://www.nature.com/articles/s41524-019-0151-x

A Semi-Formal Approach to Describing Semantics of Data Modeling Patterns

Marcin Jodłowiec[✉][iD]

Faculty of Information and Communication Technology, Wrocław University of
Science and Technology, Wrocław, Poland
marcin.jodlowiec@pwr.edu.pl

Abstract. The paper presents the method for describing data modeling
patterns. Data modeling patterns are reusable, general solutions providing the abstraction over the domain problem, which is modeled, retaining
the semantic structure of the solution. The proposed method consists of
three phases: conceptualization, specification and implementation. The
specification phase is based on the ontological concept system Conceptual Layer of Metamodels. The method has been verified by a case study,
the Generalized Network modeling pattern and its implementation in
Association-Oriented Metamodel.

Keywords: modeling patterns · specification of conceptualization ·
conceptual modeling · data modeling · association-oriented modeling ·
semi-formal methods

1   Introduction

Design patterns are general and reusable solutions for recurring problems [4,16].
The problems that arise during data modeling are repeatable. This repeatability
results from a finite number of syntactic rules that can be used in constructing
a structure, which can be assigned specific semantics in terms of the metamodel
to which the data model conforms. Of course, the structures that make up the
different models should be interpreted not only in terms of the semantics of the
metamodel, but also the semantics of the model itself. This meaning is given by
the modeler to represent a certain domain. Modeling patterns abstract from the
domain. However, they allow a significant part of the domain to be expressed
not directly in terms of the metamodel, but in terms of the pattern.

The use of patterns as a tool for expressing repeating templates of structures,
and then translating them into specific models, allows to significantly accelerate
and facilitate the process of model creation. The modeler does not have to consider how to construct a given model from scratch, but uses proven solutions.
An important issue is both the selection of a specific set of patterns, as well as
the method of their specification. Traditionally, patterns have been described
informally, i.e. by the use of natural language supported by example usage.

© The Author(s), under exclusive license to Springer Nature Singapore Pte Ltd. 2023
N. T. Nguyen et al. (Eds.): ACIIDS 2023, LNAI 13996, pp. 14–26, 2023.
https://doi.org/10.1007/978-981-99-5837-5_2

This paper proposes a method of creating and describing patterns that can be used for data modeling. The method is generic, i.e. it abstracts from a specific metamodel, using an general specification layer which can capture any data metmodel. In the proposed solution, Conceptual Layer of Metamodels (CLoM) was adopted as a kind of modeling ontology, ensuring the possibility of specifying conceptualizations common to various metamodels, while having available mappings of CLoM concepts to specific data metamodels. An example of such metamodels can be Association-Oriented Metamodel (AOM) which has been successfully conceptualized within CLoM [6]. However, other data metamodels like Extended Entity-Relationship Model (EER), Unified Modeling Language (UML), Object-Role Modeling (ORM) also are subject of such conceptualization.

This paper is structured as follows. The next chapter contains an introductory description of CLoM, a concept system which has been employed for specification. Subsequently, the method is proposed and described. The adopted approach covers three layers of different level of formalization: description, specification and implementation. Moreover, the conceptualization of the whole framework within CLoM has been described. In Sect. 4 a case study of generalized network pattern is analyzed. The pattern has been described, specified and implemented within AOM. The paper ends with conclusions.

2   Conceptual Layer of Metamodels

CLoM is a concept system which forms a semi-formal ontology. The system has been based on OMG's standard Semantics of Business Vocabulary and Business Rules (SBVR) [12]. CLoM adopted its semantics as the basis for formulating expressions. It includes a superset of concepts that perform foundational functions in the field of data modeling, their characteristics (constraints), and determines how concepts can relate to each other.

The CLoM covers the subset of concepts and relationships between them in the domain of data modeling and metamodeling. Though data metamodels have different categories, these categories share some semantics, which is common in the domain. An example could be a concepts of *association*, which is represented by the *Assoc* category in AOM and *relationship set* in EER. The concept system abstracts from a specific metamodel. The approach which has been adopted was to define a number of constraints for the modeling concepts. These constraints describe properties and semantics, which is given to a specific model element. Basic concepts in the field of metamodeling have also been defined. The conceptual framework for metamodeling platform has been based on the approach presented in [10]. It should also be added that the concept system covers only structural relationships. Therefore, some conceptual modeling techniques such as *materialization, point of view, generation* (see e.g. [3]) have not been included.

In the CLoM we have proposed a number of concepts groups which refer to a particular areas of data modeling domain (Fig. 1): *metamodeling, patterns, core, classifiers, features, categorization, associations*.

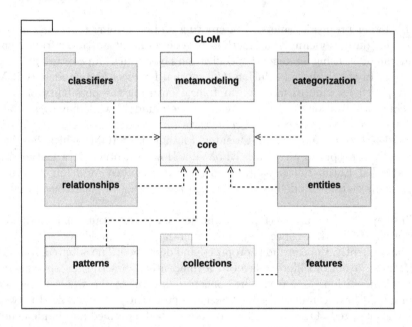

Fig. 1. UML package diagram representing CLoM concept groups

3 Method for Pattern Specification

The proposed method of modeling pattern specification is based on three layers (levels of abstraction):

1. Description Layer, informal, expressing an intent of a pattern,
2. Specification Layer, semi-formal, determining semantics of the pattern concept by the use of CLoM,
3. Implementation Layer, formal, showing how the pattern concept can be used to build actual data structures within a specific metamodel.

3.1 Description Layer

The description level of the modeling pattern describes in an informal way (in the form of a natural language) the elements that make up the idea of the pattern, its intent, the reason to its formulation, i.e. the abstracted problem that a given pattern should solve. At this stage, the participants of the pattern are also defined, i.e. the abstractions of entities that play certain roles in the pattern. Mostly, this layer has been inspired by the well-known method for describing object-oriented patterns proposed by [4]. However, it should be noted that in the case of [4] informal expressions constitute the pattern description, and in this method one of the (initial) layers, which is the starting point for further specification. Within this layer, the following should be specified:

– the name of the pattern,

- a description of the modeling problem to be solved,
- components (participants) and
- possible characteristics, i.e. set of fact sets that may exist in a specific instance of the pattern and constitute additional extensions or limitations.

3.2 Specification Layer

The specification layer has been based on the CLoM modeling ontology. The specification is based on the pattern concept and defines a series of atomic expressions. Expressions take the form of facts and rules that specify subsequent requirements that must be fulfilled in order to confirm the realization of the pattern concept. Expressions are atomic in nature, i.e. they are calculated in atomic form and are called *semantic atoms*. Thus, semantic atom concept has been defined in CLoM as follows:

semantic atom

Definition: indivisible fact or rule describing semantics

The modeling pattern specification must contain a set of necessary semantic atoms and may additionally contain a set of possible characteristics. The joint occurrence of the necessary semantic atoms is required to state the realization of the pattern concept. Additionally, the specification may include possible characteristics of a given pattern, i.e. a specification of extensions and constraints that may or may not appear in the pattern instance. The possible characteristics consist of semantic atoms. For a possible characteristic to exist, the joint occurrence of all the semantic atoms of the possible characteristics is required. The example of pattern concept specification (necessary semantic atoms) and possible characteristic has been presented in the Sect. 4.

3.3 Implementation Layer

The implementation layer describes the pattern models by presenting conceptual models in the given data metamodel that implement them together with a description of the implementation steps of individual semantic atoms, both necessary and those describing possible characteristics. The descriptions use the term implementation in its broad sense to refer to conceptual models in order to emphasize that they are real and measurable artifacts created within the adopted approach to data modeling. At this stage, for a given pattern configuration, a structure related to a given metamodel is constructed.

3.4 Patterns as a Part of a Concept System

A concept diagram representing the CLoM fragment describing the adopted approach is shown in Fig. 2. The pattern concept notion defines semantics (meaning) of a pattern, which is formed by the set of necessary semantic atoms. Moreover, this concept defines which possible characteristic are possible in terms of a pattern configuration. Each possible characteristic is formed by the set of

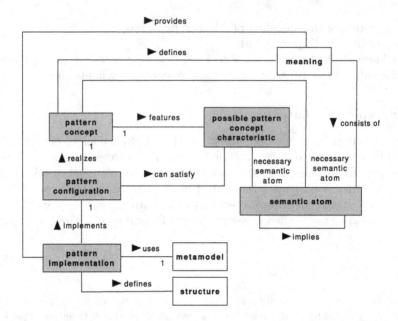

Fig. 2. CLoM concept diagram representing concepts in *patterns* concept group

semantic atoms, which jointly define it. The <u>semantic atoms</u> can imply other semantic atoms. However, an atom cannot imply itself. The <u>pattern configuration</u> realizes the assumptions of a given <u>pattern concept</u> and satisfies a number of <u>possible characteristics</u>. A <u>pattern configuration</u> can only satisfy those characteristics, which are featured by a <u>pattern concept</u>, which is realized by it. The <u>pattern implementation</u> implements a specific <u>pattern configuration</u> by the use of exactly one metamodel and definition of a specific structure in terms of this metamodel.

The CLoM termonological entries for concepts regarding *patterns* group has been semi-formally defined below.

<u>pattern concept</u>

Definition: <u>set</u> *of* concepts *defining* <u>meaning</u> *of* pattern in terms of <u>necessary semantic atoms</u> and <u>possible characteristic</u>

Note: The pattern concept is its definition framework, i.e. it defines the full semantic space used to determine which structures can meet the requirements specified within a given pattern.

<u>pattern configuration</u>

Definition: <u>configuration</u> *of* <u>necessary semantic atoms</u> and <u>subset</u> *of* <u>possible characteristics</u> featuring a specific set of properties taken into account in the process of classifying <u>structure</u> as an instance of a given pattern

<u>pattern configuration</u> **realizes** <u>pattern concept</u>

Definition: <u>pattern configuration</u>   defines <u>subset</u> *of* <u>possible characteristics</u> *featuring* <u>pattern concept</u>

Necessity: each <u>pattern configuration</u> *realizes* exactly one <u>pattern concept</u>

<u>pattern implementation</u>

Definition: mapping of <u>pattern configuration</u> to <u>structure</u> conforming to a syntax of <u>given</u> <u>metamodel</u>

pattern implementation **implements**

pattern configuration

pattern implementation **uses** metamodel

  Necessity: each pattern implementation *uses* exactly one metamodel

pattern implementation **defines** structure
  possible characteristic

  Definition: optional   property   *of* pattern concept *realized by* pattern configuration

pattern concept **features** possible characteristic

  Necessity: each possible characteristic *features* exactly one pattern concept

pattern configuration can **satisfy** possible characteristic

  Necessity: possible characteristic   cannot *feature* pattern configuration that does not realize pattern concept *featured by* given possible characteristic

pattern concept **defines** meaning

semantic atom₁ **implies** semantic atom₂

semantic atom **is part of** meaning

necessary semantic atom

  Definition: semantic aotm that *is* essential to define meaning *of* pattern concept or possible characteristic
  Concept type: role

pattern concept **has** necessary semantic atom

possible characteristic **has** necessary semantic atom

4 Generalized Network: A Case Study

As a case study of the proposed description method, the *generalized network* pattern will be considered. The problem solved by the generalized network pattern consists in the possibility of creating models based on various types of constructions, which rely on the generalization of the graph approach to data representation. The concept of the generalized network pattern is based on the application of a number of graph modeling properties noted in [7] to the problems of conceptual modeling.

4.1 Description Layer

The reason for the identification of the generalized network as a separate pattern is that the data structure stored in the instance of this pattern has been moved to a higher level of abstraction. The adopted abstraction is based on generalization of the edge and the node to the form of a linked element.

Modeling of network structures such as graphs and hypergraphs [1,14] and ubergraphs [9] is used for representation of a number of problems, structures, and issues in science and its applications [11,13,15].

4.1.1 Participants

Concepts constituting participants of the pattern:

1. CONNECTABLE ELEMENT – abstraction of VERTEX and EDGE,
2. VERTEX – abstraction of element which represents data, ,
3. EDGE – abstraction of element which represents links between data.

Concepts constituting connections between participants of the pattern:

1. CONNECTION – representation for link between VERTICES and EDGES.

4.1.2 Characteristics

Concepts constituting possible characteristics of the pattern:

1. RELATIONSHIP BINARITY – characteristic which constraints the ability for implementing hypergraph structures and allows only binary edges. In the other words, one edge can only connect two vertices.
2. VERTICAL CONNECTION CONSTRAINT – characteristic which disables creation of ultragraph structures, i.e. allows only VERTICES to be connected.
3. VERTICAL LABELABILITY – characteristic which decides whether it is possible to describe a VERTEX with an additional set of attributes.
4. EDGE LABELABILITY – characteristic which specifies the possibility to describe EDGES with additional set of attributes.
5. FIRST-CLASS PROPERTY – characteristic which decides if network is a first-class citizen within a network, i.e. networks can be elements of EDGES. This characteristic makes pattern configurations compatible with the idea of partitioned networks [2,5].
6. UNINAVIGABILITY – characteristic of a pattern which defines that each EDGE has defined direction which has to be enforced whilst traversing by the network (visiting CONNECTABLE ELEMENTS).

4.2 Specification Layer

Specification layer consists of definition of necessary semantic atoms expressed within CLoM and specification of possible characteristics.

4.2.1 Necessary Semantic Atoms of Generalized Atoms

Statements K.1 – K7 describe atoms, which are needed to be asserted in a structure in order to be classified as generalized network. The atoms introduce concepts for participants and refer to CLoM vocabulary to make implied statements about their characteristics.

K.1 generalized network *is concretization of* pattern concept

K.2 generalized network always *has* vertex that *is concretization of* entity type

K.3 generalized network always *has* edge that *is concretization of* entity type

K.4 generalized network always *has* connectable element that *is concretization of* entity type

K.5 connectable element *categorizes* classifier set *containing* node and edge

    K.5.1 this categorization *includes* polymorphism

K.6 generalized network always *has* connection that *is concretization of* association

K.6.1 __edge__ always *plays the role of* __connecting element__ *in* __connection__

K.6.2 __connectable element__ always *plays the role of* __connected element__ *in* __connection__

K.6.3 __destination__ *of* __connecting element__ *is featured by* multiplicity that *has* upper bound and __lower bound__ *equal* __1__

K.7 __connectable element__ *is featured by* non-instantiability

4.2.2   Semantics of Possible Characteristics

Statements M.1–M.6 introduce possible characteristics of generalized network pattern. The atoms which are implied describe conceptual conditions for the assertion of characteristic.

M.1 __relationship binarity__ *is concretization of* possible pattern characteristic

M.1.1 each __edge__ *plays the role of* __connecting element__ *in* exactly __2__ __connections__

M.2 __vertical connection constraint__   *is*   *concretization*   *of* possible pattern characteristic

M.2.1 each instance *classified by* __connectable element__ *playing role of* __connected element__ *in* __connection__ *is classified by* __vertex__

M.3 __vertical labelability__ *is concretization of* possible pattern characteristic

M.3.1 __vertex__ *has* at least one owned attribute

M.4 __edge labelability__ *is concretization of* possible pattern characteristic

M.4.1 __edge__ *has* at least one owned attribute

M.5 __first-class property__ *is concretization of* possible pattern characteristic

M.5.1 __generalized network__ *has* __network__ that *is concretization of* entity type

M.5.1.1 __network__ *is* *in* classifier set that *is* *categorized* *by* __connectable element__

M.5.2 __generalized network__ *has* __network's ownership__ that *is concretization of* association

M.5.2.1 __network__ *plays the role of* __owner__ *in* __network's ownership__

M.5.2.1.1 source *of* __owner__ *is featured by* lifetime dependency

M.5.2.1.2 <u>source</u> and <u>destination</u> *of* **owner** *is featured by* <u>multiplicity</u> that *has* <u>upper bound</u> and <u>lower bound</u> *equal* <u>**1**</u>

M.5.2.2 <u>**connectable element**</u> *plays the role of* <u>**network's property**</u> *in* <u>**network's ownership**</u>

M.5.2.2.1 <u>destination</u> *of* <u>**network's property**</u> *is featured by* <u>lifetime dependency</u>

M.5.2.2.2 <u>source</u> *of* <u>**network's property**</u> *is featured by* <u>multiplicity</u> that *has* <u>upper bound</u> and <u>lower bound</u> *equal* <u>**1**</u>

M.6 <u>**uninavigability**</u> *is concretization of* <u>possible pattern characteristic</u>

M.6.1 each <u>**connection**</u> *is featured by* <u>navigability</u>

M.6.2 *if* <u>source</u> *of* <u>role</u> *owned by* <u>**connection**</u> *is navigable* then <u>destination</u> *of that* <u>role</u> *is not navigable*

M.6.3 *if* <u>source</u> *of* <u>role</u> *owned by* <u>**connection**</u> *is unnavigable* then <u>destination</u> *of that* <u>role</u> *is navigable*

M.6.4 each <u>**edge**</u> that *is in* at least one <u>**connection**</u> *plays the role of* <u>**connecting element**</u> that *has* <u>source</u> that *is navigable* and each <u>**edge**</u> that *is in* at least one <u>**connection**</u> *plays the role of* <u>**connecting element**</u> that *has* <u>source</u> that *is unnavigable*

4.3   Implementation Layer

4.3.1   Implementation of Necessary Semantic Atoms

In this section it will be shown how to implement the semantics of generalized network in AOM. The semantics atoms defined in 4.2.1 are SBVR/CLoM expressions thus they abstract from the metamodel. However, if the conceptualization of metamodel is extracted, then they can be expressed in terms of the metamodel semantics. The semantics of AOM has been described within [6]. In the proces of analysis of the subsequent semantic atoms, one can deduce semantic structures, which can directly be implemented. For example, taking AOM as a metamodel for implementation, the following semantic implications can be considered for atoms K.1, K.5, K.5.1:

$$\circledast^{K.1}_{[generalized\ network]} \overset{AOM}{\models} \left\{ \ast^{[generalized\ network]}_{AOM} \right\} \tag{1}$$

$$\circledast^{K.5}_{[generalized\ network]} \overset{AOM}{\models} \left\{ \begin{array}{c} \ast^{[inheritance]}_{AOM}, \\ \odot_{[connectable\ element]} \vartriangleleft \odot_{[generalization]} \\ \odot_{[vertex\ relationship\ aspect]} \vartriangleleft \odot_{[specialization]} \\ \odot_{[vertex\ data\ aspec]} \vartriangleleft \odot_{[specialization]} \end{array} \right\} \tag{2}$$

$$\underset{[generalized\ network]}{\circledast K.5.1} \overset{\text{AOM}}{\vDash} \left\{ \overset{\text{[inheritance]}}{\underset{\circledcirc [\text{aom:nochange}]}{\ast \text{AOM}}} \lhd \circledcirc_{[\text{fulfilling roles inheritance aspect}]} \right\} \quad (3)$$

In the Fig. 3 and Fig. 4 the implementation of necessary semantic atoms of the pattern has been shown in the form of AOM diagrams. Two solutions of this problem are correct. The first one (Fig. 3) implements a pattern with a fixed, maximum connection arity. In this implementation CONNECTION has been modeled by the use of the *Role* category from AOM. The example has three particular connections: $Conn_1$, $Conn_2$, $Conn_3$. This means that structures described by this model can use at most three connections.

The other solution (Fig. 4) uses *Assoc* as a category which realizes connection. In such case it is possible to dynamically create new connections (concrete connections have been moved to model's extension). One can notice that both solutions are correct in terms of necessary semantic atoms. The K.6 atom's expression states that connection is a concretization of <u>association</u>. In AOM, both *Assoc* and *Role* are conceptually concretizations of this concept.

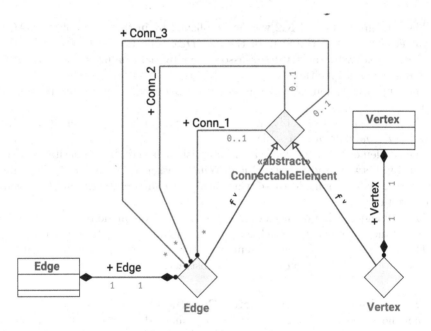

Fig. 3. Implementation of generalized network pattern in AOM. The implementation is constrained by fixed maximum connection arity

The implementation of the other model elements which are common for both implementations have been presented below.

1. The implementation of K.1 was accomplished by creating a separate model implementing this pattern.

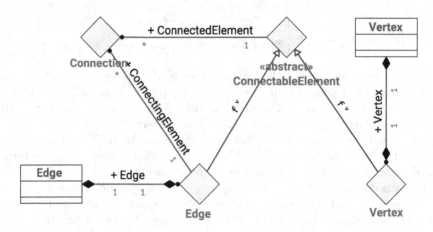

Fig. 4. Implementation of generalized network pattern in AOM. The implementation allows for unlimited connection arity

2. The implementation of K.2 was accomplished by instantiation of the BACT pattern (see [8]) in the form of the $\square Vertex$ collection associated with the $Vertex$ role with the $\lozenge Vertex$ association by a bicompositional role with equivalent multiplicities.
3. The implementation of K.3 was done analogously — by using the BACT pattern.
4. The implementation of K.4 was accomplished by creating the $\lozenge ConnectableElement$ association.
5. The implementation of K.5 was accomplished by creating inheritance relationship between the $\lozenge ConnectableElement$ association and the $\lozenge Edge$ and $\lozenge Vertex$ associations. In these relationships, $\lozenge ConnectableElement$ is a generalization.
6. The implementation of the semantic atom K.5.1 was made by specifying the inheritance mode as f^v, which enforces inheritance of rights to fulfill roles.
7. The implementation of the K.7 semantic atom was accomplished by specifying association $\lozenge ConnectableElement$ as abstract.

4.3.2   Implementation of Selected Possible Characteristics

To show how possible characteristics are implemented, VERTICAL CONNECTION CONSTRAINT has been selected. The diagram in the Fig. 5 shows an AOM model which holds this property.

This implementation assumes the realization of all necessary semantic atoms of the pattern and additionally the semantic atoms specified within this possible property.

Semantic atoms M.2 and M.2.1 has been implemented by creating a separate model and deletion of $ConnectedElement$ role of $\lozenge Connection$ association.

The implementation is based on the version with unlimited arity. The semantic atoms M.2 and M.2.1 were implemented by creating a separate model and

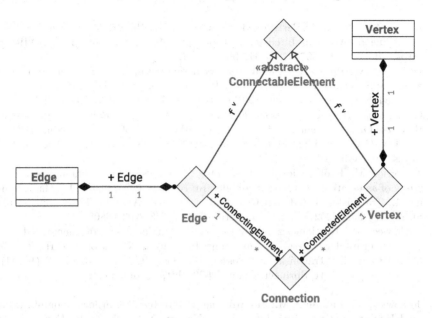

Fig. 5. Implementation of generalized network in AOM. The pattern configuration satisfies a possible characteristic: Vertical Connection Constraint

removing the *ConnectedElement* role from the ◊*Connection* association and creating a new role in the same association with the same name and the same characteristics, but specifying the purpose of the role as ◊*Vertex*.

5   Conclusions

In the paper we have presented the method for data modeling design pattern description. The method is based on semi-formal approach. It covers the issues of description, specification and implementation of data modeling patterns. The adopted approach is based on the concept system CLoM, which plays the role of data modeling ontology and provides vocabulary for describing such phenomena as i.a. metamodels, models, patterns. The method has been verified and presented by the use of generalized network pattern, a data modeling pattern based on Extended Graph Generalization concept. The pattern has been specified in CLoM and implemented in AOM.

In the future, authors plan to focus on more applications of CLoM in data semantics analysis. A very interesting and still unsolved problem is the issue of measuring and comparing semantic expressiveness and semantic capacity of metamodels.

References

1. Bretto, A.: Hypergraph theory. An introduction. Mathematical Engineering. Springer, Cham (2013). https://doi.org/10.1007/978-1-4419-9863-7_100650

2. Bundy, A., Wallen, L.: Partitioned semantic net. In: Bundy, A., Wallen, L. (eds.) Catalogue of Artificial Intelligence Tools, pp. 89–89. Springer, Berlin, (1984). https://doi.org/10.1007/978-3-642-96868-6_172

3. Dahchour, M., Pirotte, A., Zimányi, E.: Generic relationships in information modeling. In: Spaccapietra, S. (ed.) Journal on Data Semantics IV. LNCS, vol. 3730, pp. 1–34. Springer, Heidelberg (2005). https://doi.org/10.1007/11603412_1

4. Gamma, E., Helm, R., Johnson, R., Vlissides, J.M.: Design Patterns: Elements of Reusable Object-Oriented Software. Addison-Wesley Professional, 1 edn. (1994)

5. Hendrix, G.G.: Encoding knowledge in partitioned networks. In: Associative networks, pp. 51–92. Elsevier (1979)

6. Jodłowiec, M., Krótkiewicz, M.: Describing Semantics of data metamodels: a case study of association-oriented metamodel. In: Nguyen, N.T., Iliadis, L., Maglogiannis, I., Trawiński, B. (eds.) ICCCI 2021. LNCS (LNAI), vol. 12876, pp. 53–65. Springer, Cham (2021). https://doi.org/10.1007/978-3-030-88081-1_5

7. Jodłowiec, M., Krótkiewicz, M., Zabawa, P.: Fundamentals of generalized and extended graph-based structural modeling. In: Nguyen, N.T., Hoang, B.H., Huynh, C.P., Hwang, D., Trawiński, B., Vossen, G. (eds.) ICCCI 2020. LNCS (LNAI), vol. 12496, pp. 27–41. Springer, Cham (2020). https://doi.org/10.1007/978-3-030-63007-2_3

8. Jodłowiec, M., Pietranik, M.: Towards the pattern-based transformation of SBVR models to association-oriented models. In: Nguyen, N.T., Chbeir, R., Exposito, E., Aniorté, P., Trawiński, B. (eds.) ICCCI 2019. LNCS (LNAI), vol. 11683, pp. 79–90. Springer, Cham (2019). https://doi.org/10.1007/978-3-030-28377-3_7

9. Joslyn, C., Nowak, K.: Ubergraphs: a definition of a recursive hypergraph structure. arXiv preprint arXiv:1704.05547 (2017)

10. Karagiannis, D., Kühn, H.: Metamodelling platforms. In: EC-Web, vol. 2455, p. 182 (2002)

11. McQuade, S.T., Merrill, N.J., Piccoli, B.: Metabolic graphs, life method and the modeling of drug action on mycobacterium tuberculosis. arXiv preprint arXiv:2003.12400 (2020)

12. OMG: object management group, semantics of business vocabulary and rules 1.5 (2019). https://www.omg.org/spec/SBVR/1.5/

13. Qu, C., Tao, M., Yuan, R.: A hypergraph-based blockchain model and application in internet of things-enabled smart homes. Sensors **18**(9), 2784 (2018)

14. Voloshin, V.I.: Introduction to graph and hypergraph theory. Nova Science Publishers, New York (2009)

15. Wu, Z., et al.: Semantic hyper-graph-based knowledge representation architecture for complex product development. Comput. Ind. **100**, 43–56 (2018)

16. Yacoub, S.M., Ammar, H.H.: Pattern-Oriented Analysis and Design?: Composing Patterns to Design Software systems. Addison-Wesley, Boston (2004)

Obfuscating LLVM IR
with the Application of Lambda Calculus

Rei Kasuya[ID] and Noriaki Yoshiura[✉][ID]

Department of Information and Computer Sciences, Saitama University, 255,
Shimo-ookubo, Sakura-ku, Saitama, Japan
{rkasuya,yoshiura}@fmx.ics.saitama-u.ac.jp

Abstract. Software obfuscation is a method that complicates data
structures and algorithms in software to prevent software from being
analyzed. This paper proposes a method to obfuscate loop structures in
LLVM IR (LLVM is the abbreviation of "Low Level Virtual Machine"
and IR is that of "Intermediate Representation".) by applying a fixed-
point combinator in the lambda calculus. LLVM IR is an intermediate
representation in LLVM. A purpose of using intermediate representation
in this paper is to handle multiple programming languages and architectures. This paper also evaluates the proposed obfuscation method
by experiments. These experiments use loop programs, which are created artificially in this paper, and a practical program. The result of the
experiments in this paper shows that the proposed method can obfuscate
programs, but the execution times of the loop programs increase if these
programs are obfuscated by the proposed method. However, the execution times of the practical programs do not increase under the obfuscation
by the proposed method. It follows that the proposed method is available
as an obfuscation method for practical programs.

Keywords: Obfuscation · LLVM IR · Lambda Calculus

1 Introduction

The advance of reverse-engineering techniques enables to analyze or modify software [1,2,9]. Reverse-engineering techniques are, however, used for plagiarism of
intellectual properties from software. Software obfuscation is a method which
complicates data structures and algorithms in software to prevent software from
being unintentionally analyzed [8,10].

Software is classified in two types: executable files which are compiled from
source codes written in programming languages and source codes which are
executed by interpreters without being compiled. This paper focuses on the
former kinds of software and proposes a method of obfuscation for software.
Software obfuscation is classified in two types; one is to obfuscate source codes of
target software before compilation and the other is to obfuscate data or machine
codes in target software. Each of the two has disadvantages. The former method

© The Author(s), under exclusive license to Springer Nature Singapore Pte Ltd. 2023
N. T. Nguyen et al. (Eds.): ACIIDS 2023, LNAI 13996, pp. 27–39, 2023.
https://doi.org/10.1007/978-981-99-5837-5_3

requires a different obfuscation program for a different programming language used for the source codes of target software. For example, an obfuscation program for source codes written in C language cannot obfuscate source codes written in other programming languages. Similarly, the latter method requires a different obfuscation program for a different computer architecture because the syntax of machine codes are different according to computer architectures.

This paper proposes an obfuscation method using LLVM IR to resolve these problems. LLVM is a project to provide a base to develop compilers [5]. In particular, the LLVM Core library, a central subproject of LLVM, provides a unique intermediate representation which is called LLVM IR. A compiler using LLVM translates input source codes into LLVM IR codes and translates LLVM IR codes into machine codes of target architectures. *Intermediate representation* LLVM IR is a form of representation used in a translation process from source codes to machine codes. One purpose of using an intermediate representation is to optimize codes regardless of programming languages and machine architectures. Another purpose is that LLVM IR enables compliers to handle multiple programming languages and architectures.

The LLVM Core library also includes optimizers for LLVM IR and code generators for many architectures. Clang, a compiler developed in the LLVM project, can compile C, C++, and Objective-C. Third parties support translation tools from several programming languages into LLVM IR. For example, GHC, a DD facto standard compiler for Haskell, can translate Haskell codes into LLVM IR. Since LLVM IR is available for many programming languages and architectures, obfuscating LLVM IR codes can achieve obfuscating software written in multiple programming languages and compiled for multiple architectures.

This paper is organized as follows; Sect. 2 discusses the related works. Section 3 explains the purpose of this paper. Section 4 explains the obfuscation method of this paper. Section 5 describes the experiments and results. Section 6 discusses the result of the experiments. Section 7 concludes this paper.

2   Related Works

Pengwei et al. proposed an obfuscation method to replace integer comparison with functions applying lambda calculus [4]. Comparison operations often appear in conditional branches and therefore this method can significantly modify the control flow of the target software.

Pascal et al. proposed a method to obfuscate LLVM IR by complicating arithmetic operations, bitwise operations, the control flow, and function calls [3]. This method also implements tamper-proofing. Kyeonghwan et al. utilize the method of [3] to improve the resistance of Android applications against reverse-engineering [6]. Dongpeng et al. obfuscate LLVM IR with dynamic opaque predicates [11]. An *opaque predicate* is a complex tautology or contradiction. The most straightforward example is shown in Algorithm 1.

Although actual opaque predicates are more complex than the opaque predicate in Algorithm 1, one of the problems of opaque predicates is that static

---

Algorithm 1. Example opaque predicate

1: **if** $2x \bmod 2 = 0$ **then**   ▷ The control will pass through this branch regardless of the value of x.
2:    $\cdots$
3: **else**   ▷ The control will never pass through this branch, vice-versa.
4:    $\cdots$
5: **end if**

---

evaluation is possible. For example, the conditional expression $2x \bmod 2 = 0$ in Algorithm 1 is always true regardless of other program parts; thus, attackers can find the expression a tautology through static code analysis. To resolve this problem, *dynamic opaque predicates*, opaque predicates whose evaluation value might vary by each execution, have been proposed. The evaluated value of dynamic opaque predicates might vary with each execution of programs, but the programs do not change their outputs. It is difficult for attackers to detect opaque predicates.

3 Purpose

This paper proposes the method that obfuscates loop structures in LLVM IR codes by applying a fixed-point combinator in the lambda calculus. Utilizing the lambda calculus to obfuscate LLVM IR is similar to Pengwei et al [4].

Many programming languages, including C, belong to a paradigm of imperative languages. Each program has "mutable states" in imperative programming, and computers modify the states with instructions to achieve computation. LLVM IR and machine languages are also imperative languages although syntax of machine languages vary by their architecture.

Lambda calculus is a model of computation that has impacted many functional programming languages, including Haskell. The lambda calculus is calculated in a completely different way from imperative languages such as C. Thus, using the lambda calculus for obfuscation is significantly more challenging. For attackers without knowledge of the details of the proposed method, it is difficult to restore an obfuscated code to its original state.

The proposed method obfuscates loop structures in LLVM IR codes with the lambda calculus and modifies the control flow of input codes significantly. Since Pengwei et al. obfuscated integer comparison by the lambda calculus, obfuscations by the proposed method are more difficult for attackers than obfuscations by the method of Pengwei et al.

4 Method

This paper uses the lambda calculus and combinators in [7]. Because of the shortage of space, this section explains only Z-combinator. Z-combinator is widely

---

Algorithm 2. Calculate sum of 1 to 10 using fixed point combinator

1: **function** $V(F)$
2:      **return**
3:      **function** (x, i)
4:          **if** $i < 10$ **then return** $F(x + i + 1, i + 1)$ **else return** x **end if**
5:      **end function**
6: **end function**
7: FIXED_POINT_COMBINATOR$(V)(0, 0)$

---

known as an actual instance of fixed-point combinators in the lambda calculus. The definition of the Z-combinator is as follows:

$$Z = \lambda f.(\lambda x.f(\lambda y.xxy))(\lambda x.f(\lambda y.xxy))$$

The method proposed in this paper uses the Z-combinator implemented with C++. The proposed method first obfuscates input LLVM IR code and links it with the program implementing Z-combinator. In lambda calculus, this fixex-point combinator calculates loop program. For example, by using Z-combinator (fixed-point combinator), the function calculating the sum of 1 to 10 is given in Algorithm 2.

Next, we explains loop structures in LLVM IR. An LLVM IR code is described in the form of a control flow graph. Blocks and edges in the control flow graph are as follows;

Header block
Header block is an entrance block to a loop. Every node outside a loop has no edges with nodes other than a header block in the loop

Entering blocks
Entering block of a loop is a block which is outside the loop and with edges into an header block in the loop. When a loop has only one entering block and the entering block has exactly one edge, the entering block is called preheader block.

Latch blocks
Latch block is a node in a loop and has an edge into the header block of the loop.

Backedges
Backedge is an edge from a latch block into a header block.

Exiting edges
Exiting edge is an edge from a node of a loop in a node outside the loop. The source block of an exiting edge is called exiting block and the destination block of an exiting edge is called exit block.

The proposed method obfuscates only loops that have a preheader block and exactly one exit block. However only the latter condition should be satisfied because the LLVM optimizer can modify every loop so that the loop has a preheader block.

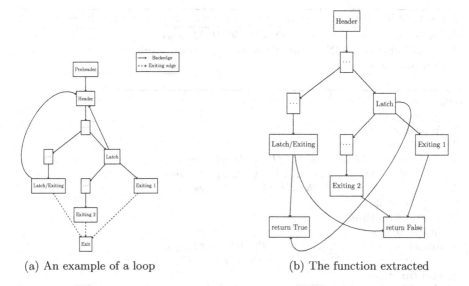

(a) An example of a loop   (b) The function extracted

Fig. 1. Obfuscating loop structure

The following explains the procedure of obfuscation by using the example loop in Fig. 1(a). Since the loop in Fig. 1 (a) has a preheader block and exactly one exit block, the loop satisfies the two conditions.

First, the procedure of obfuscation extracts instructions of the loop and defines a function for the instructions. Let this function be named "extracted." Every time the function extracted is called, it executes in-loop instructions only once and returns a truth value for representing whether the instructions should run again. More specifically, the procedure extracts blocks between the preheader block and the exit blocks, adds the blocks, "return True" and "return False" as shown in Fig. 1 (b). The created blocks consist of the function "extracted". The return value of the function "extracted" represents repeating loop or not. In the original loop structure, the control passes through a backedge when returning to the header block. Similarly, in the extracted function, the control passes through an exiting edge when exiting the loop; the return value is true when executing the loop again. The extracted function cannot refer to the variables outside the loop. If the extracted function requires to access the variables outside the loop, the variables are passed to the function extracted as arguments.

Next, the procedure of obfuscation replaces the loop by "looper function", which is described in Algorithm 3. The looper function takes, as arguments, the pointer to the extracted function and the variables required in the extracted function. The looper function, as its name suggests, executes a loop. The looper function repeatedly calls the extracted function as long as it returns "True". Algorithm 3 is the most straightforward implementation.

Algorithm 4 shows the implementation of the looper algorithm with fixed point combinator. When the loop count is enormous, a stack overflow would

---

Algorithm 3. Example implementation of looper function using while statement

```
1: function LOOPER(FUNCTION_TO_REPEAT, args ...)
2:     whileFUNCTION_TO_REPEAT(args ...) end while
3: end function
```

Algorithm 4. Example implementation of looper function using fixed point combinator

```
1: function V(F)
2:     return
3:     function (FUNCTION_TO_REPEAT, args ...)
4:         if FUNCTION_TO_REPEAT(args ...) then
5:             F(FUNCTION_TO_REPEAT, args ...)
6:         end if
7:     end function
8: end function
9: function LOOPER(FUNCTION_TO_REPEAT, args ...)
10:     Z_COMBINATOR(V)(FUNCTION_TO_REPEAT, args ...)
11: end function
```

---

happen in Algorithm 4. Therefore, in the actual implementation of the looper function with fixed point combinator, loop execution would be switched to the while-loop algorithm like Algorithm 3 when the recursion count exceeds 8192. The pseudo-code of the actual implementation is in Algorithm 5.

5   Experiment

This section evaluates the proposed method by two experiments.

Variation with the Depth of the Loop Nesting
The looper function runs per a loop in the original code. For example, when obfuscating a double-nested loop shown in Algorithm 6, two extracted functions would be created and the looper function would run twice, as shown in Algorithm 7. The number of looper function calls increases as the nest of the loop increases; the execution time would presumably increase due to the overhead of function calls. This experiment confirms that the execution time would increase, by measuring the execution time of obfuscated programs. This experiment uses single, double, triple, quadruple and sextuple nested loop codes. Figure 2 shows only double and triple nested codes because of the shortage of space.

Each source code has a loop of different nesting depths, but the total numbers of loop iterations are the same in all of the source codes. All of the source codes execute 4096 iterations of loop. For example, the total number of loop iterations in the program of a double-nested loop is $64^2 = 4096$ and the total number of loop iterations in the program of the sextuple-nested loop is also $4^6 = 4096$.

Additionally, the numbers of loop iterations in each nest level are the same in each of the source codes. For example, in Code 1 (double-nested loop) in Fig. 2, the number of loop iterations in each nest level is 64, and in Code 2 (triple-nested loop) inf Fig. 2, the number of loop iterations of each nest level is 16.

Algorithm 5. Implementation of looper function

1: **function** V(F)
2:     **return**
3:     **function** ($recursion\_count, FUNCTION\_TO\_REPEAT, args \ldots$)
4:         **if** $recursion\_count < 8192$ **then**
5:             **if** FUNCTION_TO_REPEAT($args \ldots$) **then**
6:                 F($recursion\_count + 1, FUNCTION\_TO\_REPEAT, args \ldots$)
7:             **end if**
8:         **else while** FUNCTION_TO_REPEAT($args \ldots$) **do end while**
9:         **end if**
10:     **end function**
11: **end function**
12: **function** LOOPER($FUNCTION\_TO\_REPEAT, args \ldots$)
13:     Z_COMBINATOR(V)($0, FUNCTION\_TO\_REPEAT, args \ldots$)
14: **end function**

Algorithm 6. Double-nested loop before obfuscation

1: $i \leftarrow 0, j \leftarrow 0$
2: **while** $i < 10$ **do while** $j < 10$ /* some calculations */ $j \leftarrow j + 1$ **end while**
3:     $i \leftarrow i + 1$
4: **end while**

In the experiment, the total number of loop iterations is 4096 in order to unify the total number of loop iterations in each program and the number of loop iterations in each nest level. The minimum number for the unification is 4096 because the first, second, third, fourth and sixth roots are all integers. If we use a quintuple-nested loop additionally in the experiment, the minimum number of loop iterations is 2^{60}. However, it is unfeasible to execute 2^{60} iterations. Thus, the experiment uses 4096 iterations.

The experiment consists of the three parts. The first part in the experiment measures the execution times of these loop programs which are compiled with and without the proposed obfuscation method. Every programs are run 500 times and the experimental results are the averages of 500 runs.

The second part in the experiment evaluates the performance of the proposed method quantitatively. This evaluation is based on the number of edges in the callgraph, the number of blocks and edges in the control flow graph, and the cyclomatic complexity of the control flow. The *cyclomatic complexity* of a given graph is an index of the complexity of the graph. The cyclomatic complexity is defined as $E - V + 2C$, where E is the number of edges, V is the number of vertices, and C is the number of components. Each component in a control flow graph is a graph of each individual function. The cyclomatic complexity represents the number of independent passes on the graph. The higher the cyclomatic complexity of the control flow graph of a program is, the more complex the program structure is. Cyclomatic complexity is widely used as an indicator of software complexity and maintainability.

Algorithm 7. Double-nested loop after obfuscation

1: **function** EXTRACTED1(i, j)
2:     LOOPER$(EXTRACTED2, j)$
3:     $i \leftarrow i + 1$
4:     **if** $i < 10$ **return** True **else return** False **end if**
5: **end function**
6: **function** EXTRACTED2(j)
7:     /* some calculations */
8:     $j \leftarrow j + 1$
9:     **if** $j < 10$ **return** True **else return** False **end if**
10: **end function**
11: $i \leftarrow 0, j \leftarrow 0$
12: LOOPER$(EXTRACTED1, i, j)$

Code 1.1: Double-nested loop

```
int main(void)
{
    int i, j;
    int x = 0;
    for (i = 0; i < 64; ++i) {
        for (j = 0; j < 64; ++j) {
            ++x;
        }
    }
    return 0;
}
```

Code 1.2: Triple-nested loop

```
int main(void)
{
    int i, j, k;
    int x = 0;
    for (i = 0; i < 16; ++i) {
        for (j = 0; j < 16; ++j) {
            for (k = 0; k < 16; ++k) {
                ++x;
            }
        }
    }
    return 0;
}
```

Fig. 2. Source codes of the tested loop programs

The third part measures the distribution of machine instructions in the programs with and without obfuscation. If the instruction distributions significantly differ in the programs with and without obfuscation, it might be a clue for attackers to the detail of the obfuscation method. Therefore, the desired result is no significant difference in the instruction distribution in the programs with and without obfuscation.

Performance Measurement of a Practical Program
This paper also evaluates the practicality of the proposed method by implementing a program of SHA-256, which is a well-known hash algorithm, and measuring the execution times and the instruction distributions of the SHA-256 programs with and without obfuscation. The experiment results are the average time of 500 runs of the programs on random byte sequences of 100kB as an argument.

5.1 Experimental Results

Variation with the Depth of the Loop Nesting

In Table 1, *real* refers to the elapsed time from the start to the end of the program, *user* refers to the time the CPU was running in user mode during program processing, and *sys* refers to the time the CPU was running in kernel mode during program processing.

Table 1. Execution time (ms) with the depth of loop nesting

	No Obfuscation						Obfuscation					
Depth	1	2	3	4	5	6	1	2	3	4	5	6
real	5.5	5.3	5.3	5.4		5.4	13	8.2	8.3	8.5		9.4
user	4.1	4.0	3.8	4.0		4.0	6.6	5.9	6.1	6.2		6.8
sys	1.9	1.8	1.9	1.9		1.9	6.8	2.8	2.6	2.8		3.1

Table 2. Complexity indicators with the depth of the loop nesting

	No Obfuscation						Obfuscation					
Depth	1	2	3	4	5	6	1	2	3	4	5	6
edges in the callgraph	0	0	0	0		0	67	68	69	70		72
basic blocks	5	9	13	17		25	81	84	87	90		96
edges in the control flow graph	5	10	15	20		30	51	54	57	60		66
Cyclomatic complexity	2	3	4	5		7	72	74	76	78		82

Table 1 shows that the nesting depth does not significantly affect real, user or sys without obfuscation. On the other hand, regarding obfuscation, the single-nested loop has quite long execution times for all real, user and sys, and from the double-nested loop to sextuple-nested loop, execution times gradually increase for all real, user and sys. Table 2 shows that all four indicators significantly increase and that the proposed method increases the resistance of the programs against reverse-engineering.

Figure 3 shows the instruction distributions of the loop programs. The vertical axis of each graph is the percentage of occurrences. In the graphs, vertical crosses show the data of the programs without obfuscation and diagonal crosses show the data of the programs with obfuscation. The number of the "mov" instruction significantly increases in all nesting depths. On the other hand, the numbers of some other instructions decrease vice-versa.

Performance Measurement of a Practical Program

Table 3.(a) shows that the times of real and user increase around twofold and the time of sys remain almost unchanged. Table 3.(b) shows that all four indicators increase; this is the same as the experimental result of the loop programs. Figure 4 shows that the instruction distribution does not significantly differ with and without obfuscation.

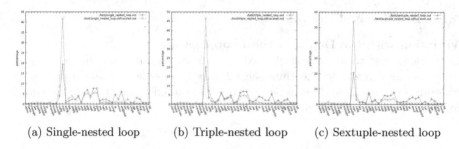

| (a) Single-nested loop | (b) Triple-nested loop | (c) Sextuple-nested loop |

Fig. 3. Variations of the instruction distributions of the loop programs

Table 3. Experiment results of the SHA-256 program

	No obfus-cation	Obfuscation
real	68 ms	140 ms
user	66 ms	130 ms
sys	23 ms	25 ms

(a) Execution time

	No obfus-cation	Obfuscation
edges in the callgraph	191	258
basic blocks	283	330
edges in the control flow graph	210	220
Cyclomatic complexity	239	296

(b) Complexity indicators

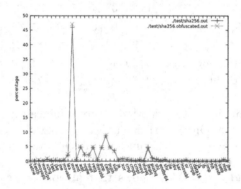

Fig. 4. Variations of the instruction distributions of the SHA-256 program

6 Discussion

6.1 Variation with the Depth of the Loop Nesting

The nesting depth did not significantly affect execution time without obfuscation in all five loop programs. However, with obfuscation, the execution time of the single-nested loop program was significantly longer than that of the other loop programs, and from the double to sextuple-nested loop program, the execution time increased gradually with the nesting depth.

We guessed that the execution time would increase as the nesting depth increased, but the measured execution time of the single-nested loop program with obfuscation was not as expected. A candidate for the causes is a difference in the depth of function calls. For example, the function call depth of the single-nested loop reaches a maximum of 4096 since the program recurses 4096 times. On the other hand, the function call depth of the double-nested loop program reaches only a maximum of $64 + 64 = 128$. This difference in function call depth might be related to the higher than expected increase in execution time.

Next, we discuss the effects and costs of obfuscation. Obfuscation significantly increases all four complexity indicators but execution time does not increase with program complexity significantly. In particular, the execution time of the single-nested loop program has more than doubled due to obfuscation and it is necessary to reduce the increase in the execution time by improving the obfuscation method. In both methods by Pascal et al. (2015) and Pengwei et al. (2017), users can specify the percentage of how much the obfuscation is to be performed, rather than obfuscating the whole code. Since the proposed method aims to prevent reverse-engineering, it is more effective to obfuscate only the parts of the program whose processing details should not be known to third parties rather than obfuscating all loops in the program. In order to obfuscate some parts of programs, we can use source code annotations and specify the parts that should be obfuscated.

Finally, we discuss the difference in the instruction distribution with and without obfuscation. The method proposed by Pengwei et al. (2017) does not make a significant difference in the instruction distribution with and without obfuscation. Making no difference is the reason why an attacker presumably cannot know the details of obfuscations. In contrast, the proposed method makes non-negligible differences in the instruction distribution; for example, the proposed method makes significant differences in the occurrence rates of "mov" instruction. These differences may provide clues for an attacker who would like to analyze programs by reverse-engineering.

6.2 Performance Measurement of a Practical Program

We discuss the result of experiment of a practical program, SHA-256 program. The four complexity indicators increased in the obfuscated SHA-256 program, but the execution time also nearly doubled. As mentioned in Sect. 6.1, specifying the parts which should be obfuscated in the programs may prevent an increase in the execution time of an obfuscated program.

Compared with the experiment of the loop programs, instruction distributions between the SHA programs with and without obfuscation are not significantly different. This result is an advantage of the proposed method.

This discuss indicates that for a small program consisting such loop programs, a difference in instruction distribution is different before and after obfuscation. However it is expected that the difference is unnoticeable when the sizes of programs exceed a certain level.

7   Conclusion

This paper proposed the method which obfuscates loop structures in LLVM IR by applying a fixed-point combinator in the lambda calculus. As a result, the complexities of the programs significantly increase, but the increase in the execution time of the program is not negligible. However specifying parts which should be obfuscated may resolve this problem.

An instruction distribution is significantly different between programs with and without obfuscation if the programs are relatively small. For programs whose sizes are relatively large, a significant difference before and after obfuscation does not appear. Since practical programs are relatively large, the proposed method does not have a disadvantage for practical programs.

Future work includes the development of more effective obfuscation methods, as described in Sect. 6, in order to reduce the increase in execution time. Another future work is to propose another obfuscation methods for other control structures such as conditional branches.

References

1. Bhardwaj, V., Kukreja, V., Sharma, C., Kansal, I., Popali, R.: Reverse engineering- a method for analyzing malicious code behavior. In: 2021 International Conference on Advances in Computing, Communication and Control (ICAC3), pp. 1–5 (2021)
2. Jain, A., Gonzalez, H., Stakhanova, N.: Enriching reverse engineering through visual exploration of Android binaries. In: Proceedings of the 5th Program Protection and Reverse Engineering Workshop (PPREW-5), pp. 1–9 (2015)
3. Junod, P., Rinaldini, J., Wehrli, J., Michielin, J.: Obfuscator-LLVM - software protection for the masses. In: Proceedings of the 2015 IEEE/ACM 1st International Workshop on Software Protection, SPRO 2015, pp. 3–9 (2015)
4. Lan, P., Wang, P., Wang, S., Wu, D.: Lambda obfuscation, security and privacy in communication networks. In: 13th International Conference, SecureComm 2017, Proceedings, pp. 206–224 (2017)
5. Lattner, C., Adve, V.: LLVM: a compilation framework for lifelong program analysis & transformation. In: Proceedings of International Symposium on Code Generation and Optimization, pp. 75–86 (2004)
6. Lim, K., et al.: An anti-reverse engineering technique using native code and obfuscator-LLVM for Android applications. In: Proceedings of the International Conference on Research in Adaptive and Convergent Systems, RACS 2017, pp. 217–221 (2017)
7. Roger Hindley J., Seldin, J.P.: Lambda Calculus and Combinators: An Introduction (2nd. ed.). Cambridge University Press, Cambridge (2008)
8. Schrittwieser, S., Katzenbeisser, S., Kinder, J., Merzdovnik, G., Weippl, E.: Protecting software through obfuscation: can it keep pace with progress in code analysis? ACM Comput. Surv. 49(1), 1–37 (2016)
9. Votipka, D., Rabin, S., Micinski, K., Foster, J.S., Mazurek, M.L.: An observational investigation of reverse engineers' processes. In: The Proceedings of 29th USENIX (2020)

10. Kang, S., Lee, S., Kim, Y., Mok, S.K., Cho, E.S.: OBFUS: an obfuscation tool for software copyright and vulnerability protection. In: Proceedings of the Eleventh ACM Conference on Data and Application Security and Privacy (CODASPY 2021), pp. 309–311 (2021)
11. Xu, D., Ming, J., Wu, D.: Generalized dynamic opaque predicates: a new control flow obfuscation method. In: 19th Internation Conference on Information Security, pp. 323–342 (2016)

Structural and Compact Latent Representation Learning on Sparse Reward Environments

Bang-Giang Le, Thi-Linh Hoang, Hai-Dang Kieu, and Viet-Cuong Ta(✉)

HMI Laboratory, VNU University of Engineering and Technology, Hanoi, Vietnam
cuongtv@vnu.edu.vn

Abstract. For the task of training RL agent in a sparse-reward, image-based observation environment, the agent should perfect both learning latent representation and having a good-exploration strategy. Standard approaches such as variational auto-encoder (VAE) could learn such representation. However, these approaches are only designed to encode the input observations into a pre-defined latent distribution and do not take into account the dynamics of the environment. To improve the training process from high-dimensional input images, we extend the standard VAE framework to learn a compact latent representation that can mimic the structures of the underlying Markov decision process. We further add an intrinsic reward based on the learned latent to encourage exploratory actions in the sparse reward environments. The intrinsic reward is designed to direct the policy to visit distant states in the latent space. Experiments on several gridworld environments with sparse rewards are carried out to demonstrate the effectiveness of our proposed approach. Compared to other baselines, our method has more stable performance and better exploration coverage by exploiting the learned latent structure property.

Keywords: Image-based Observation · Compact Latent Representation · Sparse-Reward · Exploration

1 Introduction

Reinforcement learning (RL) has been demonstrated to be an effective framework to solve highly complex, stochastic decision-making problems [1]. While achieving impressive successes, training RL agents generally requires a large number of environment samples to converge to good performance, especially in environments with high-dimensional observations, such as images, and sparse rewards environments. Having an effective algorithm learning from pixel open up greater opportunities to apply RL in practical application, while effective exploration in sparse reward environments eliminates the need to hand design the reward functions. Researches in images-based environments focus on reducing the number of high dimensions of the input by learning of a compact latent

© The Author(s), under exclusive license to Springer Nature Singapore Pte Ltd. 2023
N. T. Nguyen et al. (Eds.): ACIIDS 2023, LNAI 13996, pp. 40–51, 2023.
https://doi.org/10.1007/978-981-99-5837-5_4

representation, recent works in this approach include employing encoders by reconstruction or model learning of the environment [2,3]. On the other hand, approaches on sparse-reward environments largely involve the estimate of the novelness of states and an incentive to increase the likelihood of visiting new states.

A good latent representation can reduce the redundancy, and filter out irrelevant features and noise present in the data that can be leveraged for efficient learning. Many works tackle the representation learning of RL by using a popular idea of using reconstruction loss in unsupervised learning, which is achieved by encoder-decoder network architecture. A typical choice for learning latent representation is to use auto-encoder or variational auto-encoder (VAE) [4,5]. However, the latent learned in this way is messy and hardly contains information about the structure of the Markov Decision Process (MDP). In addition to that, it is difficult to employ VAE in the training of RL agents directly due to the stochastic nature of VAE [2]. To improve the latent representation, several works propose to impose predefined structures on the learned latent space, for example by learning a linear topology on the embedding space [6], or learning temporally structured state representations [7]. In training with sparse-reward environments, good latent representation can enhance the exploration process. Recent works in this approach include partitioning the continuous latent space into a discrete space [8] or prediction-error [9]. As such, many of the exploration methods fall under the category of using bonus rewards for intrinsic exploration [8–12].

In this work, we focus on the learning of a smooth, compact latent representation which can be used to enhance exploration in on-policy algorithms based on the reconstruction of pixel observations. For learning a compact representation, we propose to use VAE coupled with L2 regularization of adjacent states that are reachable from each other on the MDP and a procedure to prevent latent collapsing by a clipping distance trick. To improve exploration, we propose an intrinsic reward schema based on the learned embedding landscape which encourages visiting novel latent states. we conduct experiments and analyses on classical image-based, sparse reward environments to evaluate our proposed method. Among the tested baselines, our approach demonstrates that the learned latent can help to speed up the training process and solve the environment efficiently. Further analysis results highlight that by learning a proper and compact latent representation of the input observation, the agent could use to connect the exploration in the latent space and the MDP environment.

2   Related Works

There are a number of works that explore the direction of using unsupervised objectives and subtasks to learn a representation of latent space. In model-free RL, unsupervised learning has been used for obtaining latent representation to improve data efficiency. Yarats et al. [2] propose SAC-AE that is built on SAC [13] coupled with an auto-encoder to learn latent embeddings of the environment pixel observation. Nair et al. [14] employ VAE for representation learning and use it to sample goals for practice in goal-conditioned RL problems. Shelhamer [3] considers various self-supervised tasks for representation learning. Non-reconstruction loss for state representation has also been studied in [6,15]. In model-based RL, latent learning is also rendered to be an effective approach. Hafner et al. [16] learn a transition model on the latent states.

Sparse-reward problem in RL poses additional challenges for the learning agents in that they have to find high-reward state regions efficiently. A common approach for exploration is to inject a random noise term into the action of agents, popular methods in this direction include ϵ-greedy exploration [17] and entropy regularization [13,18]. In the discrete settings, the famous upper confident bound algorithm uses an intrinsic reward derived from the visitation counts of a state to encourage novel state visits [19]. Several counting-based methods have been extended to work with continuous, infinite state space and deep neural networks, including hashing [8] and pseudo-count [10]. Another line of work in exploration RL relies on the predictive errors of neural networks to exploration such as using a forward dynamic model in [9]. Other methods use a memory-based approach: by using the data collected from the past to decide on exploration strategies, both with a global memory [20] and episodic memory [21].

The structure of the latent space can be enforced in several ways. For example, Kim et al. [6] introduce the idea of imposing a linear structure on latent dynamic model, offloading the modeling burden to the encoder. Ermolov et al. [7] propose to minimize the L2 norm on the whiten latent representations of temporally close observations, effectively grouping temporally related observations latent close to each other. Raileanu et al. [22] demonstrate that the distances of the latent vectors of states can be utilized to provide information of adjacent latent vectors to motivate exploration. Our method also learns latent representation that embeds the structures of tasks, in that we encourage subsequent state representations in the MDP to be close, but our method focuses on the latent space in which the distances between state representations correlate with the number of steps taken by the agent to traverse between them.

3   Background

3.1   Fully and Partially Observable MDP

A standard definition of (Fully Observable) Markov decision process (MDP) can be described by a tuple $M = \langle S, A, P, R, \gamma \rangle$, with S the state space, A is the action space, $P(s'|s, a)$ the transition dynamic that denotes the transition probability from one state s to the other s' when a certain action a is taken, $r_t = R(s_t, a_t)$ is the reward signal at each time step t and γ the discount factor. Traditional RL objective maximizes the expected cumulative discounted rewards as $\sum_{t=0} \mathbb{E} \gamma^t r_t$.

The property of a Fully Observable MDP requires the state representations to contain all information needed to take actions. When working with image observations, the environment can be formulated as a Partially Observable Markov Decision Process (POMDP). In this framework, instead of the low-dimensional states, the agent only observes the observation o_t following conditional probabilities of s_t at each timestep. The shifting from states to observations can make the learning of RL agent more difficult, as observations can be incomplete, noisy and contain redundant information that could mislead the training algorithms. Closing the gap of performance between image-based observation and true state environments is a challenging problem and an active research area that attract a lot of attention [2,23].

3.2   Proximal Policy Optimization

PPO [18] is an on-policy, policy gradient algorithm that emulates the trusted region principle by using a surrogate objective with clipped probability ratios to perform pessimistic gradient ascend that prevents destructively large policy updates. The objective of PPO is

$$J^{\text{CLIP}}(\pi) = \mathbb{E}[\min(r\hat{A}_{\text{old}}, \text{clip}(r, 1 - \epsilon, 1 + \epsilon)\hat{A}_{\text{old}})],$$

with $\hat{A}$ is an estimator of the advantage function, r is the importance sampling ratio and the hyperparameter ϵ controls how conservative each policy update is.

3.3   Latent Encoding via AE and VAE

An autoencoder (AE) is characterized by a pair of two parameterized families of functions, the encoder, p_θ, and the decoder, q_ϕ, co-operate in a joint optimization problem in which one network attempts to encode complex, high-dimensional input data into a lower dimensional latent space while the other learns to recover the original data from the encoder's latent vectors. The reconstruction loss is

$$\mathcal{L}_{\text{AE}} = \mathbb{E}_{x \sim \mathcal{D}} c(x, q_\phi(p_\theta(x)))$$

where c is a cost metric that measures the difference between the original data and the reconstructed one, which is typically mean squared error. The observational data x is drawn from the dataset $\mathcal{D}$.

On the other hand, variational autoencoder (VAE) [4] follows a similar structure of AE, but with a different theoretical formulation. VAE aims to model the joint distribution of the original data and hidden latent variables z by maximizing the lower bound of the log-likelihood, called the Evidence Lower Bound (ELBO)

$$\mathbb{E}_{x \sim \mathcal{D}} \log p_\theta(x) \geq \mathbb{E}_{x \sim \mathcal{D}}[\mathbb{E}_{q_\phi(z|x)} \log p_\theta(x|z) - \text{KL}(q_\phi(z|x) \| \pi(z)))]. \tag{1}$$

In practice, the prior distribution $\pi(z)$ and the posterior $q(z|x)$ are both assumed to be Gaussian distributions to make the computation tractable. It is often convenient to add a coefficient to the KL term to trade off between reconstruction and regularization of VAE, which results in β-VAE [5]. When the decoding distribution is modeled as a Gaussian, the log-likelihood term in Eq. 1 reduces to the Mean Squared Error Loss.

4   Method

Our proposed method can be split into two parts (Fig. 1). Firstly, an encoder encodes environmental states to a latent space. The latent code should preserve necessary information from the original observations so that they can be recovered, which is realized via the reconstruction term $\mathcal{L}_{RC}$. For learning the structure of the latent space, we add an L2 regularization $\mathcal{L}_{ADJ}$ that pushes adjacent latent states z, z_{t+1} on the same trajectories closed together.

Secondly, based on the design of the learning latent scheme, the learned representation is expected to reflect the temporal structure of the transition dynamics, in the sense that the distances of different state encodings roughly correlate with their distances in terms of the number of timesteps needed to reach each other. Using this intuition, our exploration strategy is designed to target the remote states on the latent manifold. The agent is trained with PPO using rewards $\overline{r_t}$ combined from the intrinsic exploration bonus derived from distances between consecutive latent vectors together with the rewards r_t from the environment.

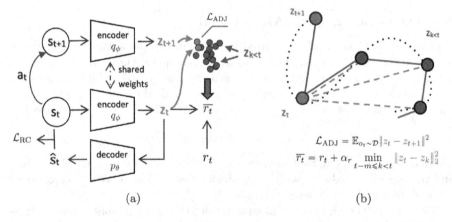

(a)   (b)

Fig. 1. Illustration of our method: a) latent representation via reconstruction loss, with an adjacent norm to pull close states together; b) contraction objective $\mathcal{L}_{ADJ}$ for adjacent states and auxiliary rewards $\overline{r_t}$ by combining extrinsic reward r_t and the intrinsic reward for exploration in latent space. Red solid lines denote latent distances of neighboring states used in $\mathcal{L}_{ADJ}$ and green dash lines indicate distances of current (yellow nodes) and past latent states (blue nodes), both within the same trajectory episode.

4.1   Learning Compact Latent Representations

When working with high-dimensional data, such as images, learning by solely relying on the gradient signal from the RL objective is usually slow and inefficient [2]. To improve efficiency, unsupervised and self-supervised learning with auxiliary tasks can be applied.

More specifically, we aim at learning an encoder q_ϕ parameterized by weights vectors ϕ. The encoder takes input as the image observation o_t and maps it to latent encoding z_t. The dimension of the encoding is generally much lower than that of the original observation. The encoding function should filter out noise and redundancy in the original observation. The training of this encoding function is unsupervised via reconstruction loss from another decoder network p_θ, which alleviates the sparsity problem of RL rewards. RL algorithms that operate on the learned latent space have an easier time learning the tasks at hand, since it has access to a dense-information representation.

To employ exploration on the latent space, we propose to learn a compact representation that closely reflects the MDP structure by enforcing adjacent states in a trajectory to be similar to the latent representations. This can be done by adding an additional constraint on the objective for encoding the latent embeddings. Concretely, we use an L2 norm regularization on the difference between the latent vectors of the two consecutive observations from the same trajectories. The training objective for learning encoder-decoder networks is as follows

$$\mathcal{L}_{\text{RC}} = \mathbb{E}_{o_t \sim \mathcal{D}} \log p_\theta(o_t|z_t) + \beta D_{\text{KL}}(q_\phi(z_t|o_t)\|\pi(z))) \tag{2}$$

$$\mathcal{L}_{\text{ADJ}} = \mathbb{E}_{o_t \sim \mathcal{D}}\|z_t - z_{t+1}\|^2 \tag{3}$$

$$z_t = q_\phi(z|o_t) \qquad \forall o_t \in \mathcal{D}$$

$$\mathcal{L}_{\text{Latent}} = \mathcal{L}_{\text{RC}} + \alpha \mathcal{L}_{\text{ADJ}} \tag{4}$$

where $\mathcal{L}_{\text{RC}}$ and $\mathcal{L}_{\text{ADJ}}$ denote reconstruction loss and adjacent loss from the encoding-decoding networks, respectively. z_t and z_{t+1} are two temporally consecutive latent states and α the coefficient that tradeoffs between reconstruction and compactness. In Eq. 2, we employ β-VAE instead of the standard VAE due to its better control on the regularization and reconstruction tradeoff. A replay buffer $\mathcal{D}$ is used to store recent observations for the training of the encoding network.

The overall training pseudocode is described in Algorithm 1. The inclusion of the value error in the objective of the encoder encourages it to learn representations that are relevant to current tasks. As suggested by Yarats et al. [2], we do not let the gradient from policy loss through the encoding network. With the assumption that our encoder q_ϕ could reflect the dynamics of the environment, an auxiliary reward $\overline{r_t}$ is added to enhance the exploration effects. The intrinsic rewards $\overline{r_t}$ is discussed in details in the next section.

Algorithm 1. Pseudocode for intrinsic reward from latent distances with PPO

Input: encoder q_ϕ, decoder p_θ, batch size M, PPO networks (policy π, value function V), buffer $\mathcal{D}$ for training encoder-decoder.

1: **for** each iteration i **do**
2:     Collect batch of transitions $B_i = (s_t, a_t, s_{t+1})_M$, append to $\mathcal{D}$
3:     Encode batch of samples (s_t, s_{t+1}) by encoder q_ϕ to obtain z_t, z_{t+1}
4:     Update the reward r_t with intrinsic reward from Eq. 5 to obtain $\overline{r_t}$
5:     Optimize $\overline{r_t}$ by PPO with on-policy batch B_i
6:     Train encoder q_ϕ and decoder networks p_θ with samples from $\mathcal{D}$
7: **end for**

4.2   Exploration on the Learned Latent Space

In the latent space, neighboring states are close to each other due to the contraction term in the objective (Eq. 3). To encourage the agent to visit novel states, the L2 norm of the latent vectors can be taken as intrinsic reward signals. The total reward at each timestep is calculated as the sum of the extrinsic rewards (the rewards from the environment) and the intrinsic rewards to the previous steps,

$$\overline{r_t} = r_t + \alpha_r \min_{t-m \le k < t} \|z_t - z_k\|_2^2, \tag{5}$$

with α_r the intrinsic reward coefficient. By taking into account the minimum distance from past states, the agent is incentivized to explore regions that are distinct from early states. When the number of timesteps grows large or the environments are infinite horizon, we can truncate it to a window of a particular size m. Also, this window size can be treated as a hyperparameter that can be used to control the locality of the exploration. The reward obtained from Eq. 5 by adding an intrinsic motivation is then optimized by PPO algorithm. Overall, the dynamics of the training process of our method forms a competing process; the encoder attempts to project the observations into a compact space, while the policy learns to reach states with distant latent encodings.

The introduction of the L2 adjacent norm on the latent states (5) combined with a powerful decoder can cause degenerate solutions to occur, in which case the latent states collapse to almost constant vectors. This makes the training of the policy and value networks, which usually have less capacity than the decoder, more difficult, as they can not differentiate states as well as the decoder. Although the intrinsic rewards in (5) could help the policy sample from lower-density regions to some extent, it does not guarantee that the L2 adjacent norm would not move to zero-constant during the training progress.

In this work, we instead employ a modification on the objective of (3). More specifically, the adjacent term only penalizes latent distances that exceed a certain threshold. Our modified adjacent norm regularization is then:

$$\mathcal{L}_{ADJ} = \mathbb{E}_{o_t \sim \mathcal{D}} \max(\|z_t - z_{t+1}\|_2 - \epsilon, 0)^2 \tag{6}$$

where ϵ is the clipping threshold that decides on how close the two consecutive representations should be. Latent states that live inside the threshold distance incur no compact loss and should spend their modeling capacity on other learning signals. Thus the value of ϵ is highly correlated with the desired density of the learned latent and the capacity of the encoder-decoder networks.

5 Experiments

5.1 Environments and Configs

Environments. For the experiments in this section, we use several environments in MiniGrid [24]. An agent has access to a partial view of the maze; at each timestep, the agent can take one in seven possible actions. Many of the environments in MiniGrid are sparse rewards; only upon reaching the goal that the agent receives a non-zero reward signal. For the purpose of the experimental analysis, we consider three environments *FourRoom*, *DoorKey8×8* and *Simple-CrossingS9N3*. All the experiments are conducted on the image observational space, input images are resized at a fixed size of $86 \times 86 \times 3$, with three channel RGB pixels. Also, since the pixel observations provide a top-down view of the map, the nature of the environment is no longer partially observable, but is still highly complex. As such, we do not use recurrent networks in our experiments, and rely solely on the encoder to extract all relevant information from raw image pixels (Fig. 2).

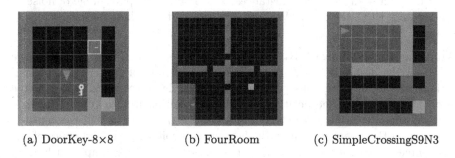

(a) DoorKey-8×8   (b) FourRoom   (c) SimpleCrossingS9N3

Fig. 2. Frames as input observation images from our three tested environments.

Model and Hyperparameters. We fix the architecture of the encoder with 3 hidden layers and 32 convolution filters for each layer. The encoder learns to encode the input RGB image observation into a 100 hidden dimension. Similarly, the decode has 3 hidden layers and 32 deconvolution filters. The number of hidden dimensions is selected following the training of SAC-AE [2]. The buffer $\mathcal{D}$ keeps the most recent 50000 samples from the sampled trajectories. For training the agent's policy, we employ the PPO algorithm. The actor and critic network of the policy are built on top of the encoder with two 64-dimension hidden layers. All of the networks are trained with 3e-4 learning rate and Adam optimizer. For learning the compact latent, we use the value $\beta = 10^{-5}$ in β-VAE (Eq. 2) and the adjacent norm coefficient $\alpha = 10^{-3}$ (Eq. 3). The reward coefficient a_t in Eq. 5 is set to 10^{-2}. For preventing latent collapse, the clipping threshold ϵ in Eq. 6 is set to 1. To improve the prediction performance, we omit the noise injection in the inference step of VAE.

5.2   Results and Analysis

We compare the performance of RL agents with and without access to a learned embedding space to verify the importance of the compact representation features on the performance of the RL algorithm. We test PPO with AE/VAE and compare the performance with PPO without any auxiliary representation learning. All experiments are limited to a fixed training budget of 1M environment samples with the same network architectures and hyper-parameters. In our experiments, we fix the random seeding of the environment at the beginning of each training run and test with 3 different seeds. The success rate of solving the tasks are reported in Fig. 3. Our method outperforms other baselines in two environments *DoorKey8×8* and *SimpleCrossingS9N3*, and is second best in *FourRoom*. The baseline of using PPO with VAE performs the best in *FourRoom* but has instable performance in the two other environments. It can be seen from the plot that our regularization objective L_{ADJ} and the auxiliary reward can improve the performances of the standard VAE. In *FourRoom* environment, we conjecture that the latent distance in some part of the environment does not represent well with the actual distances, which add some percentage of noise to the learning PPO algorithm. We plan to investigate this issue in the following section.

Analysis on the Latent Representation Effects. In Fig. 4, we study the effects of the adjacent loss $\mathcal{L}_{ADJ}$ in Eq. 3 for learning of the latent representation from input image observations. The plot illustrates the distance in the latent domains from one position to the goal position from a running seed in our experiments. From the plot, our proposed $\mathcal{L}_{ADJ}$ can be used to align the distances in the latent domain with the distances following the dynamics of the environment. Several interesting patterns can be seen. For example, in Fig. 4a, the distances between states separated by the door are distinctively different, so the sequence of actions passing through the door has a high intrinsic reward and thus will be reinforced. In Fig. 4b, the latent mapping network q_ϕ is able to map obsoleted states, which are the positions in the two separated columns on the top right, to distanced regions. However, some regions in the upper right room do not reflect well with the actual distances. We hypothesize that since the agent starts at the one room in the lower left, the upper room is not well explored enough and so the mapped distances are not accurate.

Analysis on the Exploration Effects. We examine the pure exploration aspect of the algorithm on the classical *FourRoom* environment. The agent starts at a fixed room at the bottom left of the environment and explores the rooms without any extrinsic reward signal. State visitation counts are collected each

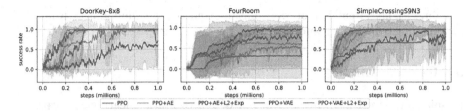

Fig. 3. Results on three MiniGrid environments, measured by success rate averaged over 3 seeds, with shaded areas indicate 1-std deviation from the mean. L2+Exp denotes experiments with **L2** adjacent norm and latent **Exp**loration.

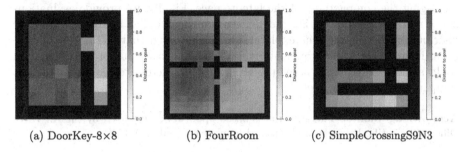

(a) DoorKey-8×8   (b) FourRoom   (c) SimpleCrossingS9N3

Fig. 4. Results on the correlation between our learned latent representation and the underlying dynamics of the environments. Green squares denote goal positions, while red denotes the starting positions of the agent. The distances are calculated between each goal and non-goal position pairs and averaged over all different views of the agent and are normalized into the range [0, 1]. The color denotes how far the distance in the latent embedding space is from each cell to the goal. (Color figure online)

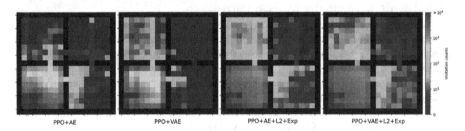

Fig. 5. Heat map distribution of PPO with AE/VAE with and without L2 adjacent regularization and latent exploration after 250k training steps on MiniGrid-Fourroom in the absence of extrinsic rewards.

time the agent interacts with the environment and the heatmap of state distributions is shown in Fig. 5 after 250k step environments. For comparison, we show the state distribution of the RL agents with and without the latent extrinsic reward bonuses. As can be seen, the state coverage of the agents trained with latent exploration is significantly improved, with most areas of the first three rooms being visited frequently and a large part of the furthest room explored.

Agents without latent exploration only depend on the randomness from the maximum entropy policy and thus cover far less state space.

6 Conclusion

In this work, we consider the problem of representation learning for sparse-reward RL environments from image observations. To learn a compact representation, we propose a method that incorporates an L2 regularization on the adjacent latent states with the reconstruction loss and prevents latent collapse by using a clipping threshold on the adjacent distance. Exploration is improved by leveraging the compactness of the latent space. Finally, we provide evaluations of our method on three environments from MiniGrid and an analysis of the state distribution to illustrate the exploration aspect. Given that learning in sparse-reward, image-based environments is a challenging task, our proposed method could achieve better and more stable performance across tested environments, compared to other baselines.

Acknowledgements. This material is based upon work supported by the Air Force Office of Scientific Research under award number FA2386-22-1-4026.

References

1. Mnih, V., et al.: Playing atari with deep reinforcement learning. arXiv preprint arXiv:1312.5602 (2013)
2. Yarats, D., Zhang, A., Kostrikov, I., Amos, B., Pineau, J., Fergus, R.: Improving sample efficiency in model-free reinforcement learning from images. Proc. AAAI Conf. Artif. Intell. **35**, 10674–10681 (2021)
3. Shelhamer, E., Mahmoudieh, P., Argus, M., Darrell, T.: Loss is its own reward: self-supervision for reinforcement learning. arXiv preprint arXiv:1612.07307 (2016)
4. Kingma, D.P., Welling, M.: Auto-encoding variational bayes. arXiv preprint arXiv:1312.6114 (2013)
5. Higgins, I., et al.: Beta-vae: learning basic visual concepts with a constrained variational framework (2016)
6. Kim, H., Kim, J., Jeong, Y., Levine, S. and Song, H.O.: Emi: exploration with mutual information. arXiv preprint arXiv:1810.01176 (2018)
7. Ermolov, A., Sebe, N.: Latent world models for intrinsically motivated exploration. Adv. Neural Inf. Process. Syst. **33**, 5565–5575 (2020)
8. Tang, H., et al.: # exploration: a study of count-based exploration for deep reinforcement learning. In: Advances in Neural Information Processing Systems, vol. 30 (2017)
9. Stadie, B.C., Levine, S., Abbeel, P.: Incentivizing exploration in reinforcement learning with deep predictive models. arXiv preprint arXiv:1507.00814 (2015)
10. Bellemare, M., Srinivasan, S., Ostrovski, G., Schaul, T., Saxton, D., Munos, R.: Unifying count-based exploration and intrinsic motivation. In: Advances in Neural Information Processing Systems, vol. 29 (2016)
11. Burda, Y., Edwards, H., Storkey, A., Klimov, O.: Exploration by random network distillation. arXiv preprint arXiv:1810.12894 (2018)

12. Zhang, T., Rashidinejad, P., Jiao, J., Tian, Y., Gonzalez, J.E., Russell, S.: Exploration via maximizing deviation from explored regions. Adv. Neural Inf. Process. Syst. **34**, 9663–9680 (2021)
13. Haarnoja, T., Zhou, A., Abbeel, P., Levine, S.: Soft actor-critic: Off-policy maximum entropy deep reinforcement learning with a stochastic actor. In: International Conference on Machine Learning, pp. 1861–1870. PMLR (2018)
14. Nachum, O., Gu, S.S., Lee, H., Levine, S.: Visual reinforcement learning with imagined goals. In: Advances in Neural Information Processing Systems, vol. 31 (2018)
15. Zhang, A., McAllister, R., Calandra, R., Gal, Y., Levine, S.: Learning invariant representations for reinforcement learning without reconstruction. arXiv preprint arXiv:2006.10742 (2020)
16. Hafner, D., et al.: Learning latent dynamics for planning from pixels. In: International Conference on Machine Learning, pp. 2555–2565. PMLR (2019)
17. Christopher, J.C.H.: Christopher JCH watkins and peter dayan: q-learning. Mach. Learn. **8**(3), 279–292 (1992)
18. Schulman, J., Wolski, F., Dhariwal, P., Radford, A., Klimov, O.: Proximal policy optimization algorithms. arXiv preprint arXiv:1707.06347 (2017)
19. Sutton, R.S., Barto, A.G.: Reinforcement learning: an introduction. MIT Press, Cambridge (2018)
20. Han, S., Sung, Y.: Diversity actor-critic: sample-aware entropy regularization for sample-efficient exploration. In: International Conference on Machine Learning, pp. 4018–4029. PMLR (2021)
21. Savinov, N., et al.: Episodic curiosity through reachability. arXiv preprint arXiv:1810.02274 (2018)
22. Raileanu, R., Rocktäschel, T.: Ride: rewarding impact-driven exploration for procedurally-generated environments. arXiv preprint arXiv:2002.12292 (2020)
23. Kostrikov, I., Yarats, D., Fergus, R.: Image augmentation is all you need: regularizing deep reinforcement learning from pixels. arXiv preprint arXiv:2004.13649 (2020)
24. Chevalier-Boisvert, M., Willems, L., Pal, S.: Minimalistic gridworld environment for gymnasium (2018)

A Survey of Explainable Artificial Intelligence Approaches for Sentiment Analysis

Bernadetta Maleszka(✉) [ID]

Faculty of Information and Communication Technology, Department of Applied Informatics, Wroclaw University of Science and Technology, Wybrzeze Wyspianskiego 27, 50-370 Wroclaw, Poland
Bernadetta.Maleszka@pwr.edu.pl

Abstract. Nowadays, the problems of sentiment analysis, opinion mining and fake news detection are very important. Artificial intelligence methods are widely used to analyze opinions in social media and to obtain the results in an efficient manner and with high accuracy. The most common approaches are ML methods using nonlinear models and complex structures, e.g. deep neural networks, SVM or random forest. These methods have only one disadvantage: they work as black-boxes so it is hard to understand how they predict the results and lowers trust to such methods. In this paper we present a survey of explainable artificial intelligence methods that are used in sentiment analysis area, analyze the differences between XAI in SA and feature selection methods and indicate trends and challenges in this area.

Keywords: Sentiment Analysis · Opinion Mining · XAI · Feature Selection

1 Introduction

Nowadays, sentiment analysis is a domain that develops rapidly. There are more and more models, methods and algorithms that help the user to form an opinion about particular topic, person, issue, service, etc. [1,5]. Developed artificial intelligence methods provide us with better and better results in this area.

To obtain better accuracy, more complicated structures of the model and more sophisticated methods are used. The problem arises when user asks how a particular result was achieved or how a particular sample has influenced the final model [34]. It is a problem of reliability of the system [35].

In this paper we present a survey of explainable artificial intelligence (XAI) methods that are used to increase user's trust to the system of sentiment analysis or opinion mining. We have provided the following sections: In Sect. 2 we present the basic concepts correlated to sentiment analysis and XAI. In Sect. 3 we describe related works for XAI methods in sentiment analysis. Some trends and challenges are provided in Sect. 4. Final remarks and summary are in Sect. 5.

N. T. Nguyen et al. (Eds.): ACIIDS 2023, LNAI 13996, pp. 52–62, 2023.
https://doi.org/10.1007/978-981-99-5837-5_5

2   Background

In this section we present the basic idea of sentiment analysis, explainable artificial intelligence and feature selection. We provide some definitions and description of the problem.

2.1   Sentiment Analysis

Nowadays, sentiment analysis is a very important issue as it can influence many aspects of everyday life. Before a user decides to buy or order a product or service, he or she tries to find the best offer but more and more often he or she looks for opinion from other users about the product or service.

In the last few years, the development of e-commerce systems and social networks has allowed the user to share his or her opinion easily [31]. On the other hand, the user can find a huge amount of e.g. product reviews, so that it is impossible to manage out all information. Many systems offer recommendations or decision support algorithms to improve user experience. Using sentiment analysis techniques allows to additionally enrich the accuracy of recommendations as they reflect users opinions.

The most popular tasks based on sentiment analysis are as follows: opinion mining [31], fake news detection [28,29] and stance detection [10]. The main contribution of sentiment analysis is to extract opinions from different modalities, e.g. text, image, video, etc. and usually combine them to obtain a final polarity. There arises a problem of opinion veracity and credibility which lead us to the fake news detection issues. It is possible to use sentiment analysis approaches to judge if a news is true or fake. The stance detection problem is correlated to users attitude toward a situation or an event. The user can agree or disagree with statements of other users.

According to Phan et al. [29] "Sentiment is the feeling, attitude, evaluation, or emotion of users toward specific aspects of topics or for the topics". The set of possible values of sentiment can be defined in many ways, e.g. [15,17,29]:

- $s = \{positive, \ negative\}$.
- $s = \{positive, \ neutral, \ negative\}$.
- $s = \{positive, \ neutral, \ negative, \ mixed\}$.
- $s = \{strong/very \ positive, \ positive, \ neutral, \ negative, \ strong/very \ negative\}$.
- $s = \{very \ very \ negative, \ very \ negative, \ negative, \ somewhat \ negative, \ neutral, \ somewhat \ positive, \ positive, \ very \ positive, \ and \ very \ very \ positive\}$.

Sentiment analysis is a "process used to determine the sentiment orientation in opinions" [29]. The process can be treated as a classification problem: classify a given opinion o toward specific aspect or topic into one sentiment polarity from set s [6]. Sentiment analysis can be divided into three levels: document level (when we judge the polarity of the final conclusions of some report), sentence level (polarity of each sentence) and aspect level (polarity towards particular aspect).

Phan et al. [29] defines the problem in a wider way: sentiment is an attitude of a particular user u in a timestamp t towards a given topic p. The user u delivers an opinion about the topic p and the task is to judge whether the opinion is positive or negative.

The most popular methods for sentiment analysis are those based on machine learning approaches or those based on lexicon approaches (e.g. corpus or dictionary based approaches) [22,29,39]. In the first group one can use supervised methods (e.g. probabilistic classifiers: Naive Bayesian, Bayesian network, maximum entropy, linear classifier: SVM, neural network, decision tree, rule-based methods, etc.), semi-supervised methods (e.g. self-training, graph-based, generative models), unsupervised methods (k-means, fuzzy c-means, agglomerative and divisive algorithms) or deep learning methods (RNN, CNN, LSTM, GNN, GCN, etc.) [37]. We can also find many hybrid approaches that combine machine learning with lexicon-based approaches, especially deep neural networks and lexicon-based methods. Usually, methods from the last group obtain the best results.

To judge the efficiency of the method we can use typical efficient metrics, such as precision, recall, F-measure and accuracy. Usually more complicated methods obtain better results than linear or simple methods. On the other hand, these methods are hard to explain. It is not obvious how single opinion or statement affects the final result. This is the reason for the popularity of developing explainable methods.

2.2 Explainable Artificial Intelligence

Artificial intelligence has appeared in many aspects of our life, e.g. medicine, transport, e-commerce, intelligent houses, etc. The systems can help the doctor to analyze X-ray or magnetic resonance images [13], support car drivers [16], recommend us some personalized products or services [25], allows us to "talk" with ChatGPT [9], etc.

The main contribution of XAI is to increase user's trust in AI systems. User confidence is crucial in many situation, especially when the results of these systems affect our health or even life.

The main idea of the XAI is to explain why the system obtained the particular result. It can be illustrated with the Albert Einstein's quote: *"If you can't explain it simply, you don't understand it well enough"*. It is an important aspect of many deep algorithms were it is not obvious what information does the network contain or why does this particular input lead to that particular output [14].

The most frequent division of XAI approaches is into two groups: visualization methods and post-hoc analysis. In the first group, there exists a few algorithms that do not need any explanation as they are transparent enough. They are: linear or logistic regression, decision trees, kNN, rule based learners, general additive model, or Bayesian models [3]. The category of post-hoc analysis contains more sophisticated methods that do not allow to easily explain why a particular case was classified into particular class. They are e.g. tree ensembles, SVM, deep neural networks: multi-layer, convolutional or recurrent neural networks. Usually, the following techniques are used for explaining how they work:

model simplification, feature relevance, local explanations or visualization in the post-hoc step.

Athira et al. [4] differentiate two concepts: interpretability and explainability. In the first case, we have a simple structure and it can be used to interpret or explain how the method works (e.g. linear model, decision trees, association rules). It assumes that the used algorithms or methods are transparent and does not need any explanation. It can be also called model-based explainability, or explainability by design [24]. The category of post-hoc explanations tries to explain how a black box (an algorithm or a method) works based on the final results [38]. It is crucial for such models that are non-linear: ensemble methods or neural networks (e.g. CNN, RNN [2]).

Arrieta et al. [3] and Ding et al. [8] have defined more aspects of explainability: understandability – user can understand how the algorithm works without any additional explanation about the internal structure; comprehensibility – the result of the learning algorithm should be understandable for human, it is also connected with the model complexity; transparency – the model by itself is understandable.

Another division of XAI models is into global and local explanation [38]. The global one aims to explain how the input variables influence the model. The local explanation focuses on how each feature influences the result (e.g. SHAP algorithm [20]).

Dazeley et al. [7] claim that full XAI system should implement two processes: social and cognitive. The first process should take into account interactions with other actors like people, animals, other agents, etc. The cognitive process should identify general causes and counterfactuals [11].

The authors have proposed the following levels of explanations according to the factors of user beliefs and motivations [7]:

- Reactive: it is an explanation of an agent's reaction to immediately perceived inputs – like instinctive behaviour of animals in dangerous situation.
- Disposition: it is an explanation of an agent's underlying internal disposition towards the environment and other actors that motivated a particular decision – the agent's decision is based on its beliefs or desires.
- Social: it is an explanation of a decision based on an awareness or belief of its own or other actors' mental states.
- Cultural: it is an explanation of a decision made by the agent based on what it has determined is expected of it culturally, separate from its primary objective, by other actors.
- Reflective: it is an explanation detailing the process and factors that were used to generate, infer or select an explanation.

The first four levels are object-level explanations based on decisions or arguments and the last meta-explanation is based on the scenario structure or historical decisions or justifications.

In the literature one can find many methods for XAI but the majority of them can be classified to the lower levels: reactive, disposition or social.

In the next part of this section we present the most popular approaches to XAI. The methods in the group of visualization are based on visual form of explanation, like highlighted text in natural language processing [23] or explicit visualization of the results according some subsets of features [33]. The post-hoc explanations' aim is to find feature relevance, model simplification, text explanation or explanation by example [3]. In many cases the post-hoc methods also use visualization approaches.

Visualization. Nowadays, there is more and more methods to train the model but it is hard to explain why we obtained any specific final results, what was an impact of particular set of features or cases during the training process and how they have influenced the final prediction mechanism [33].

Visualization approach allows us to take a look inside the data in a simpler way than using analytical methods. It can provide us with some intuitions about data distribution or differences between some subsets of cases.

So et al. [33] claim that the basics of explanation is the set of features that can be visualized. They differentiate the following aspects:

– feature importance – it calculates how the feature of all observations impacts the prediction. The most popular method is SHAP (SHapley Additive exPlanations) [20] or counterfactual explanations [11];
– additive variable attributions – it estimates which instances of the dataset are outliers;
– what-if analysis – one can use ceteris-paribus plot to analyze a relationship between features and response.

One of the most effective algorithms for sentiment analysis uses CNN architecture. Souza et al. [36] proposed five different PIV (particle image velocimetry) techniques to visualize the flow of the method. They are as follows: guided backpropagation (GBP), saliency (SAL), integrated gradients (IGR), input × gradients (IXG) and DeepLIFT (DLF).

Post-hoc XAI Methods. An input for a post-hoc XAI methods is a trained model. An expected output of the method is an approximate model that explains how the original model works [24]. It can also reflect decision logic or generate some representation of the model that is understandable, e.g. set of rules, feature importance score or heatmaps.

Most of the XAI methods dedicated for the text processing are model-specific approaches [3].

Some exemplary methods are described below [24]:

– LIME (Local Interpretable Model-agnostic Explanations) – the algorithm introduces some perturbations to real samples and provides observations about the output of the model;
– If-then rules – they should reflect the dependencies between the features. The generated rules should represent the original black-box model; determining the optimal set of rules is an optimization task.

The results obtained from post-hoc XAI methods that have found some dependencies between features, can be used for the feature selection methods. The main aim of feature selection methods is to reduce the dimensionality of the dataset and the complexity of the solution. It is possible because a lot of data is redundant [21,32]. The task is to delete (or omit) some data as it does not significantly change the result of the algorithm.

The methods and techniques of Explainable Artificial Intelligence presented above focus on the feature – how a particular feature influences the result. They take care about the form of explanation, use a subset of features to obtain the result and they are separated from the model [11].

These methods also have disadvantages: they cannot show us, e.g. what is a minimal set of samples or instances that guarantees the obtained results [41] and using these methods, is not clear which input instance has determined the final result.

3 Explainable AI in Sentiment Analysis

Sentiment analysis is an area where transparency is a crucial feature of the user's trust in the system [2]. Before a user makes a purchase decision or decides to use the service, he or she may decide to check the opinion about the topic, product or delivered service, etc.

Explainable artificial intelligence techniques allow us to better understand prediction of the model [12]. More and more methods and models in this area are predictive – to increase user's confidence in the system, it should provide transparent and trustworthy results. As the authors claim, more effective algorithms mean less transparency.

The main objective of the XAI methods in sentiment analysis area is to answer the query: "How can XAI methods reveal potential bias in trained machine learning models for the prediction of product ratings?" [34].

In this section we present a classification of the existing solution for XAI in sentiment analysis domain. Most commonly used methods focus on the following aspects [34]:

- Feature importance – it approximates the global relevance of the feature in the model. It depends on the model, e.g. for models based on trees it can split the tree and for linear models it is correlated with regression coefficient.
- Local attributions – this approach allows to visualize the impact of a single feature's variance as it can be missed by the analysis of global feature importance.
- Partial dependency plot – it presents how each feature or several features can impact the final result.

Above mentioned methods are based on the visualization of the results. They can be used both to explain how the model works and to improve it: a feature can be not used in the model when it is not important, it has too high variance or it has weak relationship with other attributes.

The improvement of the interpretability or explainability can be achieved mostly by high transparency of the model that can be developed from structure of the network, feature importance, local gradient information, redistribution the function's value on the input variables, specific propagation rules for neural networks [2].

In Table 1 we summarize existing papers focused on XAI in SA. Each paper is analyzed according to the main problem, feature and techniques used for sentiment analysis and type of explainability.

All these papers developed models for predicting sentence polarity. Most of them work on text reviews of documents or movies or simply tweets using a wide range of possible models of the data and methods for sentiment analysis: naive Bayes, decision trees, random forests, LSTM, softmax attention, neural networks (CNN, RNN, etc.).

The most popular approach to provide explanations of the results is a visualization method: SHAP, BertViz [12], LIME [12,35], feature importance [19,26,40], local feature attributions and partial dependency plots [34], contextual importance and utility [30].

4   Trends and Challenges

The area of XAI methods is more and more developed to ensure more transparent and confident results that user can trust them. There are still many aspects that should be taken into account.

The challenges of XAI methods in sentiment analysis are correlated with development of new methods for SA, especially deep neural network approaches. As they become more and more popular and are used by wider and wider group of people (sometimes they even use them without thought or awareness how they work), it is important to take care about the responsibility of the results. Arrieta et al. [3] highlighted the need of preparing and using a set of principles that should be satisfied. They called this trend as responsible AI – it should include the following issues: fairness, privacy, accountability, ethics, transparency, security and safety.

XAI algorithms used in presented papers focus on visualization approaches. It allows us to see the impact of a feature or set of features on the final result. It increases the transparency and it can help to reduce the dimensionality of the problem.

There appears more and more sophisticated algorithms that take into account more information and obtain more accurate results. Unfortunately, they do not focus on the interpretability.

Most of responsible AI aspects are still not introduced to SA methods. The users would like to have trustworthy methods for analyzing opinion mining so explainable sentiment analysis is a promising investigation area. Due to the wide variety of the SA methods, better explainable algorithms should be also created.

Table 1. Summary of the XAI methods in SA problems.

No. Paper	Main problem	Feature used	Techniques used	Type of explainability
[18]	generating opinions summary (aspect-based SA)	review content – new dataset of opinions about one entity in the restaurant domain	aspect extraction, sentences grouping, rules of interest extraction	discovering subgroup of features using statistical methods; generating the rules of classification to subgroups; developing quality measures that are easy to understand for humans
[12]	comparing effectiveness of selected NLP models	tweets	explainable FEs (EFEs); pre-trained DL FEs that do not require training on task-specific data; and trainable DL FEs that require training on task-specific data	local interpretable model-agnostic explanations (LIME), the variant called submodular pick LIME (SP-LIME); Shapley additive explanations (SHAP); BertViz, designed specifically for transformer LMs
[34]	sentiment analysis of online reviews; extracting features for product rating prediction	online reviews	knn, support vector machines, random, forests, gradient boosting machines, XGBoost	local feature attributions and partial dependency plots
[30]	model for sentence polarity for the Italian language	sentence content	BERT model, Long-Short Term Memory (LSTM) and the WMAL-based text representation module	Lexicon-driven classification explanation; contextual importance and utility; explanatory and WMAL attention
[26]	investigation of the capability of an attention mechanism to "attend to" semantically meaningful words	text from videos of Stanford Emotional Narratives Dataset (SEND) - dataset consisting of videos of people narrating emotional events in their lives	Window-Based Attention (WBA) consisting of a hierarchical, two-level long short-term memory (LSTM) with softmax attention	attention based explanation: word deletion experiments and visualizations of results
[40]	leveraging a sentiment knowledge graph to better capture the sentiment relations between aspects and sentiment terms	online learner review dataset	knowledge-enabled language representation model BERT for aspect-based sentiment analysis	knowledge-enabled BERT model delivers explainable information to boost performance
[27]	examination of interpretable HMMs methods performance under various architectures, parameters, orders and ensembles	annotated datasets of documents or movies reviews	interpretable Hidden Markov Models (HMM)-based methods for recognizing sentiments in text	visual interpretation of the HMM
[19]	attention-based multi-feature fusion method for intention recognition	movie comments	fusing features extracted from frequency-inverse document frequency (TF-IDF), convolutional neural networks (CNNs), long short term memory (LSTM)	attention mechanism for measuring feature importance
[35]	securing the reliability of machine learning-based sentiment analysis and prediction	movie reviews	multinomial naïve Bayes, random forest, random boosting, decision trees	LIME - visualization

5   Summary

In this paper we have presented the explainable artificial intelligence methods that are used in sentiment analysis. We have described definitions and an overview of the existing methodologies used in SA. The second part focuses on explainable methods that are more and more popular in general area of artificial intelligence. And finally, we presented exemplary research articles that use XAI methods in the opinion mining.

Most of presented paper uses only visualization methods to help the user to interpret the result so it is still a potential research domain.

References

1. Alsaif, H.F., Aldosssari, H.D.: Review of stance detection for rumor verification in social media. Eng. Appl. Artif. Intell. **119**, 105801 (2023)
2. Arras, L., Montavon, G., Muller, K.R., Samek, W.: Explaining recurrent neural network predictions in sentiment analysis. In: Proceedings of the 8th Workshop on Computational Approaches to Subjectivity, Sentiment and Social Media Analysis, pp. 159–168 (2017)
3. Arrieta, A.B., et al.: Explainable Artificial Intelligence (XAI): concepts, taxonomies, opportunities and challenges toward responsible AI. Inf. Fusion **58**(2020), 82–115 (2020)
4. Athira, A.B., Kumar, S.D.M., Chacko, A.M.: A systematic survey on explainable AI applied to fake news detection. Eng. Appl. Artif. Intell. **122**, 106087 (2023)
5. Birjali, M., Kasri, M., Beni-Hssane, A.: A comprehensive survey on sentiment analysis: approaches, challenges and trends. Knowl.-Based Syst. **226**(2021), 107–134 (2021)
6. Chaturvedi, I., Satapathy, R., Cavallari, S., Cambria, E.: Fuzzy commonsense reasoning for multimodal sentiment analysis. Pattern Recogn. Lett. **125**(2019), 264–270 (2019)
7. Dazeley, R., Vamplew, P., Foale, C., Young, Ch., Aryal, S., Cruz, F.: Levels of explainable artificial intelligence for human-aligned conversational explanations. Artif. Intell. **299**, 103525 (2021)
8. Ding, W., Abdel-Basset, M., Hawash, H., Ali, A.M.: Explainability of artificial intelligence methods, applications and challenges: a comprehensive survey. Inf. Sci. **615**(2022), 238–292 (2022)
9. Dwivedi, Y.K., Kshetri, N., et al.: "So what if ChatGPT wrote it?" Multidisciplinary perspectives on opportunities, challenges and implications of generative conversational AI for research, practice and policy. Int. J. Inf. Manage. 71, 102642 (2023). https://doi.org/10.1016/j.ijinfomgt.2023.102642. ISSN 0268–4012
10. Esuli, A., Sebastiani, F.: SentiWordNet - a publicly available lexical resource for opinion mining. In: Proceedings of the 5th Conference on Language Resources and Evaluation (LREC 2006), pp. 417–422 (2006)
11. Fernandez, C., Provost, F., Han, X.: Explaining data-driven decisions made by AI systems: the counterfactual approach (2020). arXiv:2001.07417v1. Accessed 5 Mar 2023
12. Fiok, K., Karwowski, W., Gutierrez, E., Wilamowski, M.: Twitter account: comparison of model performance and explainability of predictions. Expert Syst. Appl. **186**, 115771 (2021)
13. Fuhrman, J.D., Gorre, N., Hu, Q., Li, H., El Naqa, I., Giger, M.L.: A review of explainable and interpretable AI with applications in COVID-19 imaging. Med. Phys. **49**(1), 1–14 (2022). https://aapm.onlinelibrary.wiley.com/doi/10.1002/mp.15359

14. Gilpin, L.H., Bau, D., Yuan, B.Z., Bajwa, A., Specter, M., Kagal, L.: Explaining explanations: an overview of interpretability of machine learning (2019). arXiv:1806.00069v3. Accessed 18 Mar 2023

15. Gutierrez-Batista, K., Vila, M.-A., Martin-Bautista, M.J.: Building a fuzzy sentiment dimension for multidimensional analysis in social networks. Appl. Soft Comput. **108**, 107390 (2021)

16. Hacohen, S., Medina, O., Shoval, S.: Autonomous driving: a survey of technological gaps using google scholar and web of science trend analysis. IEEE Trans. Intell. Transp. Syst. **23**(11), 21241–21258 (2022)

17. Hussein, D.M.E.D.M.: A survey on sentiment analysis challenges. J. King Saud Univ. Eng. Sci. **2018**(30), 330–338 (2018)

18. López, M., Martínez-Cámara, E., Luzón, V., Herrera, F.: ADOPS: Aspect Discovery OPinion Summarisation Methodology based on deep learning and subgroup discovery for generating explainable opinion summaries. Knowl.-Based Syst. **231**, 107455 (2021)

19. Liu, C., Xu, X.: AMFF: a new attention-based multi-feature fusion method for intention recognition. Knowl.-Based Syst. **233**, 107525 (2021)

20. Lundberg, S.M., Lee, S.-I.: A unified approach to interpreting model predictions. In: NIPS 2017: Proceedings of the 31st International Conference on Neural Information Processing Systems, pp. 4768–4777 (2017)

21. Lötsch, J., Ultsch, A.: Enhancing explainable machine learning by reconsidering initially unselected items in feature selection for classification. Biomedinformatics **2**, 701–714 (2022). https://doi.org/10.3390/biomedinformatics2040047

22. Medhat, W., Hassan, A., Korashy, H.: Sentiment analysis algorithms and applications: a survey. Ain Shams Eng. J. **5**, 1093–1113 (2014)

23. Montavon, G., Samek, W., Muller, K.R.: Methods for interpreting and understanding deep neural networks (2017). https://arxiv.org/pdf/1706.07979.pdf. Accessed 21 Mar 2023

24. Moradi, M., Samwald, M.: Post-hoc explanation of black-box classifiers using confident itemsets. Expert Syst. Appl. **165**, 113941 (2021)

25. Nabizadeh, A.H., Leal, J.P., Rafsanjani, H.N., Shah, R.R.: Learning path personalization and recommendation methods: a survey of the state-of-the-art. Expert Syst. Appl. **159**, 113596 (2020)

26. Nguyen, T.-S., Wu, Z., Ong, D.C.: Attention uncovers task-relevant semantics in emotional narrative understanding. Knowl.-Based Syst. **226**, 107162 (2021)

27. Perikos, I., Kardakis, S., Hatzilygeroudis, I.: Sentiment analysis using novel and interpretable architectures of Hidden Markov Models. Knowl.-Based Syst. **229**, 107332 (2021)

28. Phan, H.T., Nguyen, N.T., Hwang, D.: Fake news detection: a survey of graph neural network methods. Appl. Soft Comput. **139**, 110235 (2023)

29. Phan, H.T., Nguyen, N.T., Hwang, D.: Sentiment analysis for opinions on social media: a survey. J. Comput. Sci. Cybern. **37**(4), 403–428 (2021)

30. Polignano, M., Basile, V., Basile, P., Gabrieli, G., Vassallo, M., Bosco, C.: A hybrid lexicon-based and neural approach for explainable polarity detection. Inf. Process. Manage. **59**, 103058 (2022)

31. Serrano-Guerrero, J., Romero, F.P., Olivias, J.A.: Fuzzy logic applied to opinion mining: a review. Knowl.-Based Syst. **222**, 107018 (2021)

32. da Silva, M.P.: Feature Selection using SHAP: an Explainable AI approach. University of Brasilia. Doctoral thesis (2021)

33. So, Ch.: Understanding the prediction mechanism of sentiments by XAI visualization. In: 4th International Conference on Natural Language Processing and Information Retrieval, Sejong, South Korea, 18–20 December 2020. ACM (2020)

34. So, C.: What emotions make one or five stars? Understanding ratings of online product reviews by sentiment analysis and XAI. In: Degen, H., Reinerman-Jones, L. (eds.) HCII 2020. LNCS, vol. 12217, pp. 412–421. Springer, Cham (2020). https://doi.org/10.1007/978-3-030-50334-5_28

35. Song, M.H.: A study on explainable artificial intelligence-based sentimental analysis system model. Int. J. Internet Broadcast. Commun. **14**(1), 142–151 (2022). https://doi.org/10.7236/IJIBC.2022.1.142

36. de Souza Jr., L.A., et al.: Convolutional Neural Networks for the evaluation of cancer in Barrett's esophagus: explainable AI to lighten up the black-box. Comput. Biol. Med. **135**, 104578 (2021)

37. Ventura, F., Greco, S., Apiletti, D., Cerquitelli, T.: Explaining the Deep Natural Language Processing by Mining Textual Interpretable Features (2021). https://arxiv.org/abs/2106.06697. Accessed 31 Mar 2023

38. Zacharias, J., von Zahn, M., Chen, J., Hinz, O.: Designing a feature selection method based on explainable artificial intelligence. Electron. Mark. **32**, 2159–2184 (2022). https://doi.org/10.1007/s12525-022-00608-1

39. Zhang, L., Wang, S., Liu, B.: Deep learning for sentiment analysis: a survey (2018). https://doi.org/10.1002/widm.1253. Accessed 11 Mar 2023

40. Zhao, A., Yu, Y.: Knowledge-enabled BERT for aspect-based sentiment analysis. Knowl.-Based Syst. **227**, 107220 (2021)

41. https://elula.ai/feature-importances-are-not-good-enough/. Accessed 10 Mar 2023

Social Collective Model of Non-internet Social Networks

Marcin Maleszka[1](✉) and Sinh Van Nguyen[2,3]

[1] Wroclaw University of Science and Technology,
st. Wyspianskiego 27, 50-370 Wroclaw, Poland
`marcin.maleszka@pwr.edu.pl`
[2] School of Computer Science and Engineering, International University,
Ho Chi Minh City, Vietnam
`nvsinh@hcmiu.edu.vn`
[3] Vietnam National University, Ho Chi Minh City, Vietnam

Abstract. Social collective is one of possible models of a group for tasks of knowledge diffussion and opinion formation. Previous papers have described multiple possibilities for behaviors of singular agents, with only two basic models of their interaction. In this paper the focus is on describing various structures of group and methods of communication, with specific focus on non-internet social networks. We describe traditional mail, phone communication, group learning and discussion in a physical location. We compare them in terms of stability of collectives, based on the measure of drift.

Keywords: Collective intelligence · Group modeling · Multi-agent simulation · Non-internet social network

1 Introduction and Related Works

Collective Intelligence is a broad term describing various approaches to model the behavior of groups and make use of those models. Computational collective intelligence considers the issue from the point of view of certain computer science tools, including multi-agent systems. In our research we have worked to develop a specific new tool for group modeling, called social collective. It is a multi-agent simulation of a social network, with focus on internal workings of the agent, instead of a more common global view of the group. Such intelligent collective could be used to work with problems of group dynamics, knowledge diffusion, opinion formation, etc.

The concept of a social collective was initially based on the notion of asynchronous communication between a group of identical non-cognitive agents. In subsequent research we have expanded upon this model to develop the model representing a single social network [13], and then modified parts of it for a different one [14]. Due to this approach the final model is highly modular, with four main parts that can be exchanged. For the internal working of social agents in

N. T. Nguyen et al. (Eds.): ACIIDS 2023, LNAI 13996, pp. 63–73, 2023.
https://doi.org/10.1007/978-981-99-5837-5_6

the collective we can swap (1) the structure of their internal knowledge, (2) the method they use to internalize new knowledge received in messages from other agents, and (3) how their internal knowledge changes over time. For a group of agents we can also determine the (4) how they are interconnected and what type channels of communication exist.

Our previous papers focused on internal workings of social agents, with only Twitter-like and Facebook-like structures of the social collective being considered. In this paper we focus solely on the group part and study different non-internet methods of communication: traditional mail, telephones (cellphones), learning (and group learning), and a simple group in the same physical location. Some of those social networks are rarely used in computer science literature, due to difficulty in studying them. They occur more often in social sciences literature. In consequence we base part of the proposed methods on past research conducted in cooperation between computer science and social science teams [11]. We provide detailed description that allows for building a model in an agent simulation environment and conduct some basic social collective tests on those social networks.

This paper is organized as follows: in Sect. 2 we describe other research relevant to social collective; in Sect. 3 we describe the general overview of the social collective model with brief description of each of its four modules; in Sect. 4 we provide a detailed description of the communication module and all its variants; in Sect. 5 a short description of evaluation method for the variants and the results of evaluation is given; finally in Sect. 6 we give some concluding remarks and present further planned extensions to the social collective model.

2   Related Works

The underlying concepts of the social collective come from several research areas. First are the consensus-based approach of data and knowledge integration, especially common with the nontion of time dependent consensus. These ideas could be related to multi-agents systems with continuous-time consensus as used in autonomous robots and network systems [20], which has application to flocking, attitude alignment and formation control in aerial drones, negotiations between communicating agents, etc. There is also overlap with finite-time consensus research, as the notion of agent system stability [2] is similar to the notion of draft used in our research – both consider the changes in agent state in terms of its convergence. Another factor taken into account is the presence of a leader (as opposed to supervisor) in a multi-agent system, in these terms our approach has similarities to leaderless finite-time consensus simulation, where agents will reach a common result in a given time if no disturbances are present. With them, only pairs of agents may reach consensus [10]. The impact of the number of agents on the finite-time consensus is also a factor that we have considered as part of our research, similar to [17], where it was determined that in group decision making larger groups lead to better results with an additional, increasing cost – there is a point, where increasing the number of participants further is too costly. The mathematical basis for the consensus approach that we use is derived from the median calculations and the generic topic of consensus theory, which operates on the postulate basis, for example: reliability, unanimity, simplification, quasi-unanimity, consistency, Condorcet consistency, general consistency, proportion, 1-optimality and 2-optimality [5, 16]. Both the algorithmic ones (optimality variants) and generic ones (reliability, unanimity) are taken into account in the knowledge integration strategies of the social collective model. In the above approaches, as well as ours, the agents are mostly uniform, which has been observed to be detrimental to the quality of results [8]. On the other hand we extend the details in the structure of communication channels, creating some of different importance levels, which provides positive effect on quality [15].

Another important underlying concept of the social collective is in its potential applications in knowledge diffusion or opinion formation tasks. It was not initially developed for these tasks, but in recent iterations of the model, it has become a variant of a predictive influence maximization approach, with elements of explanatory SIRS model included. The explanatory models are mostly derived from epidemic ones (and recently used again for their original purpose). They are mostly graph representations of a population, where each node is a person and an edge is a contact (communication) between them. In a SIR type of model, a person is initially Susceptible to new information, may become Infected with it, with some probability, then develops Resistance to new information, and after time may become Susceptible again [9]. There exist more complex variants, for example S-SEIR model [22] requires acknowledging received knowledge only depending on their actual content. Explanatory influence models work on a more individual level, where main focus is determining nodes that are most influential on the rest of the network, for example by bridging information from one group to another. Influence maximization models in particular work to determine all key nodes that would facilitate knowledge diffusion, possibly by using network parameters, user attributes [6], or mathematical models like Markov chains [23].

3   Overall Model of the Social Collective

Following our previous research [13, 14], the model of the social collective consists of identical social agents and the structure facilitating their communication. Each agent have some internal knowledge or opinions. Over time, they may chose to send parts of it to random other agents. Upon receiving the message, agents may act differently to integrate it with their own previous knowledge (opinion), changing it fully, partially, or not at all. Additionally, agents may forget or change parts of their own knowledge over time.

The main focus of this paper are the various modes of communication associated with different types of real world and Internet-based collectives. Communication and social structure are the focus of one of exchangeable modules in the model of social collective. The other modules focus on internal structure of the agent (their knowledge), the processing of incoming messages and integrating them with that knowledge, and with possible changes to agent knowledge over time (i.e. forgetting knowledge).

Common knowledge structure is an important aspect of the collective, as without it, sharing any knowledge becomes impossible. In the social aspect it corresponds to its internal knowledge, but in the overall model, this module also deals with selecting parts of the knowledge for communication and on the algorithmic part of merging two knowledge structures (when to do it, and which parts to merge, is the role of the integration module). We have conducted simulations with knowledge structured in vectors, weighted vectors, fuzzy weighted vectors, labeled trees, and fuzzy labeled trees. An ontology based structure is also compatible with the model of social collective, but we have not tested them due to computational limitations.

Knowledge induction module is closely related to the previous one. It describes when and how is the received knowledge integrated with the internal knowledge of the social agent. This is presented as formalization of various social science approaches to the process of induction (being influenced by received messages). The classic research [7] determines that the receiver may be fully resistant to that external knowledge, partially resistant, have no resistance, or the resistance may be mixed (different depending on the source of knowledge). In our research we also studied approaches based on consensus theory [16], persuasion, leadership, and teaching [21].

Knowledge modification module that deals with changes of social agents internal knowledge that are not caused by any external factor. These may express different situations, for example observing the environment and self-learning. In our previous research we have studied how social agents may forget their knowledge: by forgetting random parts of it, last learned, newest, least *interesting*, etc.

4   Communication Module

The first model of communication in a social collective we have considered was initially generic, but bearing similarities to the Twitter social network. In consequence we have brought it more in line to that well known social network [13]. In this platform, the messages are sent out with no guarantee of being read by specific receivers. Instead a reader may follow a general feed (observe some specific tag) or messages of a specific account (follower). Messages are short and consider a single issue and by definition they are in one direction only.

Twitter-like communication in a social collective could be described as:

- Communication modes are one issue, one direction. In first mode, message is sent either to the whole population of receivers; in second, only to its subset determined by neigbourhood graph.
- The same message could be addressed to multiple receivers, mixing both communication modes.

In subsequent research, we have proposed an approach based on the Facebook social network [14]. The structure in this network is built on the basis of bi-directional relations between friends and communication may only occur between them. It can be bidirectional and immediate, as in case of chat function, or one direction and delayed, as in case of *wall* posts. In turn those posts could be forwarded (*liked*) or commented upon. Most communication is one issue only, but chat may be multi-issue.

Facebook-like communication in a social collective could be described as:

- For simplicity all communication covers one issue at a time only. Multi-issue discussions via chat are modelled by a communication attempts following each other.
- All communication is done only on a basis of a bi-directional neighbourhood graph.
- Four modes of communication exist: bi-directional chat, one direction post, one direction like (forward post to own neighbourhood graph), one direction comment (forward post to own neighbourhood graph, but may change parts of it).

In the *current* phase of research, we expanded the available communication modules for the social collective by the following approaches.

Traditional paper mailing is can be used to form a form of basic social collective with severe limits on communication. Until more modern modes of communication became standard, a message sent in this form could be very complex, covering multiple different issues and opinions. It would also occur only very rarely and be only one-directional (the response time could be measured in days, weeks, or longer). On the other hand the influence of the message would be much stronger, as other points of view would not be taken into account. These days, this form of communication is used less often. Some elements of it could be found in advertisement mail, which are focused on a single issue, but have much less influence on the receiver.

Traditional mail communication in terms of a social collective could be described as:

- One mode of communication, but much lower probability of sending any message in a given time moment.
- Communication is one-direction only, but in the neighborhood graph all connections are bidirectional (i.e. receiver of the message knows the address of the sender).
- Any number of knowledge issues can be considered in a single message.

Telephone communication could also be used as a basis of a social collective. In this case we will consider it for modern smartphone communication, which mostly occurs in three modes: classic phone call, text message (including multimedia text message) and chat communication (WhatsApp, Viber, etc.). Different groups have a preference for distinct modes, and some of them even avoid using a specific mode at all. For an anecdotal example, an older person may prefer calling, may use text, but cannot use chat; a slightly younger person prefers to use text messages, but uses all forms of communication if necessary; and a very young person uses almost only chat, never texts (chat has the same function for them) and rarely phone (and even then: using the same chat application). Functionally phone call and chat are similar, as they are a long bidirectional communication considering multiple issues; but call may have a slightly larger influence on the change of opinions (it would need to be experimentally verified if it is significant, if the communication was one-direction, but in this case the effects cancel each other out). Text communication is slightly different, as it is one direction only and considers one issue per message.

In terms of a social collective, phone communication could be described as follows:

- Two distinct modes of communication. Call and chat are aggregated into one mode: bidirectional, multiple issues per message. Text is the second mode: one-direction, one issue per message.
- Neighborhood graph has bidirectional vertices, but text mode may be to a random recipient in a full graph.
- The probabilities for different modes vary depending on specific groups, but call/chat always has non-zero probability.
- In phone communication we assume possibility of misunderstanding the message, with probability P_m.

Communication occurring during learning process can also be expressed in terms of a social collective. In this case we consider three different modes: direct teaching in a classroom, where a single teacher covers a single issue and multiple students learn by extending their knowledge; tutoring between a singular teacher and a singular student, where a single issue is considered, but some information may flow back to the teacher ("learning by teaching"); and groupwork of students, where a small group combines their knowledge with a series of exchanged messages on a single issue. Additionally, outside of communication, students may learn on their own and gradually change their knowledge over time (technically, this could also be considered a very delayed one-directional communication from the author of teaching materials). While only knowledge increase may be considered in a basic approach, many real cases show that students may also decrease their knowledge (for example by learning wrong things), therefore we do not consider any specific ("directed") structure of knowledge here.

In terms of a social collective, we consider learning as follows:

- Three different modes of communication, all considering a single issue only: one-direction teaching, bi-direction tutoring, multi-direction group work.
- Collective members may change their knowledge over time on their own.
- Communication mode is not determined by probabilities, but a schedule of types of communication.
- Neighborhood graph has three distinct types of connections (teacher-student, tutor-student, student-student).

We also consider a specific case of spreading gossip, as presented in [11], in terms of a social collective. The paper mentions a situation occurring in queues, mostly in medical facilities, where people discuss the topic with each other, but due to limitations of physical structure (corridor), the messages spread linearly and mostly on a neighbor to neighbor basis. There are no alternative modes of communication, but it is bidirectional and multiple issues may be considered (in most cases only one). The original research suggested possibility of the messages being subject to random changes when spread, on the basis of messages not being received properly.

In terms of social collective, we consider the corridor model as follows:

- A single mode of communication, multi-issue, bidirectional, but with possibility of random change in received message (similar to mutation in Genetic Algorithms).
- Physical limits of the *corridor* (a rectangle) determine the limits of the social graph. Agents are located in two lines along longer edges of the rectangle.
- We limit the model to a maximum of 20 social agents, 10 along each edge of the rectangle, following [11].
- The distance is standarized, so that between two neigbours on one edge it is 1. The width of the corridor is l (usually $l > 1$). Manhattan distance is used, in this environment the effect is identical to Euclidean distance, but calculations are simplified.
- Probability of communication (discussion or overhearing) is $(P_c)^d$, where d is standarized distance between agents.
- Because messages could be misunderstood, the probability of changes to received message is $(P_m)^{\frac{1}{d}}$.

5   Evaluation of the Model

Evaluation of social collective models is done in terms of drift and stability, following their introduction in our previous research [13]. This is also similar approach to finite-state consensus [2], approached with modifications derived from social sciences research area.

Definition 1. k_i-**drift** is the absolute value of the average change of weights describing the knowledge statement k_i in the whole collective over one iteration.

Definition 2. Collective Drift is the average of k_i-drifts about every knowledge statement k_i possible in the Closed World interpretation.

Definition 3. ϵ, τ-**stability.** A collective is ϵ, τ-stable, if the average weights describing knowledge of its members change over time τ by no more than ϵ.

Definition 4. Stable collective. A collective is called stable if it is drift is no larger than 0.1. Otherwise, it is called unstable.

The notion of drift, and collective stability based on it, could be also used in opinion formation models. In a stable collective a common opinion will be formed (the smaller the drift, the faster it will be) and it will represent the group opinion quite well. In an unstable collective the common opinion may not be formed at all, and if it is – it will be distant from the initial opinion of individual agents (this may be a positive effect in some situations).

In our previous research we have created a simulation environment for social collectives and determined a set of parameters that allow models with Twitter and Facebook communication modules to approximate the behavior of relevant social networks. On that basis we have further tuned the parameters for each new variant of the social collective model presented in this paper. While all structures of the social network representing each collective are appropriate, in order to make all variants comparable we did not limit them to appropriate size. Specifically, we simulate a group of 1000 social agents for each experiment. This size of group is sufficient to simulate much larger social networks like Twitter, Facebook, traditional mail or phone communication. However, learning environments are much smaller (for example 20–30 students), and a group spreading gossip on some corridor should also be limited to a maximum of 20 agents in a realistic situation. The basic experiment calls for generating the group of agents with a uniform distribution of opinions, then observing the changes in their opinions over a long period of time. The drift is measured, and if it exceeds the threshold, the setup is determined as unstable. The results of these simulations are shown in Table 1.

Table 1. Collective Drift of social collectives with various integration strategies and different communication modules. Drift over 1 represents an unstable collective. Note (1): group without teacher; for learning a larger drift is desirable if the direction is correct. Note (2): size of a group for corridor not comparable practical cases.

Integration Strategy	Substitute	Imm. Cons.	Del. Cons.	Persuasion	Polarization
Twitter	**2**	0.69	0.66	0.12	**2.32**
Facebook	0.36	0.25	0.04	0.05	0.27
Trad. Mail	0.05	0.07	0.06	0.02	0.07
Phone	0.41	0.05	0.10	0.07	0.29
Learning (1)	0.39	0.06	0.03	0.06	0.31
Gossip (2)	0.61	0.03	0.06	0.09	0.59

The communication modules following the approaches of the Internet-based social networks were already described in detail in our previous papers [13,14]. The results in these series of experiments do not deviate significantly from the previous ones: Substitution and Polarization strategies for knowledge integration lead to unstable collectives, where every small deviation from average is significantly increased in subsequent steps; and Twitter-based social network is more easily influenced than Facebook-based one (with the assumed parameters).

The communication module modelling phone communication behaves in a matter similar to Facebook-based one. The underlying structure of connections between group members is very similar, and the only difference in specific methods of communication is their probability. Communication modeling traditional mail is the most stable. A part of this effect comes from slowest communication speed compared to other methods. Additionally, any influence on social agents internal knowledge is spread among multiple issues in each single message. Additional experiments show that over extended timeframe the results are similar to the previous communication modules (the change is slower, but has similar characteristics).

Learning-based module may be distinguished from the others, as the desirable outcome is a change (increase) of the knowledge. In opposition to other approaches, the collective should be unstable. In the experiment with parameters aligned to match the other models such collective is instead stable. However, introducing even a single outside source of knowledge that constantly provides identical messages leads to quick unstability (learning), as described in detail in [12].

The last new communication module we consider is based on a concept of a talking group with knowledge (*gossip*) being spread in linear manner (*along a corridor*). In the initial simulation, the number of group members is unrealistically high, in order to better compare the approach to the previous ones. In such situation it is somewhat less stable than Facebook-based communication. Similar values of drift are calculated also for realistic size of group (5–50 members).

Several more simulations were conducted with different setups. All communication modules are subject to increase of drift due to an outside source repeatedly repeating the same knowledge to any random members of the collective, as discussed in detail in [12].

6   Conclusions

This paper details several communication modules as part of a social collective model. This includes both structure of the collective and types of messages. We have considered several types of groups and the communication withing them, and compared them using the measure of drift adapted from social sciences. We specifically did not test the approach in opinion formation or knowledge diffusion terms. While the social collective model can be applied as such, it is first a tool for modeling a group without the influence of external factors.

Our previous papers discussed other modules that build the social collective: the structure of knowledge used in the collective, how new knowledge is integrated, and how old knowledge is changed or forgotten. This concludes our research into developing the specific details of the model. Consequently, future research concerning the social collective model will be focused on applying it in specific situations, in order to better organize groups. Some initial work in this directions has already been done regarding organizing student groups in order to increase their knowledge acquisition due to exchange of individual knowledge.

References

1. Barbieri, N., Bonchi, F., Manco, G.: Topic-aware social influence propagation models. Knowl. Inf. Syst. **37**, 555–584 (2012)
2. Bhat, S.P., Bernstein, D.S.: Finite-time stability of continuous autonomous systems. SIAM J. Control. Optim. **38**(3), 751–766 (2000)

3. Chen, B., Tang, X., Yu, L., Liu, Y.: Identifying method for opinion leaders in social network based on competency model. J. Commun. **35**, 12–22 (2014)
4. Cheung, C.M.K., Lee, M.K.O.: A theoretical model of intentional social action in online social networks. Decis. Support Syst. **49**, 24–30 (2010)
5. Dubois D., Liu W., Ma J., Prade H.: The basic principles of uncertain information fusion. An organised review of merging rules in different representation frameworks. Inf. Fus. **32**, 12–39 (2016)
6. Fan, X.H., Zhao, J., Fang, B.X., Li, Y.X.: Influence diffusion probability model and utilizing it to identify network opinion leader. Chin. J. Comput. **36**, 360–367 (2013)
7. Kelman, H.C.: Interests, relationships, identities: three central issues for individuals and groups in negotiating their social environment. Annu. Rev. Psychol. **57**, 1–26 (2006)
8. Jiang, A., Marcolino, L.S., Procaccia, A.D., Sandholm, T., Shah, N., Tambe, M.: Diverse randomized agents vote to win. Adv. Neural Inf. Process. Syst. **27**, 2573–2581 (2014)
9. Jin, Y., Wang, W., Xiao, S.: An sirs model with a nonlinear incidence rate. Chaos Solit. Fractals **34**, 1482–1497 (2007)
10. Li, S., Dua, H., Lin, X.: Finite-time consensus algorithm for multi-agent systems with double-integrator dynamics. Automatica **47**, 1706–1712 (2011)
11. Maleszka, M., Nguyen, N.T., Urbanek, A., Wawrzak-Chodaczek, M.: Building educational and marketing models of diffusion in knowledge and opinion transmission. In: Hwang, D., Jung, J.J., Nguyen, N.-T. (eds.) ICCCI 2014. LNCS (LNAI), vol. 8733, pp. 164–174. Springer, Cham (2014). https://doi.org/10.1007/978-3-319-11289-3_17
12. Maleszka, M.: The increasing bias of non-uniform collectives. In: Nguyen, N.T., Pimenidis, E., Khan, Z., Trawiński, B. (eds.) ICCCI 2018. LNCS (LNAI), vol. 11055, pp. 23–30. Springer, Cham (2018). https://doi.org/10.1007/978-3-319-98443-8_3
13. Maleszka, M.: Application of collective knowledge diffusion in a social network environment. Enterp. Inf. Syst. **13**(7–8), 1120–1142 (2019)
14. Maleszka, M.: An intelligent social collective with Facebook-based communication. In: Paszynski, M., Kranzlmüller, D., Krzhizhanovskaya, V.V., Dongarra, J.J., Sloot, P.M.A. (eds.) ICCS 2021. LNCS, vol. 12744, pp. 428–439. Springer, Cham (2021). https://doi.org/10.1007/978-3-030-77967-2_36
15. De Montjoye, Y.-A., Stopczynski, A., Shmueli, E., Pentland, A., Lehmann, S.: The Strength of the Strongest Ties in Collaborative Problem Solving. Scientific reports 4, Nature Publishing Group (2014)
16. Nguyen, N.T.: Inconsistency of knowledge and collective intelligence. Cybern. Syst. Int. J. **39**(6), 542–562 (2008)
17. Nguyen, V.D., Nguyen, N.T.: An influence analysis of the inconsistency degree on the quality of collective knowledge for objective case. In: Nguyen, N.T., Trawiński, B., Fujita, H., Hong, T.-P. (eds.) ACIIDS 2016. LNCS (LNAI), vol. 9621, pp. 23–32. Springer, Heidelberg (2016). https://doi.org/10.1007/978-3-662-49381-6_3
18. Saito, K., Nakano, R., Kimura, M.: Prediction of information diffusion probabilities for independent cascade model. In: International Conference on Knowledge-Based and Intelligent Information and Engineering Systems, pp. 67–75 (2008)
19. Pratkanis, A.R. (ed.): Social Influence Analysis: An Index of Tactics. In Frontiers of Social Psychology. The Science of Social Influence: Advances and Future Progress, pp. 17–82. Psychology Press (2007)

20. Ren W., Beard R.W., Atkins E.M.: A survey of consensus problems in multi-agent coordination. In: American Control Conference. Proceedings of the 2005, pp. 1859–1864. IEEE (2005)

21. Wedrychowicz, B., Maleszka, M., Sinh, N.V.: Agent based model of elementary school group learning. In: Szczerbicki, E., Wojtkiewicz, K., Nguyen, S.V., Pietranik, M., Krótkiewicz, M. (eds.) Recent Challenges in Intelligent Information and Database Systems, ACIIDS 2022. CCIS, vol. 1716. Springer, Singapore (2022). https://doi.org/10.1007/978-981-19-8234-7_54

22. Xu, R., Li, H., Xing, C.: Research on information dissemination model for social networking services. Int. J. Comput. Sci. Appl. **2**, 1–6 (2013)

23. Zhu, T., Wang, B., Wu, B., Zhu, C.: Maximizing the spread of influence ranking in social networks. Inf. Sci. **278**, 535–544 (2014)

A Data-Driven Scheduling Strategy for Mobile Air Quality Monitoring Devices

Giang Nguyen[1], Thi Ha Ly Dinh[1(✉)], Thanh Hung Nguyen[1], Kien Nguyen[2,3], and Phi Le Nguyen[1]

[1] School of Information and Communication Technology,
Hanoi University of Science and Technology, Hanoi, Vietnam
`ly.dinhthiha@hust.edu.vn`
[2] Institute for Advanced Academic Research, Chiba University, Chiba, Japan
[3] Graduate School of Engineering, Chiba University, Chiba, Japan

Abstract. Along with the process of urbanization and mechanization, environmental pollution is becoming more and more serious all over the world. To this end, numerous efforts are directed toward evolving effective monitoring operations, including both stationary and mobile systems. Compared to traditional fixed monitoring stations, mobile air quality monitoring systems offer a more flexible and cost-effective way to achieve a fine-grained air quality map. However, as mobile air quality monitoring systems typically rely on lightweight devices with low-capacity batteries, conserving the energy of these devices becomes a significant challenge. A trivial method to reduce devices' energy is to set a low-frequency sampling rate and allow the device to go to sleep during the idle period. Nonetheless, this method may diminish the temporal granularity of the gathered data. To this end, in this paper, we propose a deep learning-based approach that adaptively regulates the activities of devices to simultaneously accomplish two goals: energy conservation and data quality assurance. The primary idea is to allow devices to go to sleep when the fluctuations of air quality indicators are minimal and to use a predictive model to forecast the air quality during the idle period. We evaluate our proposal on Fi-Mi, an actual bus-based air monitoring system in Hanoi, Vietnam. Experiment results indicate that our proposed method saves approximately 42% of the device's energy consumption with an air monitoring error of only 3%.

Keywords: Mobile air monitoring · Fi-Mi · Energy efficiency · Multi-task deep learning · Prediction model · Scheduling

1 Introduction

Due to accelerated industrialization and urbanization, air pollution is becoming an increasingly severe environmental concern. Consequently, people have raised

N. T. Nguyen et al. (Eds.): ACIIDS 2023, LNAI 13996, pp. 74–86, 2023.
https://doi.org/10.1007/978-981-99-5837-5_7

significant concerns regarding air quality, especially in crowded urban areas. Under these circumstances, a comprehensive solution for monitoring air quality and predicting future air conditions on a large scale is essential for assisting individuals in protecting their health and the government in prompt policy planning.

Traditionally, air quality indices are monitored and collected at stationary monitoring stations. For instance, in [1], the authors introduced a four-step strategy for nitrogen dioxide and ozone monitoring stations based on the sampling methodology. By leveraging IoT techniques and low power wide area (LPWA) technology, [2] deployed battery-powered sensor nodes over a broad geographic region to observe and provide real-time air quality to end-users via their website and mobile application. Despite their widespread use, the stationary air quality monitoring approach suffers from an inherent drawback: the limitation of the monitoring areas. Indeed, because of being stationary, fixed stations can only capture information around their locations; thus, many monitoring terminals are required to achieve a fine-grained air quality map. However, the significant deployment and maintenance expenses hinder such large-scale installation.

In recent years, mobile air quality monitoring has emerged as a viable alternative. In mobile air monitoring systems, compact monitoring devices are mounted on vehicles [3] or carried by individuals [4], thereby widening the monitoring range with a limited number of devices. Typically, mobile air quality monitoring devices rely on low-capacity batteries that frequently run out of power after a few hours of operation. Therefore, one of the critical issues in handling mobile air quality monitoring systems is extending the lifetime of the devices. To conserve energy, the most straightforward approach is to set a low sampling rate that enables devices to perform monitoring tasks at a low frequency and enter a sleep state during the inactive period. Reducing the sampling rate, however, diminishes the temporal granularity of the collected data, creating a trade-off between energy efficiency and data quality. To this end, we propose an adaptive scheduling solution based on the fluctuations of the air quality indicators. Intuitively, when the air quality indicators have little variation over time, we put the device to a sleep state to save energy and use a deep learning model to predict the data during the device's idle period. Also, based on that prediction, we will estimate the variability of the air quality indicators and utilize our proposed heuristic algorithm to decide whether to turn on the device or not in the next time slot. All these algorithms are implemented on the server side.

The major contributions of this paper are listed as follows.

1. We propose a mechanism that combines a deep learning-based air quality prediction and a heuristic algorithm for adaptively regulating the activities of air quality monitoring devices to simultaneously accomplish two goals: energy conservation and data quality assurance. The predictive model is designed using the multi-task learning approach so that it can accurately predict the data of multiple devices simultaneously.
2. The proposed algorithm is implemented on an actual vehicle-based mobile air monitoring system named Fi-Mi [7]. Experimental results indicate that our proposal can save 42% of energy while ensuring a 3% monitoring error.

2   Related Work

Mobile Air Quality Monitoring Systems. In the literature, there is a great deal of research focusing on issues related to mobile air quality monitoring systems. The authors [5] proposed a concept of Bus as a Sensor (BaaS) that leverages buses to carry air quality monitoring sensors to achieve a high-resolution air quality map. Reference [6] designed, implemented and evaluated UbiAir, an ambient environment-aware system that achieves fine-grained monitoring of urban air quality via a bicycle-based mobile crowdsensing system.

Another mobile air quality system based on buses was proposed in [4]. The authors focused on the susceptibility of moving vehicles to airflow disturbances. To achieve this, they developed Mosaic-Nodes with a novel design for constructive airflow disturbance based on a precisely tuned airflow structure and a GPS-assisted filtering method.

Energy Optimization for Sensor Node. Many studies have worked on optimizing energy for sensor nodes to extend the system's life cycle. Among them, [9] presented a method for adjusting the working period of sensor nodes. Namely, to obtain a given task performance, the sensors are alternated between active and sleep states by harvesting theory such that a required minimum active duration is guaranteed. Considering the sensor nodes equipped with both a rechargeable battery and an energy harvesting unit, [10] introduced Lazy Scheduling Algorithms (LSA) for energy and time scheduling to arriving tasks, i.e., the amount of energy and time should be assigned to complete a task such that all a given set of tasks are finished before the deadline and energy exhaustion.

Time Series Forecasting. Air quality index prediction is a time series prediction problem that is commonly addressed by leveraging recurrent neural networks (RNNs), including the long short-term memory (LSTM) technique [11]. As shown in [12], the LSTM-based model gains an advantage in the air pollution prediction problem. Also, [13] designed an RNN-based method incorporating LSTM network to predict air quality in Taiwan based on historical data from 77 fixed air quality monitor stations, which effectively predicted the PM2.5 value, but only for the next 4 h. Reference [14] introduced a Deep Multi-output LSTM (DM-LSTM) model integrating with three deep learning algorithms, including mini-batch gradient descent, dropout neuron, and L2 regularization, to derive key features of complex space-time relationships. The results showed that DM-LSTM improved the space-time stability and the accuracy of air quality predictions significantly. In [15], LSTM was combined with the convolutional graph network (GCN) to predict the PM2.5 index for up to 72 h with adequate recall and correlation coefficient. Reference [16] adapted a convolution and LSTM neural network model by introducing K-nearest neighbor algorithm to exploit space-time information for air pollution prediction, which was proved to be efficient for diverse time predictions at several regional scales.

Unlike the existing works that focus on only enhancing the precision of air quality index prediction, this work puts more effort to reduce sensors' energy consumption while guaranteeing the quality of prediction as detailed next.

3   Proposed Framework

3.1   Overview of the Air Quality Monitoring System

We consider a mobile air quality monitoring system composed of M mobile sensing devices and a processing server as depicted in Fig. 1. Mobile devices collect the air quality indices during their journeys and send these monitored data to a processing server which performs calibration and forecasting.

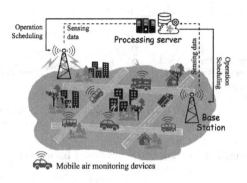

Fig. 1. System model

In particular, each mobile sensing device has two operating states: active and sleep. When device m is active at time slot t, it consumes an amount of energy W_{active} to collect air quality indices and to send observed results x_t^m to the server through an API. On the contrary, if device m is in sleep mode, no data collection and API interaction are operated, resulting in a less consumed power $W_{\text{sleep}} < W_{\text{active}}$. For this idle slot of device m, the missing monitoring data is then predicted by the server, denoted as y_t^m.

Obviously, the more y_t^m the server predicts, the more energy the mobile device can save by sleeping but at the cost of a large difference $|x_t^m - y_t^m|$ between the prediction and the real data due to the spatial and temporal variance of the air condition. Therefore, our proposal is first to enhance the quality of y_t^m prediction to keep the device in a longer idle state and by tracking the prediction quality, the server then decides when the device is active to gather ground-truth data if the prediction error exceeds a specific threshold. In particular, instead of independent prediction and decision for each device, the server enables to perform this process for all M terminals simultaneously as described next.

3.2   Overview of the Framework

In this section, we introduce a novel framework for an efficient mobile air monitoring system in both terms of energy and quality. Our framework is constructed with two parts: a multitask learning-based air quality prediction model and an adaptive scheduler as illustrated in Fig. 2. The prediction model takes historical air metrics and returns forecast values for the next time slots. Based on these values, the server performs the heuristic scheduling algorithm to determine the operation states of terminals and sends this decision to the devices.

3.3   Multitask Learning-Based Air Quality Prediction Model

Motivated by the aforementioned success of deep LSTM-based models in time series forecasting problems, we here also design a deep learning model leveraging the LSTM technique for air quality indices prediction. Moreover, given the number M of mobile air monitoring devices, it will be computational time and resource-consuming for training M prediction models independently. To handle this issue, we propose to build a deep multitask learning-based model to predict air metrics for multiple devices simultaneously.

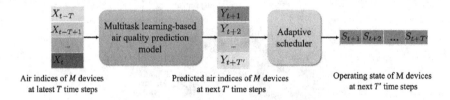

Fig. 2. Overview of the proposed framework

Namely, based on the Encoder-Decoder model introduced in seminal paper [20] for sequence-to-sequence processing, we design our predictive model for forecasting air indices when the monitoring devices are idle, as depicted in Fig. 3. The Encoder is to apprehend the historical data of devices and then compress them into a unique context vector describing the current state of the whole system. Based on this encoded vector, the Decoder predicts the air metrics of all devices for several next time-steps. Both Encoder and Decoder parts perform their missions by leveraging LSTM neural networks to process the different lengths of inputs as detailed next.

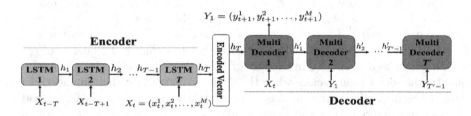

Fig. 3. Architecture of multitask learning-based air quality prediction model

The Encoder: The Encoder is composed of a stack of T LSTM blocks, where T is the input length representing the number of latest air quality values used to predict future values. At the time step t, the model receives two input vectors: the input data X_t and the previous hidden state h_{t-1}. Each input data $X_t = \{x_t^1, x_t^2, \ldots, x_t^M\}$ consists of air indices of M monitoring devices, where x_t^i denotes the value observed by the device i^{th} at the time step t. As such, our framework enables multi-task learning by considering the data of M devices simultaneously.

After capturing all T latest information of all devices in the system, the Encoder encodes it into a context vector C. The context vector C then becomes the hidden state h_T of the last time step T and carries the characteristic information of M devices (tasks) through T time steps.

The Decoder: The Decoder encompasses T' multi-decoder blocks for air quality indices prediction of all devices in T' next time steps. Namely, we design each multi-decoder block composed of M distinct sub-blocks corresponding to M monitoring devices in the system.

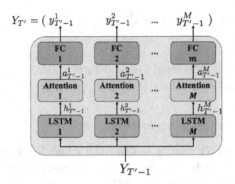

Fig. 4. A Multi-Decoder block in the Decoder

In particular, each sub-block is built with three layers as depicted in Fig. 4. The first one is an LSTM layer, followed by a scaled dot-product attention layer and the last one is a fully connected layer. By considering the historical and current data of all devices, the LSTM layer decides which current information to serve as a hidden state of the next layer. Denoting $h_{t'}^i$ as the hidden state of the sub-block i at the time step t', we calculate $h_{t'}^i$ by

$$
h_{t'}^i = \begin{cases} f'(x_T^i, C) & \text{if } t' = t+1, \\ f'(y_{t'}^i, y_{t'-1}^i) & \text{if } t' \neq t+1, \end{cases}
\tag{1}
$$

where $i = \{1, 2, ..., M\}$, $t' = \{t+1, t+2, ..., t+T'\}$ and f' is the transformation function of hidden states in the LSTM layer.

To take the impact of these data into account, the second layer leverages the scaled dot-product attention scheme [18], which assigns different weights (so-called attention weight) to the input data for different devices. In detail, at time step t', each output $h_{t'}^i$ of the LSTM layer in a sub-block i will be multiplied with three matrices W_q^i, W_k^i, W_v^i to obtain the query Q_i, the key K_i and the value V_i as the inputs of the i-th attention layer. Then by considering the query and the key have the same size d_k, the scaled dot-product attention weight of key K_i in Q_i is computed by following equation, where K' is the transposed matrix of K.

$$
a_i(Q_i, K_i, V_i) = Softmax(\frac{Q_i K_i'}{\sqrt{d_k} V}),
\tag{2}
$$

Finally, given the output of the attention layer, the predicted air indices $Y_{t'}$ of all M devices at time step t' are computed by a fully connected layer as

$$Y_{t'} = ReLu(W_1 A + b_1)W_2 + b_2, \tag{3}$$

where $A = (a_1, a_2, \ldots, a_M)$ is the output of the attention layer of all sub-blocks, W_1, W_2 are the weight matrices and b_1, b_2 are the bias factors.

Loss Function: During training M tasks simultaneously, the loss L of our model is calculated as a weighted sum over M loss components of M tasks, by

$$L = \frac{1}{M} \sum_{i=1}^{M} (\lambda_i \times L_i). \tag{4}$$

Here, λ_i is the weight of individual loss and L_i is a Mean Square Error between the actual and the predicted values at device i, given by $L_i = \frac{1}{T'} \sum_{i=1}^{T'} (y_j - \hat{y}_j)^2$.

3.4 Adaptive Scheduling Algorithm

Based on the predicted air quality indices, the scheduler decides to turn on/off the monitoring devices such that their consumed energy is minimized while the air monitoring quality is maintained. For that, we propose an adaptive scheduling algorithm based on prediction error. That is, the strategy keeps the devices in their idle states until the prediction error surpasses the predefined threshold α. In other words, the devices need to be active to observe the ground-truth data for prediction enhancement at each following time slot.

Algorithm 1: Adaptive scheduling algorithm based on prediction error

---

    Input: Historical data $X = (X_{t-T}, X_{t-T+1}, \ldots, X_t)$
1 **while** *True* **do**
2   mode = Sleep;
3   **for** $t' = t+1, \ldots, t+T'$ **do**
4   Y = Prediction model (X);
5   mode = Active ;
6   **for** $t' = t+T'+1, \ldots, t+T'+k$ **do**
7   Collect real data $x_{t'}$;
8      $t' \leftarrow t+T'+k+1$;
9   **while** *True* **do**
10         $X' = (X'_{t'-T-1}, X'_{t'-T}, \ldots, X'_{t'-1})$;
11   Y = Prediction model (X');
12   **if** $MAPE(Y_1, X'_{t'}) < \alpha$ **then** break

The pseudo-code of proposed adaptive scheduling is given in Algorithm 1. In detail, it is supposed that the sensing device is off from slot t to $t + T'$ (Lines

2–4). After T' time steps, its operating mode is switched to active for the next k time steps, where k is a small arbitrary number (Lines 5–7). Then, from time step $t + T' + k + 1$, we keep the device in the active state until its prediction error $MAPE$ is less than threshold α (Lines 9–13), where $MAPE$ is defined by Eq. (5) in Sect. 4.4.

4   Performance Evaluation

4.1   Hardware Implementation

As a use-case, we experiment with the proposed framework on a practical bus-mounted air quality monitoring system, namely Fi-Mi [7]. Fi-Mi is a system of small sensor devices attached to Hanoi internal buses, forming mobile air quality monitors. Generally, a Fi-Mi device is composed of four function blocks: a sensor block for collecting air quality indices; a microcontroller unit for pre-processing sensed data; a communication block for reporting these data to the server via Wifi or LTE; and a power supply. At the server with a GPU of 24 GB RAM NVIDIA GeForce RTX 2080 Ti, we implement our framework using Pytorch.

4.2   Experimental Setup

After deploying the proposed framework on the Fi-Mi system, we utilize the dataset composed of 8 subsets, each of which contains 40000 records of PM2.5 values collected by a Fi-Mi device on its route [7] for training and evaluation.

We set up the parameters of our framework as follows. For the prediction part, the Encoder uses $T = 10$ LSTM layers, each of which has 32 hidden nodes. The Decoder predicts PM2.5 values for $T' = 10$ next time-steps by 10 Multi-Decoder blocks. A Multi-Decode block contains $M = 8$ identical sub-blocks, each of which is composed of an LSTM layer with a size of 32 hidden nodes, an attention layer and a fully connected (FC) layer with the same size. We train the Predictive model with a batch size of 32 and apply Adam optimization to update the weights. The initial learning rate (γ) is set to 10^{-3} and revised during the training process by $\gamma = \gamma * 0.9^{\lfloor \frac{e}{25} \rfloor}$, where e is the current epoch number.

For the adaptive scheduling algorithm, its parameters are tuned manually through a preliminary assessment over varying values of α and k. We found that $\alpha = 5, k = 0$ is the best setting for the proposed adaptive scheduler.

4.3   Benchmark Algorithms

To evaluate the prediction performance of the proposed deep multitask learning-based method, we compare our model with the single-task one where the reference model trains and predicts the air indices for each device based on their own historical knowledge independently, not being noised by the others' data.

We also make comparisons between our proposed adaptive scheduling algorithms and two baselines strategies, namely

- **Cyclic strategy** (*Ref. 1*): This scheme turns on/off the sensing devices periodically, i.e., a period of τ active and then T' inactive time steps are iterated during the device's operation.
- **Prediction variability-based strategy** (*Ref. 2*): This strategy adjusts the devices' operating states based on the variability of predicted values. Namely, if there are two consecutive predictions at time i and $i + 1$, whose difference exceeds a predefined threshold β, the device is turned on to gather the ground truth for the next l time steps since slot $i + 1$. If no such situation happens, the device is set to be inactive for T' time steps as in *Ref. 1*.

For fair evaluation, the parameters of two reference algorithms are also tuned manually. Namely, we set $\tau = 5$ for the reference cyclic strategy, and $l = 2, \beta = 0.5$ for the reference prediction variability-based strategy.

4.4   Evaluation Metrics

We evaluate the performance of all algorithms on two statistical metrics commonly used for regression methods, including Mean Absolute Percentage Error (MAPE) and Pearson Correlation (r) [21], given by

$$MAPE = \frac{1}{n} \sum_{i=1}^{n} \frac{|x_i - y_i|}{x_i}, \tag{5}$$

$$r = \frac{\sum_{i}^{n} (x_i - \bar{x})(y_i - \bar{y})}{\sqrt{\sum_{i}^{n} (x_i - \bar{x}) \sum_{j}^{n} (y_j - \bar{y})}}, \tag{6}$$

where x_i, y_i are the actual and predicted values at time step i, respectively. Following that, $\bar{x}, \bar{y}$ denote the average of n actual and predicted values.

In addition, we define a metric named *Saving* to assess the level of energy saving of the proposed strategies, which is a ratio between the energy saved by going into sleep mode for some intervals and the total energy consumed by a non-sleep device, given by

$$Saving = 1 - \frac{W_{\text{sleep}} * t_{\text{sleep}} + W_{\text{active}} * t_{\text{active}}}{W_{\text{active}} * (t_{\text{active}} + t_{\text{sleep}})} * 100\%, \tag{7}$$

where $W_{\text{sleep}} = 0.617$ and $W_{\text{active}} = 3.185$ Watts are the power consumption of a Fi-Mi device in sleep and active modes [7], respectively.

4.5   Numerical Results

4.5.1   Performance of Deep Multitask Learning-Based Prediction Model

We evaluate the effectiveness of the proposed multitask learning-based air quality prediction by comparing it with the single-task one as mentioned in Sect. 4.3.

Table 1. The number of model parameters and the running time of two methods

Method	Data set	No. of parameters	Running time (s)
Single-task	Sensor 1	14273	1.66
	Sensor 2	14273	1.65
	Sensor 3	14273	1.65
	Sensor 4	14273	1.65
	Sensor 5	14273	1.66
	Sensor 6	14273	1.65
	Sensor 7	14273	1.65
	Sensor 8	14273	1.65
	Total	114184	13.22
Multi-task	8 sensors	83720	12.08

First, we show the number of parameters and running times for both single and multi-task learning-based prediction methods in Table 1. For the single-task method, Fi-Mi devices have the same number of parameters and running time due to their identical structures and the same training data size. However, since the training process is performed independently for each device, the single-task scheme needs to train 114184 parameters in 13.22 s[1] in total. Meanwhile, by merging the data from devices into one input and predicting their next PM2.5 values simultaneously, the multi-task method reduces the number of training parameters down to 83720, i.e., about 26.68% lower than that of the single-task learning model, and takes 9% faster of running time.

In spite of such a lower number of parameters, the multi-task learning-based prediction method still keeps a similar performance as the single-task one in both terms of $MAPE$ and Pearson Correlation r, as presented in Fig. 5. We observe that the multi-task model achieves a slightly better $MAPE$ and r for some devices such as Fi-Mi 1 and Fi-Mi 6, whereas the same or slightly worse performance is produced for the other devices. The reason is due to the (either good or bad) influence among data of devices during the training process in the multi-task model. However, by averaging over the whole dataset, the single-task learning-based prediction method achieves a $MAPE$ of 2.29% and a Pearson correlation r of 95.60%, whereas that are respectively 2.40% and 95.67% for the multi-task learning-based prediction. This indicates that the multi-task model provides a better trade-off between the performance and the complexity, especially in large-scale systems with a large number of sensing devices.

---

[1] Obviously, this running time could be reduced by parallel training processes, which is out of the scope of this study.

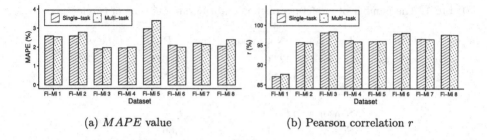

(a) MAPE value   (b) Pearson correlation r

Fig. 5. Performance of single- and multi-task learning-based prediction methods

Table 2. Performances of proposed sampling frequency regulation algorithms

Dataset	Ref. 1		Ref. 2		Proposed method	
	MAPE (%)	Saving (%)	MAPE (%)	Saving (%)	MAPE (%)	Saving (%)
Fi-Mi 1	3.64	37.91	3.64	37.91	**3.38**	**41.35**
Fi-Mi 2	3.49	37.91	3.22	35.61	**2.99**	**41.78**
Fi-Mi 3	2.40	37.91	**2.29**	37.69	2.33	**43.32**
Fi-Mi 4	2.59	37.91	2.43	36.81	**2.42**	**43.51**
Fi-Mi 5	5.40	37.91	**3.96**	32.71	4.33	**39.75**
Fi-Mi 6	2.40	37.91	2.39	36.60	**2.37**	**43.26**
Fi-Mi 7	2.81	37.91	2.70	36.62	**2.49**	**42.52**
Fi-Mi 8	2.73	37.91	2.60	37.60	**2.45**	**43.02**
Average	3.18	37.91	2.90	36.44	**2.85**	**42.31**

4.5.2   Performance of Adaptive Scheduling Algorithm

Next, we compare and evaluate the performance of the proposed adaptive scheduling algorithm in comparison with two baseline methods as presented in Sect. 4.3. Their detailed results are given in Table 2 in terms of $MAPE$ and *Saving*. It can be observed that the proposed method outperforms the cyclic algorithm (*Ref. 1*) overall dataset in both metrics, namely about 10.4% lower of $MAPE$ and 11.6% higher energy efficiency on average. Compared to the prediction variability-based strategy (*Ref. 2*), the proposed algorithm also provides better $MAPE$ and *Saving* for almost devices, except in the case of Fi-Mi 3 and 5 with higher $MAPE$ values.

Ref. 1 does not work well because the on-off state is just switched periodically without any consideration of prediction quality. In other words, the algorithm possibly falls into the case that Fi-Mi device is turned on while the predictive model gives a good prediction and vice versa. Also, *Ref. 2* may face such an issue, since the variability bias does not give us a sight of whether the predicted value is good or not. Meanwhile, the proposed method takes prediction error into account for the sampling frequency adjustment, thereby overcoming the weakness of the two others. It regulates the device off when the model's prediction error is less than an acceptable threshold α and otherwise, turn the device on, i.e., the prediction quality should be enhanced by the actual data. This explains why the prediction error-based strategy obtains the best performance.

5 Conclusion

In this paper, we investigated the issue of efficient scheduling for mobile air quality monitoring devices, which not only conserve the energy of moving sensing terminals but also guarantee high monitoring performance. For that, we proposed a framework composed of two main parts: a deep multi-task learning-based prediction model and an adaptive scheduling algorithm that automatically adjusts the device's operating state based on the prediction error. Experimental results on the practical mobile air quality monitoring system, Fi-Mi, showed that the multi-task learning method for prediction has improved the calculation speed by 9% faster compared to the single-task method and reduced 27% the total number of model parameters. The results also depict the efficiency of our scheduling algorithm with the energy saving level of 42.31% while the error is only 2.85%.

In the future, we will increase the depth of the model, and apply advanced techniques related to multi-task learning to enhance the accuracy and reduce the computational time of the deep learning model. Besides, improving the auto-tuning algorithm is also an important task for us to consider.

Acknowledgments. This work was supported in part by the Japan Society for the Promotion of Science (JSPS) under Grant 20H0417, 23H03377. This research is also funded by Hanoi University of Science and Technology (HUST) under grant number T2022-PC-044, and partially supported by NAVER Corporation within the framework of collaboration with the International Research Center for Artificial Intelligence (BKAI), School of Information and Communication Technology, HUST under project NAVER.2022.DA01.
This work was also supported by Vingroup Joint Stock Company (Vingroup JSC), Vingroup and supported by Vingroup Innovation Foundation (VINIF) under project code VINIF.2020.DA09.

References

1. Lozano, A., et al.: Air quality monitoring network design to control nitrogen dioxide and ozone applied, in Malaga, Spain. Microchem. J. **93**(2), 164–172 (2009)
2. Zheng, K., et al.: Design and implementation of LPWA-based air quality monitoring system. IEEE Access **4**, 3238–3245 (2016)
3. Xu, X.: Deployment of a vehicle-based environmental sensing system. In: Proceedings of the 14th ACM Conference on Embedded Network Sensor Systems CD-ROM, pp. 376–377 (2016)
4. Gao, Y., et al.: Mosaic: a low-cost mobile sensing system for urban air quality monitoring. In: INFOCOM, pp. 1–9 (2016)

5. Arfire, A., Marjovi, A., Martinoli, A.: Enhancing measurement quality through active sampling in mobile air quality monitoring sensor networks. In: IEEE International Conference on Advanced Intelligent Mechatronics, pp. 1022–1027 (2016)
6. Wu, D., et al.: When sharing economy meets IoT: towards fine-grained urban air quality monitoring through mobile crowdsensing on bike-share system. Proc. ACM Interact. Mob. Wearable Ubiquit. Technol. 4(2) (2020). Article 61
7. Nguyen, V.A., et al.: Realizing mobile air quality monitoring system: architectural concept and device prototype. In: 26th IEEE APCC, pp. 115–120 (2021)
8. Kortoçi, P., et al.: Air pollution exposure monitoring using portable low-cost air quality sensors. Smart Health 23, 100241 (2022)
9. Kansal, A., et al.: Performance aware tasking for Environmentally Powered Sensor Networks. ACM SIGMETRICS Perform. Eval. Rev. 32, 223–234 (2004)
10. Moser, C., et al.: Real-time scheduling with Regenerative Energy. In: ECRTS, p. 10 (2006)
11. Hochreiter, S., Schmidhuber, J.: Long short-term memory. Neural Comput. 9(8), 1735–1780 (1997)
12. Kök, I., et al.: A deep learning model for air quality prediction in smart cities. In: IEEE Big Data, pp. 1983–1990 (2017)
13. Tsai, Y.T., et al.: Air pollution forecasting using RNN with LSTM. In: IEEE 16th International Conference on Dependable, Autonomic and Secure Computing, In: 16th International Conference on Pervasive Intelligence and Computing, 4th International Conference on Big Data Intelligence and Computing and Cyber Science and Technology Congress, pp. 1074–1079 (2018)
14. Zhou, Y., et al.: Explore a deep learning multi-output neural network for regional multi-step-ahead air quality forecasts. J. Clean. Prod. 209, 134–145 (2019)
15. Qi, Y., et al.: A hybrid model for spatiotemporal forecasting of PM2. 5 based on graph convolutional neural network and long short-term memory. Sci. Total Environ. 664, 1–10 (2019)
16. Wen, C., et al.: A novel spatiotemporal convolutional long short-term neural network for air pollution prediction. Sci. Total Environ. 654, 1091–1099 (2019)
17. Caruna, R.: Multitask learning: a knowledge-based source of inductive bias. In: Machine Learning: Proceedings of the 10th International Conference, pp. 41–48 (1993)
18. Vaswani, A., et al.: Attention is all you need. Adv. Neural Inf. Process. Syst. 30 (2017)
19. Roundy, S., et al.: Power sources for wireless sensor networks. In: European Workshop on Wireless Sensor Networks, pp. 1–17 (2004)
20. Sutskever, I., et al.: Sequence to sequence learning with neural networks. Adv. Neural Inf. Process. Syst. 27 (2004)
21. Plevris, V., et al.: Investigation of performance metrics in regression analysis and machine learning-based prediction models (2022)

Hybrid Approaches to Sentiment Analysis of Social Media Data

Thanh Luan Nguyen[1], Thi Thanh Sang Nguyen[1]([✉])(iD), and Adrianna Kozierkiewicz[2](iD)

[1] School of Computer Science and Engineering, International University, VNU-HCMC, Vietnam National University, Ho Chi Minh City, Vietnam
`itdsiu18041@student.hcmiu.edu.vn, nttsang@hcmiu.edu.vn`
[2] Faculty of Computer Science and Management, Wroclaw University of Science and Technology, Wroclaw, Poland
`adrianna.kozierkiewicz@pwr.edu.pl`

Abstract. Today, social media is accessible all around the world. It has become an online place for reviewing products or services. Social media is an appealing resource for enterprises looking to monitor user attitudes because of the massive volume of posts. It has gathered much attention in the field of sentiment analysis. Sentiment analysis models could be classified into three categories: Lexicon-based, classical Machine Learning, and Deep Learning model. The classification methods have received a lot of attention in the research area. Although supervised learning algorithms are crucial components, the preprocessing step greatly impacts the accuracy of the sentiment analysis task. In preprocessing, the negation handling method needs to get more attention as the most popular technique creates redundant features. This research proposes a negation handling method that utilizes the lexicon-model VADER (Valence Aware Dictionary for sEntiment Reasoning) for concentrating on high-sentiment words. The proposed negation handling method improves the accuracy of the classical machine learning algorithms Logistic Regression and Naive Bayes. In addition, the research proposes an approach to handling elongated words and emoji characters instead of removing them. An experiment is conducted for a comparison of the proposed methods and the chosen base techniques.

Keywords: Sentiment Analysis · Negation Handling · Preprocessing

1 Introduction

Sentiment analysis or opinion mining is the process of extracting polarities such as positive, neutral, or negative of a document, a sentence, or a word. In other words, sentiment analysis is a text classification problem. For instance, the sentence "The staffs were rude, I'll never go shopping there" could be classified as negative. Extracting sentiment from posts on social media such as Twitter has earned a concentration in different research areas and applications of business, medical, behavioral, and political studies. In disaster recovery, sentiment analysis techniques are used to propose a system

N. T. Nguyen et al. (Eds.): ACIIDS 2023, LNAI 13996, pp. 87–98, 2023.
https://doi.org/10.1007/978-981-99-5837-5_8

that classifies three Twitter datasets which are the Nepal earthquake, the Italy earthquake, and Covid-19 into two classes "resource needs" and "resource availability" to support resource management during a disaster [1]. Another similar application is from the work of Ragini et al. [2] in which they proposed an end-to-end machine learning system that extracts tweets from Twitter and classifies them into five categories "food", "water", "shelter", "medical emergency" and "electricity" for understanding the most basic needs of the people during a disaster. Another application in the political area is to predict the winner of an election. In the work of Somula et al. [3], a lexical-based model is used to classify the tweets into "positive" or "negative" in each state of the United States to predict the next president in the 2016 election between two candidates Hillary Clinton and Donald Trump. Sentiment analysis has appeared in several research and industry areas, however, mining polarity from social media posts and comments is difficult because of the messy unstructured nature of text data. Social media also contains many wrong spelling words, wrong grammar, and teenage language. Nevertheless, social media is an excellent big data source for sentiment analysis if we have an appropriate mining approach depending on our objective.

Different types of models have been employed in sentiment analysis tasks, and they can be categorized into three approaches include rule-based, traditional machine learning, deep learning, and hybrid approach. The performance of a sentiment analysis model is determined by various factors, including the domain, the machine learning method, and the preprocessing step. The preprocessing process is a crucial step in sentiment analysis, especially for classical machine learning methods [4]. For supervised learning, training data has a great impact on the result. The preprocessing step often is different between the type, domain, and quality of data. The preprocessing step may remove unnecessary and noisy features, and it may create new features, for example, the n-gram model, to improve the result of sentiment analysis.

In preprocessing text data for sentiment analysis, one of the problems is the way of handling negation clauses. However, the most currently used negation handling method [5] creates redundant noisy features. The methods append a prefix "NEG_" to every word in a clause if a negation word is present. For example, in the sentence "This laptop is not really a good one, you should not buy it.", the negation handled sentence will be "This laptop is not NEG_really NEG_a NEG_good NEG_one NEG_, you should not NEG_buy NEG_it.". Words like "really", "one", and "it" are not required to apply negation handling. Although a preprocessing step may remove the words "one" and "it" as stopwords, a negation handling method is more efficient when applied to only high-sentiment tokens. *The main focus of this paper is to propose a more efficient negation handling method that focuses on high-sentiment words; therefore, it does not produce unnecessary features.* The new negation handling method utilizes the lexicon-based model VADER [6] to focus on high-sentiment tokens. The proposed negation handling method works more efficiently with classical machine learning algorithms which are Logistic Regression, Bernoulli Naive, and Multinomial Naive Bayes. In addition, *an efficient pipeline of preprocessing text data is provided*, which is specialized for social media text data in English. It is based on the preprocessing pipeline in [2] as a baseline, and the difference is that *an approach for elongated words and emoticons is proposed instead of removing these useful sentiment features.*

The rest of the paper is organized as follows. The next section describes background and related work. Section 3 contains the description of our methodology and developed method. We describe the results of the experiments in Sect. 4. Finally, we present the conclusions in the last section.

2 Literature Review

2.1 Preprocessing

The impact of preprocessing text data before classification has been proved through several research. For instance, by removing emoticons and applying bigrams, the accuracy of Naive Bayes increased by 8.12% in the work of Alam et al. [4]. In the research of S. Pradha et al. [7], techniques like stemming, lemmatization, and spelling correction are compared, and the result is that spelling correction consumes much processing time compared with the other two techniques, yet it provides the worst accuracy. Different from spelling correction both stemming, and lemmatization aim to reduce a word's inflection and convert it into its base form. For example, "am" and "is" to "be" or "cars" to "car". However, lemmatization and stemming work differently. Stemming is a primitive heuristic procedure that cuts off the ends of words in the expectation of attaining its goal, and it frequently involves the removal of derivational affixes [8]. Lemmatization prefers to work correctly by using a vocabulary and morphological analysis of words, with the goal of removing only inflectional ends and returning the base form of a word, known as the lemma [8]. In another research, Duong et al. [9] provided various processing techniques for Vietnamese sentiment analysis and data augmentation methods that improve accuracy on different datasets and algorithms. Some preprocessing methods include elongated characters removal, abbreviations or wrong-spelling lexicons replacement, emoticon replacement, punctuation removal, numbers removal, and negation handling. Nonetheless, elongated words are common in social media posts, and removing them will lose useful features for sentiment analysis. *An approach for normalizing elongated words is proposed in this paper to keep their presence but reduce noise from them.* The preprocessing methods are various for different supervised learning algorithms. The research of Cannannore et al. [10] pointed out that a preprocessing pipeline significantly reduces the accuracy of traditional classifiers like Logistic Regression and Naïve Bayes while the same preprocessing techniques increased the accuracy of the Convolution Neural Network (CNN) model.

2.2 Vectorization

Vectorization in sentiment analysis is a step to convert text data to its numerical representation because most machine learning algorithms require numeric values for their own computation. Bag-of-Words and Term Frequency-Inverse Document Frequency (TF-IDF) are the most popular and basic vectorization methods in natural language processing. Each of them is suitable for different machine learning algorithms. For example, Bernoulli Naive Bayes requires Bag-of-Words (BoW) with binary representation for the calculation of the algorithm's prior probability. Word embedding is another more advanced technique that is usually suitable for deep learning methods. In the experiment, BoW and TF-IDF are used in the models for evaluating the proposed methods.

2.3 Negation Handling

In sentiment analysis, negation is a problem since its presence can reverse the sentiment of a sentence while the negation word is not enough to dominate machine learning algorithms to realize the opposite attitude of the negation corpus. For instance, the sentence "His favorite phone is not good for you." could be more likely to be classified as "positive" while its ground-truth label is "negative" because of the positive words like "favorite" and "good". Those positive words frequently appear in positive documents, so a machine learning algorithm will increase the "positive" weight on the above sentence. As a result, supervised learning algorithms notice those positive words and classify the given sentence as "positive".

One of the questions is whether negation words are associated with negativity. The correlation between negation and negativity is analyzed in the research of Christopher [11]. The research points out that words in the IMDB dataset like "great", "amazing", etc., which are positive scalar modifiers, are frequent at high rating reviews, on the other hand, negative scalar modifiers such as "bad", "terrible", etc., are frequent at bad rating reviews. He also provides evidence that negation and negativity are strongly correlated across datasets in different contexts. Nonetheless, the presence of negation in positive documents still exists. Isaac et al. [12] present a negation detection method based on a conditional random field modeled using features from an English dependency parser. As the authors mention linguistic negation exists in many forms including the use of negating words like "not" or "no", and morphological negation where negating prefixes such as "dis", "un", and "none" or suffix like "less". Negation forms can also be categorized into denials, rejections, imperative, questions, support, and repetition; each represents unique challenges and makes a negation detection method more complicated. With the lexicon negation detection model, their sentiment analysis system shows a significant increase in precision and recall. Another negation handling method is proposed by Itisha [13] to avoid misclassifying the double negation clauses and rhetoric questions. For example, the sentence "I am not sure about #DeMonetisation but in such rhetoric, there is no one as good as him." has a double negation presence, and the rhetorical question like "How do you solve a problem like Maria?".

2.4 Lexicon-Based Model

Lexicon-based or rule-based models in sentiment analysis contain defined rules or lexicons that can classify a document into, for example, "positive" or "negative". A sentiment lexicon is a collection of lexical elements, such as words, that are often classified as either positive or negative depending on their semantic orientation. With added variance, distance calculations, negation rules, and numerous other rules, the lexicon can be exceedingly complicated. Some popular lexicon-based models include SentiWordNet, SenticNet, and VADER [6]. Mayu et al. [14] proposed a formula for calculating the co-occurrence correlations between emotion words and emojis. In their formula, the computed scores range from -1 to 1 where -1 is the most negative score and 1 is the highest positive score. For example, the emoji with a cake has a score of 0.95 and the emoji of emergency lights has a score of -0.21. Although they did not propose a complete lexicon model for sentiment analysis, their result can be combined with the current lexicon-based model in future research or application.

Valence Aware Dictionary for sEntiment Reasoning (VADER) [6] uses a combination of qualitative and quantitative methods to construct sentiment lexicons that are especially attuned to microblog-like contexts. These lexical features are combined with consideration for five generalizable rules that embody grammatical and syntactical conventions humans use when expressing or emphasizing sentiment intensity. By incorporating these heuristics, VADER improves the accuracy of the sentiment analysis engine across several domain contexts (social media text, movie reviews, and product reviews). However, the VADER lexicon performs exceptionally well in the social media domain. In the research of Nemes et al. [15], multiple models were used to analyze sentiments of COVID-19-related tweets to observe the emotion of people during the pandemic. They used the baseline model, i.e., Bidirectional Encoder Representations from Transformers - BERT, which was considered the outstanding model due to its transformer mechanism, to evaluate other models' performance compared to it. Their results showed that the VADER lexicon-model did not provide an approximated outcome with BERT compared to the RNN (Recurrent Neural Network) model; however, it outperformed TextBlob. This paper uses VADER as a crucial element in improving our proposed negation handling method.

3 Methodology

3.1 Modeling

The diagram in Fig. 1 describes the modeling process proposed for later experiments. Feature selection is crucial in text feature engineering as they will select the best features for classification, however, it is not included in the modeling pipeline of the experiments to clearly see the performance of the preprocessing steps. Because feature selection methods may select the same features for two preprocessing pipelines, we could not notice the impact of the preprocessing steps. Moreover, the presence or absence of the feature selection step does not affect the created features after the preprocessing step, but those produced features could strongly impact the final sentiment results. The produced features are vectorized in order to be input into classification algorithms to obtain the learned models. Thus, the preprocessing step is performed before splitting the data into training and testing sets. The following presents the details of proposed solutions in the preprocessing pipeline which can improve the performance of the learning process. Testing the learned models are evaluated later.

3.2 Preprocessing Pipeline

A pipeline of preprocessing steps, named BasePP for convenience, referenced from the research of S. Pradha et al. [7], is used as the baseline to test the proposed preprocessing framework. The base preprocessing pipeline includes lowercase, tokenization, lemmatization, and removal of usernames, hashtags, hyperlinks, encoded characters, stopwords, and punctuations; however, this pipeline can be improved by keeping emoticons and elongated words.

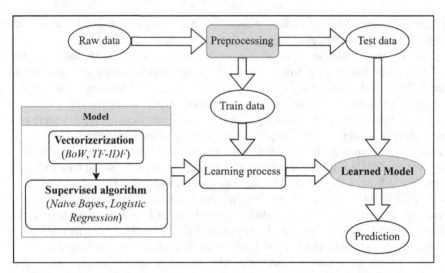

Fig. 1. The diagram depicts the modeling process.

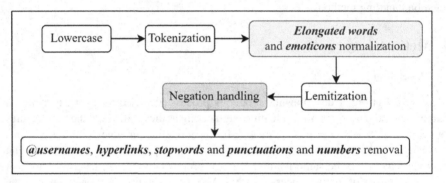

Fig. 2. The proposed preprocessing pipeline.

In the proposed preprocessing pipeline (Fig. 2), which is notated as *PP*, techniques include lowercase, tokenization, elongated words normalization, emojis normalization, lemmatization, negation handling, and the removal of username, stopwords, punctuations, and number. The difference between PP from BasePP is the elongated words normalization, emoticon normalization, and negation handling method. The base preprocessing method prefers to remove the noisy tokens like elongated words and emojis, while the proposed preprocessing method favors "normalizing" them by converting them to less noisy tokens. For instance, instead of eliminating elongated tokens in a tweet, the word "loooovvvveeeee" or "looovvvvee" is converted to "loovvee". The approach removes all the same consecutive characters of a length greater than two. The sequence "oooo" becomes "oo", and "vvvvvvv" is converted to "vv". For emoji characters, two lists of some positive emojis and negative emojis are defined as shown in Table 1. Positive emoticons are converted to "happy" and negative emoticons are transformed to "sad".

Table 1. Some samples of emoji characters are converted to their correspondent token.

Emoticon	Sentiment
:-), :), ;), :o), :], :3, :c), :>, =], 8), =), :}	Positive
:L, :-/, >:/, :S, >:[, :@, :-(, :[, :-|, =L, :<	Negative

3.3 Negation Handling with VADER

Table 2. The pseudocode of the proposed negation algorithm NEG-VADER.

FUNCTION NEGATE (document, N_w)
1   words = Tokenize (document)
2   is_negation = **False**
3   **for** i **from** 0 **to** N_w:
4   **if** words[i] **is** a negation word
5   **if** is_negation **is False**
6   is_negation = **True**
7   **Continue**
8   **else if** IS_FEATURE_WORD (threshold, word[i]) **is True**
9   words[i] += "_NEG"
10   **else if** is_negation **is True and** words[i] **is** a punctuation
11   is_negation = **not** is_negation
12   **else if** is_negation **is True and** words[i] **is not** a punctuation
13   **if** IS_FEATURE_WORD (threshold, word[i]) **is True**
14   words[i] += "_NEG"
15   new_document = Concat (words)
16   **return** new_document

Appending a distinguished prefix or suffix to words in a negation handling task for marking words in a negation clause different from the words in a non-negation sentence. The negation handling algorithm described in Narayanan et al. [5] is used as the baseline, namely NEG, for comparison to the proposed negation handling methods. The negation handling method NEG appends a suffix "_NEG" to each word if there is a presence of a negation word like "not" or "n't" in a clause. On the other hand, the proposed negation handling technique, namely NEG-VADER, combined with the lexicon-based model VADER [6] to produce a result that only focuses on high sentiment words. For instance, the comment "How can you not like this video, its explanation is the best." is labeled as positive. The baseline method NEG will result in the output (1) "How can you not_NEG like_NEG this_NEG video_NEG, its explanation is the best." while NEG-VADER will produce (2) "How can you not like_NEG this video, its explanation is the best.". Comparing outputs (1) and (2), a classifier like Naive Bayes would be more likely to classify the first one more negatively than the second because more words with the suffix "_NEG" in (1) than in (2). Furthermore, the proposed method will reduce the features used in the training process. Every word with the suffix "_NEG" is a new word that is treated differently from its original form, so the machine learning algorithms distinguish the word "best" and "best_NEG". As a result, a significant amount of vocabulary for training could be decreased with the proposed method (Tables 2 and 3).

Table 3. The pseudocode of the function IS_FEATURE_WORD in Table 2.

FUNCTION IS_FEATURE_WORD (threshold, word)
1   result = **False**
2   neg_prob = **VADER.get_neg_prob(**word)
3   pos_prob = **VADER.get_pos_prob(**word)
4   **if** neg_prob >= threshold **or** pos_prob >= threshold
5 | result = **True**
6   **return** result

4 Experiments

4.1 Experiment Setup

The whole training and testing phase is processed with five trials, and 30% is taken on the whole dataset for testing each time. Each train-test epoch has a different random seed; therefore, the training set and testing set are sampled randomly for each trial. The 5-fold cross validation is also applied to each trial.

The two popular datasets used for testing are Sentiment140 [16] and T4SA [17]. The original dataset T4SA contains three labels "negative", "neutral", and "positive", so the dataset is filtered to get only examples with negative and positive labels. The sizes of the Sentiment140 and T4SA for this project are 1,600,000 and 358,100 examples, respectively. For each trial, the number of random examples of Sentiment140 is randomly sampled to 400,000, while for T4SA all data is loaded to train for each trial. The distribution of the positive and negative in the two datasets are equal, and the training and testing sets are also stratified. Thus, the problem of imbalanced data is not a concern. Therefore, the performance of training models is evaluated using *accuracy* which is defined as the number of correct predictions divided by the number of instances in the testing data in each trial. Then, the average testing accuracies are calculated by taking the average testing accuracies of five trials. Tables 4 and 5 present the notations of the models and preprocessing pipelines used in the experiments, respectively.

4.2 Results

Tables 6 and 7 show the testing accuracy in each trial in detail, and they provide the average accuracy for each combination of the defined models and preprocessing pipelines. The number of total features after each preprocessing pipeline and trial are shown in Fig. 3 and Fig. 4 for each dataset.

Table 4. Notation for the models used in the experiments. Bigram features are created before vectorization. The Laplace smooth of Naïve Bayes is set to 0.1.

Vectorization	Supervised algorithm	Notation
Frequency BoW	Multinomial NB	**FreBoW-MultiNB**
TF-IDF	Multinomial NB	**TFIDF-MultiNB**
TF-IDF	Logistic Regression	**TFIDF-LogReg**
Binary BoW	Bernoulli NB	**BinBoW-BerNB**

Table 5. Description and notation of the preprocessing pipe used in the experiments.

Preprocessing pipeline	Notation
The base processing pipeline	BasePP
The proposed preprocessing pipeline with the base negation handling method	NEG & PP
The proposed preprocessing pipeline with the proposed negation handling method	NEG-VADER & PP

Table 6. Average testing accuracy on the dataset Sentiment140 in five trials. The notation of the models and the preprocessing pipelines are described in Tables 4 and 5.

	BasePP	NEG & PP	NEG-VADER & PP
FreBoW-MultiNB	0.7731	0.7893	**0.7941**
TFIDF-MultiNB	0.7751	0.7929	**0.797**
TFIDF-LogReg	0.7882	0.8105	**0.8106**
BinBow-BerNB	0.7729	0.7783	**0.7923**

Table 7. Accuracy on the dataset T4SA in five trials. The notation of the models and the preprocessing pipelines are described in Tables 4 and 5.

	BasePP	NEG & PP	NEG-VADER & PP
FreBoW-MultiNB	0.9572	0.9511	**0.9597**
TFIDF-MultiNB	0.954	0.9493	**0.9576**
TFIDF-LogReg	0.9652	0.9658	**0.9666**
BinBow-BerNB	0.9581	0.9509	**0.9598**

Overall, Table 6 and Table 7 show that the proposed negation handling method NEG-VADER is more effective than the base one NEG. The NEG & PP improved the accuracy by around 2% compared to the BasePP, which has no negation handling, on the Sentiment140; however, for the data T4SA, the NEG & PP is less effective by 0.4% than the BasePP. The preprocessing pipeline with negation handling NEG creates the most features, which include worthless words with the suffix "_NEG", compared the other preprocessing pipeline. The number of features after the preprocessing of NEG-VADER & PP is less than 4% and 6% compared to that of BasePP and NEG-PP, respectively, on the dataset Sentiment140 (see Fig. 3), and it is 3% and 5%, respectively, on the T4SA dataset (see Fig. 4), because the NEG-VADER only focuses on words with high polarity while NEG spans to all words in a clause. As a result, the overall accuracy is improved by paying attention to only high-polarity words. Another reason the NEG-VADER & PP provides the least features is the proposed preprocessing pipeline PP has normalized emoticons, and elongated words besides removing unnecessary tokens like usernames, hyperlinks, stopwords, and punctuations.

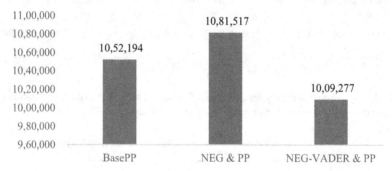

Fig. 3. The average number of features for each preprocessing pipeline in the dataset Sentiment140 in five trials.

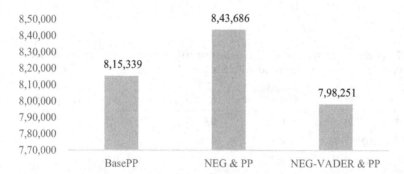

Fig. 4. The average number of features of each preprocessing pipeline on the dataset T4SA in five trials.

5   Conclusions

In this research, a preprocessing pipeline for social media with an approach for handling elongated words and emoticons for social media data has been proposed. The experiments have shown that the preprocessing pipeline PP combined with the new negation handling method NEG-VADER is more efficient than the base one BasePP. Importantly, by taking advantage of the lexicon models VADER, NEG-VADER only focuses on the highly polarized words in documents. The results show that the negation handling method NEG-VADER reduces the number of features and improves the final accuracy of binary sentiment analysis. However, the preprocessing pipeline PP could be improved with an efficient algorithm by eliminating the names of objects in documents to reduce noisy features. Moreover, the handled emojis in the proposed preprocessing pipeline does not include encoding emojis, while they can be presented in byte on many social media platforms. In addition, the scope of this paper only covers binary sentiment analysis, but the sentiment label contains more than two classes.

References

1. Behl, S., Rao, A., Aggarwal, S., Chadha, S., Pannu, H.: Twitter for disaster relief through sentiment analysis for COVID-19 and natural hazard crises. Int. J. Disaster Risk Reduct. **55**, 102101 (2021)
2. Ragini, J.R., Anand, P.R., Bhaskar, V.: Big data analytics for disaster response and recovery through sentiment analysis. Int. J. Inf. Manag. **42**, 13–24 (2018)
3. Somula, R., Dinesh Kumar, K., Aravindharamanan, S., Govinda, K.: Twitter sentiment analysis based on US presidential election 2016. In: Satapathy, S., Bhateja, V., Mohanty, J., Udgata, S. (eds.) Smart Intelligent Computing and Applications. SIST, vol. 159, pp. 363–373. Springer, Singapore (2020). https://doi.org/10.1007/978-981-13-9282-5_34
4. Alam, S., Yao, N.: The impact of preprocessing steps on the accuracy of machine learning algorithms in sentiment analysis. Comput. Math. Organ. Theory **25**, 319–335 (2019)
5. Narayanan, V., Arora, I., Bhatia, A.: Fast and accurate sentiment classification using an enhanced Naive Bayes model. In: Yin, H. (et al.) IDEAL 2013. LNCS, vol. 8206, pp. 194–201. Springer, Heidelberg (2013). https://doi.org/10.1007/978-3-642-41278-3_24
6. Hutto, C., Gilbert, E.: VADER: a parsimonious rule-based model for sentiment analysis of social media text. In: Proceedings of the International AAAI Conference on Web and Social Media, vol. 8, no. 1, pp. 216–225 (2014)
7. Pradha, S., Halgamuge, M.N., Tran Quoc Vinh, N.: Effective text data pre-processing technique for sentiment analysis in social media data. In: 2019 11th International Conference on Knowledge and Systems Engineering (KSE), pp. 1–8 (2019)
8. Manning, C.D., Raghavan, P., Schütze, H.: Introduction to Information Retrieval. Cambridge University Press. Section. 2.2.4 (2008)
9. Duong, H.T., Nguyen-Thi, T.A.: A review: preprocessing techniques and data augmentation for sentiment analysis. Comput. Soc. Netw. **8**, 1 (2021)
10. Kamath, C.N., Bukhari, S.S., Dengel, A.: Comparative study between traditional machine learning and deep learning approaches for text classification. In: Proceedings of the ACM Symposium on Document Engineering 2018 (DocEng 2018), pp. 1–11. Association for Computing Machinery, New York, Article no 14 (2018)
11. Potts, C.: On the negativity of negation. Semant. Linguist. Theory **20** (2011)

12. Councill, I., McDonald, R., Velikovich, L.: What's great and what's not: learning to classify the scope of negation for improved sentiment analysis. In: Proceedings of the Workshop on Negation and Speculation in Natural Language Processing, Uppsala, Sweden, pp. 51–59. University of Antwerp (2010)
13. Gupta, I., Joshi, N.: Feature-based Twitter sentiment analysis with improved negation handling. IEEE Trans. Comput. Soc. Syst. **8**(4), 917–927 (2021)
14. Kimura, M., Katsurai, M.: Automatic construction of an emoji sentiment lexicon. In: Proceedings of the 2017 IEEE/ACM International Conference on Advances in Social Networks Analysis and Mining 2017 (ASONAM 2017), pp. 1033–1036. Association for Computing Machinery, New York (2017)
15. Nemes, L., Kiss, A.: Information extraction and named entity recognition supported social media sentiment analysis during the COVID-19 pandemic. Appl. Sci. **11**(22), 11017 (2021)
16. Go, A., Bhayani, R., Huang, L.: Twitter sentiment classification using distant supervision. Processing 150 (2009)
17. Vadicamo, L., et al.: Cross-media learning for image sentiment analysis in the wild. In: 2017 IEEE International Conference on Computer Vision Workshops (ICCVW), pp. 308–317. Vadica-mo_2017_ICCVW, October 2017

Integrating Ontology-Based Knowledge to Improve Biomedical Multi-Document Summarization Model

Quoc-An Nguyen[1], Khanh-Vinh Nguyen[1], Hoang Quynh Le[1], Duy-Cat Can[1],
Tam Doan-Thanh[1,2], Trung-Hieu Do[3], and Mai-Vu Tran[1(✉)]

[1] University of Engineering and Technology,
Vietnam National University, Hanoi, Vietnam
{annq,22025025,lhquynh,catcd,vutm}@vnu.edu.vn
[2] Viettel Group, Hanoi, Vietnam
tamdt9@viettel.com.vn
[3] Hanoi Medical University, Hanoi, Vietnam
dotrunghieu05220161@daihocyhanoi.edu.vn

Abstract. Most existing extractive summarization models use the original text's internal information and calculate each sentence's importance individually. When applied to specific domains (such as verbal text, biomedical literature, etc.), these models have some drawbacks: the variety of synonym terms, unknown words or terminologies, and the intra-document and inter-document relations between sentences or terms. In this work, we proposed an ontology-based summarization model that leverages many knowledge bases to understand the input documents. Our proposed model was built with an integrated ontology and a signal transmission-based method for extending domain knowledge such as related terms, and relationships between terms and sentences. The proposed model has been proven effective with the highest `ROUGE-2 F1` score in the test dataset of the MEDIQA 2021 MAS shared tasks.

Keywords: extractive summarization · multi-document summarization · query-based summarization · ontology construction

1 Introduction

In an age of overwhelming information, human beings have to distill a tremendous amount of data, which leads to time-consuming, costly, and heavy workloads. One effective solution to this problem is applying text summarization. The general idea of text summarization is receiving the document(s) as input, then generating a concise and informative summary [8]. Its goal is to assist users to capture the crucial information of the original documents without the need to

---

Q.-A. Nguyen and K.-V. Nguyen—Contributed Equally.

read the entire one. Text summarization is a challenging topic in the field of natural language processing and information retrieval. Researchers are constantly exploring new techniques to improve the quality of summaries.

Biomedical summarization is a potential area of research in text summarization. However, there are several distinctive challenges that make it become a tough task. Firstly, biomedical documents are highly scientific and domain-specific, with plenty of jargon and definitions such as drugs, diseases, and symptoms. Moreover, a term can have divergent meanings in different contexts, and different terms can be used to describe the same definition. This can lead to ambiguity and variability in the language used in biomedical literature, making it challenging to accurately summarize the content. Secondly, new biomedical terms can not be updated in pre-processing models. As the result, pre-processing models label plenty of unknown words. Thirdly, answering some questions requires biomedical relations such as chemical-disease, disease-gene, and disease-symptoms. The identification of biological relationships plays an important role in the query-based summary problem. From these challenges, in this work, we have proposed the ontology-based extractive multi-document summarization model in which ontology is used to leverage many knowledge bases to understand the input documents.

This research has two vital missions: *Ontology construction* and *Building extractive multi-answer summarization model*. Ontology construction tasks focus on building integrated ontology, which is leveraged to extend biological knowledge. WordNet also is used for enhancing common sense knowledge. After that, a summarization model is built based on scoring methods to select important sentences in original answers. Our paper contributions are:

- Proposing an ontology integration process and application scenario to augment question sense.
- Improving sentence scoring methods by injecting question keywords' weights.
- Building a summarization model which has better performance than the published model on MEDIQA 2021 MAS shared tasks dataset[1].

2   Related Work

There are two main approaches to text summarization - extractive and abstractive summarization. Extractive summarization generates synopsis by selecting remarkable sentences or expressions from the original content, while abstractive strategies reword and rebuild sentences to form the summary [1]. Because the advantage of extractive summarization is rapid, we decided to utilize it for complicated data like biomedical ones. There are various extractive text summarization methods have been researched and developed over an extended period. Frequency-based methods are the earliest one, it aims to estimate the importance of components based on the frequency of words or sentences [7]. Graph-based

---

[1] https://sites.google.com/view/mediqa2021.

methods commonly use sentence-based graphs to represent a document or cluster. LexRank/TexRank builds a graph from sentences/words and uses degree centrality as the score for ranking [9]. Besides, machine learning methods convert the summarization problem to a supervised classification problem at the sentence level. Some supervised models implemented are Naive Bayes, Decision tree, and Support Vector Machine to deep learning models [6]. These models depend on training data, they sometimes overfit these data but do not generalize well to novel ones, which leads to poor performance in real-life applications.

In some cases, models use additional user queries to extract summaries that only focus on answering these queries. Some question-driven scoring methods are used to estimate the relation between the query and answer components. Weighted-relaxed word mover's distance (wRWMD) calculates the shortest distance between the query and answer sentences [18]. The Hierarchical Sentence Ordering (HSO) method is used for filtering important sentence which has tight relation to questions and other sentences [19]. Because the user's queries tend to be short and have insufficient information to understand, query sense should be improved by using some external databases. Wikipedia is a sizeable unstructured database for searching keywords in Wikipedia articles and detecting related keywords [16]. ConceptNet and WordNet provide pre-existing relationships between words [19]. For an extended biological sense, the Medical Subject Headings database can be used in query expansion method [10].

Ontologies have been proven to be the most effective way for humans and machines to communicate and share information [17]. In this paper, ontology is manipulated to extend medical knowledge. There is a large number of biological ontologies nowadays suc as Disease Ontology[2], Gene Ontology[3], MeSH[4]. However, there are no comprehensive ontologies, because actual ontologies focus on particular aspects of health. That makes it hard to apply to the summary problem because the user questions are diverse in many aspects. Ontology integration and ontology population are used to handle these obstacles [12]. Numerous ontology integration and population process are published [3,13]. Although they have high performance and can be applied to many different data types including unstructured text, these approaches require much time and labour.

3   Proposed Model

The proposed architecture is shown in Fig. 1. It contains three main phases: ontology construction phase, pre-processing and ontology augmentation phase, and summarization phase. The ontology construction phase focus on creating an integrated ontology from the existing biomedical database. Pre-processing and ontology augmentation phase is used to generate a structured representation of the question and answers. Then, the summarization phase is used to manipulate

---

[2] https://disease-ontology.org/.

[3] http://geneontology.org/.

[4] https://www.ncbi.nlm.nih.gov/mesh/.

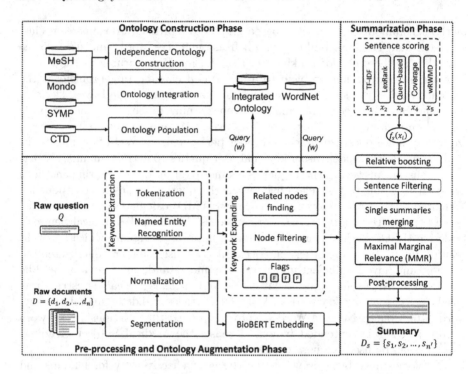

Fig. 1. Overview of proposed model

to create the extractive summary containing key information answering the given question.

3.1   Ontology Construction Phase

The biomedical text contains various term types (synonyms, keyword variation, related hypernym, related hyponym) and term relations (chemical-disease, disease-gene,disease-symptoms). The ontology can show relationships between terms about medical meaning. In a study focusing on solving the summarization problem, we propose a simple process that builds a comprehensive ontology based on existing structured databases. The pipeline for ontology construction has four main parts. Firstly, four biomedical databases are crawled:

– **Medical Subject Headings (MeSH):** MeSH[5] is used to index book collections. In MeSH, the descriptors or subject headings follow a hierarchy, which helps build the relationship structure between terms.

   However, MeSH provides an incomplete amount of keywords on essential topics such as diseases, chemicals, and genes. There are two problems to solve:

---

[5] https://www.ncbi.nlm.nih.gov/mesh/.

pruning unimportant trees to reduce noise and adding more in-depth data sources.

- **Mondo Disease Database (Mondo):** Mondo[6] is a semi-automated database that merges diseases from existing sources into a consolidated database. It has a tabular structure, providing IDs from original databases. It supports adding disease knowledge to the summary model.
- **Symptom Ontology (SYMP)** Symptom Ontology[7] includes a patient's reported changes in feelings or appearance that are suggestive of disease. SYMP only provides symptoms knowledge without mapping to related diseases. Mapping SYMP to other diseases database is a great difficulty.
- **Comparative Toxicogenomics Database (CTD)** CTD[8] contains relations information such as chemical-gene chemical-disease, and gene-disease, which are created by human knowledge. It is an important data set in the ontology population phase.

Since databases have different structures and access methods, independent ontology is used to create a unified ontology structure from crawled data. Ontologies have a taxonomy structure, in which records are grouped or classified in hierarchical architecture. Each node will consist of the following attributes:

- `ID`: A unique sequence of characters is used to retrieve records.
- `terms`: Contains terms that refer to the same concept.
- `stem_from`: Storing name of original databases, serving for integration phase.
- `define`: Saving node's definition, which supports ontology population phase.
- `edge`: Storing pointers to related nodes (**hypernym**, **hyponym**) and related relations. They help optimise speed access and computation.

In which, terms and definitions are normalized like raw texts. The data fields of MeSH, Mondo and SYMP are extracted and filtered to build independent ontologies respectively.

Because of overlapping information, independent ontology is entered into the ontology integration step. For indexing articles target, MeSH has a clear structure and a variety of topics, about 16 categories. MeSH is used as a backbone and Mondo ontology is integrated with MeSH based on cross references. There are two main cases:

- **Existence of cross-references between pair nodes:** Mondo's node (A) is integrated into MeSH node (B) with two steps, (i) fusing A's terms and B's terms, (ii) cutting the descendant of A and converting it to the descendant nodes of B.
- **No cross-references between the pair nodes:** Mondo's node is integrated into a shallow position

---

[6] https://mondo.monarchinitiative.org.

[7] http://symptomontologywiki.igs.umaryland.edu.

[8] http://ctdbase.org.

Finally, we use an ontology population process to add new semantic relations to integrated ontology. There are two ways to extract biomedical relationships. Firstly, we extract $1,048,547$ chemical-disease and $1,048,576$ gene-disease relations from CTD. Moreover, one method is used based on the term's definition with three steps: (i) getting keywords from the term's definition, (ii) searching keywords in ontology, (iii) if keywords exist, the relation is added to the integrated ontology [14]. As a result, chemical-disease, symptom-disease and gene-disease relations are added to the ontology.

3.2 Pre-processing and Ontology Augmentation Phase

Pre-processing. The pre-processing phase takes the original documents and a raw question, as the input. Firstly, documents are segmented into sentences. After that, the normalization method is used to remove noise from the raw documents (HTML tags, duplicate spacing, etc.). Both processed documents and the raw question are put into the keyword extraction phase to extract keywords including tokens and NERs. The question's keywords are expanding by Ontology Augmentation. Finally, BioBERT is used for creating word embedding vectors and sentence embedding vectors from the text.

Ontology Augmentation. Ontology is leveraged to extend biological knowledge while WordNet is used for enhancing common sense knowledge. With ontology, when getting a keyword, the method searches key nodes which contain this keyword base on the longest common sub-sequence and sets an initial weight for them. Then, this method expands to related nodes (having relation to key nodes) and calculates their weight based on the spreading activation function. Spreading activation relies on signal transmission ideals when the strength of the signal gets weaker the farther it travels [18]. In the tth expansion, the weights of the extended words are as follows:

$$weight_{w_t} = (1 - decay) * \frac{weight_{w_{t-1}}}{\left|Out_{w_{t-1}}\right|} \tag{1}$$

where $weight_{w_t}$ is weight of w_t and $\left|Out_{w_{t-1}}\right|$ is their number of outputs in $(t-1)^{th}$ expansion. The spreading process begins to find related nodes and weight until $weight_t < threshold$. The method also provides some flags which allow setting which related term types are chosen. When stopping, this method returns keywords and their node's weight in pair format. The spreading process in WordNet is similar to that in Ontology, in which a synset as a node.

3.3 Summarization Phase

Using information from the pre-processing and ontology augmentation state, the summarization phase generates the summary for each sample. Our summarization model tries to predict which sentences are important to the document by sentence scoring. This phase has four main parts: sentence scoring, relative boosting, removing duplicate semantic sentences and post-processing.

Sentence Scoring. We use five different types of scores to determine the importance of sentences. Besides, this paper proposed some customizations to inject the question keywords' weight.

TF-IDF score is a probabilistic approach that represents the importance of words in sentences [7]. The TF-IDF score of a word w located in document d of corpus D is defined as $TF - IDF(w, d, D)$. In this score, we add the weight of the maximum keywords max_w to the original formula as follows:

$$TF - IDF(w, d, D) = \max_w \times TF(w, d) \times IDF(w, D) \tag{2}$$

A sentence's TF-IDF score is the sum of the TF-IDF scores of its component words.

LexRank score is used for detecting essential sentences in the answer based on the document's graphs [5]. In LexRank, the graph's nodes present sentences, and a graph's weighted edge refers to the similarity score of node pairs. The score of sentences presenting the centrality of the sentence in the answer is calculated as follows:

$$p(e) = \frac{\alpha}{n} + (1 - \alpha) \sum_{r \in adj_e} \frac{sim(e, r)}{\sum_{z \in adj_r} sim(z, r)} p(r) \tag{3}$$

where adj_e is the set of nodes adjacent to e, n is the number of nodes, and α is the damping factor, sim is the Cosine similarity between vectors.

Coverage-Based Score. We used the longest common sub-sequence to calculate the keyword's ratio per sentence as follows:

$$kw(s) = \frac{\left| \left\{ k : k \in q \Big| \frac{\max_{i \in s} lcs(k,i)}{\text{length}(k)} \geq \text{thres} \right\} \right|}{|s|} \tag{4}$$

where q is the question's keywords and extending keywords, s is the sentence's keywords, and thres is the threshold.

Query-Based Score. This score of sentence s focuses on calculating the similarity between the question vector v_s and sentence vector v_q by Cosine similarity as follows:

$$score(s) = sim(v_s, v_q) \tag{5}$$

Weighted-relaxed word mover's distance (wRWMD) is used to estimate the "distance" between question and sentence [18], which is as follows:

$$wRWMD(q, s) = \frac{\sum_{i \in q} w_i \times \max_{j \in s} sim(v_i, v_j)}{\sum_{i \in q} w_i} \tag{6}$$

where q is the question's keywords and extended keywords which exist in the answers, s is the sentence keywords set, w_i is normalized IDF. In the model, the question's keyword weight is added to the wRWMD method with w_i' as follows:

$$w_i' = weight_i \times w_i \tag{7}$$

Scores Combination. All scores are normalized by the min-max normalization method. Then, the final score is calculated in a weighted sum.

Relative Boosting. Because adjacent sentences often complement each other about semantic information (e.g. explaining or summarizing the previous sentences). As a result, answers frequently contain continuous sentences as a cluster in the paragraph. Relative boosting increases scores for some sentences close to high-scoring sentences. The updated score of sentence i-th is as follows:

$$score'_i = \max_{j=\max(i-n_l+1,1)}^{\min(i+n_r-1,n)} score_j \tag{8}$$

where n is the total number of sentences, n_l and n_r are the number of sentences in two directions: left and right of the current sentence i. After ranking and filtering the top highest-score sentences, we restore the sentence's position and concatenate all selected sentences to create single-answer extractive summaries.

Removing Duplicate Ideas. The single-answer summaries are concatenated and then, passed into the maximal marginal relevance (MMR) [2]. MMR is used to remove overlap semantic and question-irrelevant sentences as follows:

$$\text{MMR} = \text{argmax}_{s_i \in A \backslash R} \left[\lambda \left(\text{sim}\,(s_i, q) - (1 - \lambda) \max_{j \neq i}(\text{sim}(s_i, s_j)) \right) \right] \tag{9}$$

where R is the result sentence set, and A is the remaining sentences in the answer. s is the sentence in A, and q is the question. The output of this method also is multi-answer extractive summarization. The Cosine distance is followed to calculate the similarities.

Post-processing. This method removes noises from the summary as questionable sentences, example sentences, long sentences, and duplicated information.

4   Experiments and Results

4.1   Dataset

This model uses MEDIQA-AnS Dataset as training set [20] which comprises 156 health-related questions, corresponding answers, and expert-created summaries of these answers. It includes single-answer and multi-answer summaries for extract and abstract strategies. The validation and test set are the datasets provided by the MEDIQA 2021 MAS shared tasks. The validation and test sets are summaries written by experts based on the original answers from the question-answering system CHiQA[9]. The validation set includes 50 questions, as well as answers and multi-answer summaries to these questions. There are 80 questions and related answers in the test set. Because extract relates to choosing sentences and abstract rewritten sentences, Table 1 provides statistics on given datasets to extract based on the sentence level.

---

[9] https://chiqa.nlm.nih.gov.

Table 1. The statistics of dataset

Statistic aspects	Training		Validation	Test
	Article	Section		
Average				
Answers per Question	3.54	3.54	3.85	3.8
Sentences per Answer	84.93	29.07	14.50	13.03
Sentences per Single-answer Summary	6.31	6.31	–	–
Sentences per Multi-answer Summary	10.30	10.30	11.06	–
Compression ratio				
Single-answer Summary	0.12	0.49	–	–
Multi-answer Summary	0.06	0.18	0.33	–

4.2   Experiments and Results

Figure 2 indicates the statistic of nodes and terms in three independent ontologies and in the integrated ontology. In three independent ontologies, MeSH has the most quantity of terms and nodes with 29,917 nodes and 245,918 terms in 16 topics. Mondo contains 24,409 nodes and 113,034 terms that focus on human diseases. SYMP has 944 nodes and 1179 terms about symptoms. In integrated ontology, there are 51,227 nodes and 346,098 terms.

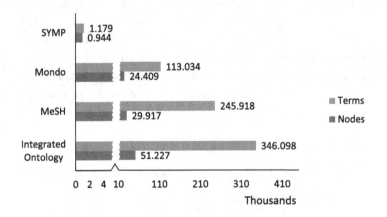

Fig. 2. The statistic of nodes and terms

Table 2 shows the number of relations after the population process. Extracting the CTD database allows adding 115,632 relations while extracting the term's definition allows adding 42,152 relations. As a result, a total of 157,784 semantic relations were extracted in the ontology population part.

ROUGE scores are used to evaluate automatic summarization and machine translation technologies [11]. The metric compares the generated summary to a

Table 2. The statistic of extracted relations in ontology population

Quantity	CTD database	Extracting term's definition
Chemical-disease relations	67,456	24,726
Symptom-disease relations	–	17,425
Gene-disease relations	48,176	–

human reference(s), the higher scores show the closer relations between them. Following MEDIQA 2021 MAS shared tasks, some ROUGE scores such as ROUGE-1, ROUGE-2, and ROUGE-L are used to evaluate the automatic summarization model in this paper, where F1 scores are main evaluation scores.

Table 3 shows the performances of the proposed model with some published models in this dataset [4,15,21,22]. The ontology-based summarization model has the best performance on all three F1 scores. The proposed model has the best performance score when cutting integrated ontology where nodes about the drug, disease, symptoms, drug-disease relations and symptom-disease relations are focused on, and other related topics and gene-disease relations are disabled.

Table 3. Comparison model's results

Model name	ROUGE-1			ROUGE-2			ROUGE-L F1
	P	R	F1	P	R	F1	
Fine-tuning RoBERTa	0.475	0.878	0.585	0.407	0.767	0.508	0.435
BART	**0.616**	0.672	0.607	**0.473**	0.531	0.472	0.429
Fine-tuning T5	0.420	**0.899**	0.547	0.358	**0.774**	0.468	0.328
Prosper-thy-neighbor	0.528	0.814	0.611	0.432	0.680	0.504	0.441
Ontology-based model	0.530	0.821	**0.616**	0.437	0.687	**0.511**	**0.446**

The highest results in each column are highlighted in bold.

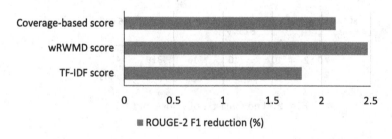

Fig. 3. Reduction of ROGUE-2 $F1$ per each scoring method

Figure 3 shows the reduction of ROGUE-2 $F1$ per each scoring method when disabling the Ontology Augmentation phase. Keyword-based score and wRWMD have significant gains when using ontology.

5   Conclusions

This paper presents the ontology-based extractive multi-document summarization model in which ontology is used to leverage many knowledge bases to understand the input documents. Because biomedical ontologies focus on particular aspects of health, we proposed a process to build a comprehensive ontology from some existing databases. After that, question keywords are through an ontology augmentation phase to find related keywords, and calculate their weight. In the summarization phase, we combined and optimized several scoring criteria such as TF-IDF, LexRank, Textrank, Coverage-based, Query-based and MMR scores. We propose some customizations to inject the question keywords' weight into sentence scoring. As the result, an integrated ontology includes 16 aspects and 5 biomedical relations. The summarization model performs better with 0.511 ROUGE-2 F1 than other published models in MEDIQUA 2021 MAS.

In the future, this model can be extended to build an ontology in more biomedical topics, using text data sources to infer related terms and biomedical relations. Moreover, this model can be implemented in Vietnamese to meet the needs for medical knowledge in my country.

Acknowledgments. This research has been done under the research project QG "Research and Development of Vietnamese Multi-document Summarization Based on Advanced Language Models" of Vietnam National University, Hanoi (Code: QG.22.61). Quoc-An Nguyen was funded by the Master, PhD Scholarship Programme of Vingroup Innovation Foundation (VINIF), code VINIF.2022.ThS.001.

References

1. Abualigah, L., Bashabsheh, M.Q., Alabool, H., Shehab, M.: Text summarization: a brief review. In: Recent Advances in NLP: The Case of Arabic Language, pp. 1–15 (2020)
2. Bennani-Smires, K., Musat, C., Hossmann, A., Baeriswyl, M., Jaggi, M.: Simple unsupervised keyphrase extraction using sentence embeddings. In: Proceedings of the 22nd Conference on Computational Natural Language Learning, pp. 221–229 (2018)
3. Blomqvist, E.: OntoCase-automatic ontology enrichment based on ontology design patterns. In: Bernstein, A., et al. (eds.) ISWC 2009. LNCS, vol. 5823, pp. 65–80. Springer, Heidelberg (2009). https://doi.org/10.1007/978-3-642-04930-9_5
4. Can, D.C., et al.: UETrice at MEDIQA 2021: a prosper-thy-neighbour extractive multi-document summarization model. In: Proceedings of the 20th Workshop on Biomedical Language Processing, pp. 311–319 (2021)
5. Erkan, G., Radev, D.R.: LexRank: graph-based lexical centrality as salience in text summarization. J. Artif. Intell. Res. **22**, 457–479 (2004)
6. Gambhir, M., Gupta, V.: Recent automatic text summarization techniques: a survey. Artif. Intell. Rev. **47**(1), 1–66 (2017)
7. Hovy, E., Lin, C.Y., et al.: Automated text summarization in SUMMARIST. In: Advances in Automatic Text Summarization, vol. 14, pp. 81–94. MIT press Cambridge, MA (1999)

8. Ježek, K., Steinberger, J.: Automatic text summarization (the state of the art 2007 and new challenges). In: Proceedings of Znalosti, pp. 1–12. Citeseer (2008)
9. Kaynar, O., Görmez, Y., Işık, Y.E., Demirkoparan, F.: Comparison of graph-based document summarization method. In: 2017 International Conference on Computer Science and Engineering (UBMK), pp. 598–603. IEEE (2017)
10. Kogilavani, A., Balasubramanie, P.: Ontology enhanced clustering based summarization of medical documents. Int. J. Recent Trends Eng. 1(1), 546 (2009)
11. Lin, C.Y.: Rouge: a package for automatic evaluation of summaries. In: Text Summarization Branches Out, pp. 74–81 (2004)
12. Lubani, M., Noah, S.A.M., Mahmud, R.: Ontology population: approaches and design aspects. J. Inf. Sci. 45(4), 502–515 (2019)
13. Mitra, P., Noy, N.F., Jaiswal, A.R.: OMEN: a probabilistic ontology mapping tool. In: Gil, Y., Motta, E., Benjamins, V.R., Musen, M.A. (eds.) ISWC 2005. LNCS, vol. 3729, pp. 537–547. Springer, Heidelberg (2005). https://doi.org/10.1007/11574620_39
14. Mohammed, O., Benlamri, R., Fong, S.: Building a diseases symptoms ontology for medical diagnosis: an integrative approach. In: The First International Conference on Future Generation Communication Technologies, pp. 104–108. IEEE (2012)
15. Mrini, K., et al.: UCSD-adobe at MEDIQA 2021: transfer learning and answer sentence selection for medical summarization. In: Proceedings of the 20th Workshop on Biomedical Language Processing, pp. 257–262 (2021)
16. Nastase, V.: Topic-driven multi-document summarization with encyclopedic knowledge and spreading activation. In: Proceedings of the 2008 Conference on Empirical Methods in Natural Language Processing, pp. 763–772 (2008)
17. Osman, I., Yahia, S.B., Diallo, G.: Ontology integration: approaches and challenging issues. Inf. Fusion 71, 38–63 (2021)
18. Ozyurt, I.B., Bandrowski, A., Grethe, J.S.: Bio-AnswerFinder: a system to find answers to questions from biomedical texts. Database 2020 (2020)
19. Rahman, N., Borah, B.: Improvement of query-based text summarization using word sense disambiguation. Complex Intell. Syst. 6(1), 75–85 (2020)
20. Savery, M., Abacha, A.B., Gayen, S., Demner-Fushman, D.: Question-driven summarization of answers to consumer health questions. Sci. Data 7(1), 1–9 (2020)
21. Yadav, S., Sarrouti, M., Gupta, D.: NLM at MEDIQA 2021: transfer learning-based approaches for consumer question and multi-answer summarization. In: Proceedings of the 20th Workshop on Biomedical Language Processing, pp. 291–301 (2021)
22. Zhu, W., et al.: paht_nlp@ MEDIQA 2021: multi-grained query focused multi-answer summarization. In: Proceedings of the 20th Workshop on Biomedical Language Processing, pp. 96–102 (2021)

An Improvement of Diachronic Embedding for Temporal Knowledge Graph Completion

Thuy-Anh Nguyen Thi[1]([✉])[ID], Viet-Phuong Ta[2][ID], Xuan Hieu Phan[2][ID], and Quang Thuy Ha[2][ID]

[1] Banking Academy of Vietnam, Hanoi, Vietnam
anhntt@hvnh.edu.vn
[2] VNU-University of Engineering and Technology (UET), Vietnam National University (VNU), Hanoi, Vietnam
{hieupx,thuyhq}@vnu.edu.vn

Abstract. Knowledge graph completion (KGC) aims to predict missing information in a knowledge graph (KG). Knowledge embedding approaches learn the representations of entities and relations in a static knowledge graph, playing an essential role in knowledge graph completion. However, the previous methods have been designed for static knowledge graphs. A temporal knowledge graph (TKG), on th other hand, contains timestamp-based facts indicating relationships between entities at different times. R. Goel et al. (2020) introduced novel models for temporal knowledge graph completion through existing static models with a diachronic entity embedding function, which are able to provide characteristics of entities at any point in time. Moreover, their method is flexible since it can be potentially combined with any existing static model. In this paper, we propose two models, - (DE-RotatE and DE-RotatE-sinc) -, to combine the diachronic entity embedding with RotatE - the model is introduced by Z. Sun et al. (2019). Through experiments, we show that the results of our proposed models are better than those in the work by R. Goel at al. Our source code is available on Github, https://github.com/anhntt1202/DE-RotatE.

Keywords: Temporal Knowledge graph completion · Diachronic Embedding · RotatE model

1 Introduction

A knowledge graph (KG) is a collection of triple facts (h, r, t), where h, t are head and tail entity, and r is a relationship, representing a relation between h and t. For instance, *(Hanoi, capitalOf, Vietnam)* means Hanoi is the capital of Vietnam. Clearly, Hanoi and Vietnam are two entities and captitalOf is the relation between them. KGs play an essential role in representing factual knowledge and are the core of many downstream tasks, e.g., recommender system [1], and

© The Author(s), under exclusive license to Springer Nature Singapore Pte Ltd. 2023
N. T. Nguyen et al. (Eds.): ACIIDS 2023, LNAI 13996, pp. 111–120, 2023.
https://doi.org/10.1007/978-981-99-5837-5_10

natural language processing [2]. There are a lot of real-world knowledge graphs, including Freebase, Yago, and Wordnet, but they are usually incomplete [3]. Thus, a critical problem for KGs is predicting the missing links. The study on KGs is attracting a growing interest in both academia and industry communities. Many studies have been proposed to learn a lower-dimensional representation of entities and relations for missing link prediction such as knowledge graph embedding (KGE). The KGE approaches are shown to be scalable and effective.

However, typical KGs, normally represent knowledge facts without incorporating temporal information, and they are called static knowledge graphs (SKGs). Consequently, some events are valid only in certain moments or in a range period of time. For example, the fact (Steve Ballmer, CEO-of, Microsoft) is valid from January 13, 2000 to February 4, 2014. The time information makes the facts in KGs more accurate and specific. The KGs that meet this demand are called temporal knowledge graphs (TKGs). Recently, there have been some TKGs, including ICEWS [4], GDELT [5], and Wikidata [6]. They are used for representing temporarily valid events. Especially, triples in TKGs are presented with their corresponding time-stamps at the same time. Thus, a triple fact now becomes a quadruple of a time-sensitive fact in TKGs as (h, r, t, τ), where h, t are still the head and tail entities, r is a relationship and τ is the temporal information.

There exist many large-scale TKGs. However, SKGs, those TKGs remain remarkably incomplete. One of the tasks of knowledge graph completion is called link prediction. It is a significant task that predicts the missing entities for incomplete queries in the form of $(h, r, ?, \tau)$ or $(?, r, t, \tau)$ in TKGs.

Recently, some studies have dealt with the problem of link prediction in temporal knowledge graph completion. Most of them consider time information as embeddings to extend conventional KGE models. One of the directions is to improve the embeddings of time-stamps. To get the time series features by using a long short-term memory, TA-TransE and TA-DistMult [7] use the embeddings of time information to TransE [8] and DistMult [9], respectively. HyTE [10], an extension of TransH [11], projects the entities and relations to a time-specific hyper-plane.

R. Goel et al. [12] introduced a novel model for temporal knowledge graph completion (TKGC) based on an intuitive assumption to provide a score for a quadruple. For example, *(Real Madrid, championOf, UEFA Champions League, 2022)*, one needs to know Real Madrid's and UEFA Champions League's features in 2022. Of course, providing a score based on different times may be misleading because the champion of the UEFA Champions League may have been quite different in 2022 compared to now. To obtain entity features at any given time, Goel et al. [12] defined entity embedding as a function that takes an entity and a time-stamp as the input. The output is a hidden representation of the entity at that time. Their proposed embedding is called diachronic embedding (DE). Moreover, they showed that DE is model-agnostic that means any SKG embedding model can be potentially extended to TKGC by leveraging DE. R. Goel et al. indicated that combining DE with SimplE [13], denoted by

DE-SimplE, is the best result for TKGC models. It is the motivation for us to improve the result of TKGC models.

In [3], Z. Sun et al. proposed a model for static knowledge graph embedding. It is called *RotatE*. Their motivation is from Euler's identity $e^{i\theta} = cos(\theta) + i \sin \theta$, which indicates that a unitary complex number can be regarded as a rotation in the complex plane. Specifically, the RotatE model maps the entities and relations to the complex vector space and defines each relation as a rotation from the head entity to the tail entity.

In summary, our contributions are briefly as follows.

- Motivated by the results from [3,12], we propose new models for temporal knowledge graph completion by combining diachronic embedding and RotatE. We denote our proposed models as the DE-RotatE and the DE-RotatE-sinc models.
- As the same approaches in [12], we also show the merit of our proposed models on subsets of ICEWS [4] and GDELT [5] datasets. By the experiments, we show that the results of our proposed models are better than the results in [12].

The remainder of this article is organized as follows. In the next section, we introduce the concepts of temporal knowledge graph completion (TKGC), diachronic embedding. Section 3 describes the DE-SimplE model and our proposed models in detail. Experiment and results, related works, and conclusion will be given in Sect. 4, Sect. 5, and Sect. 6, respectively.

2   Temporal Knowledge Graph Completion

Firstly, we would like to give the concept of temporal knowledge graph completion that is an extension of the concept of static knowledge graph completion.

Temporal Knowledge Graph Completion: Let $\mathcal{E}$ and $\mathcal{R}$ be a finite set of entities and a finite set of relations, respectively, and $\mathcal{T}$ be a finite set of time-stamps. Let $\mathcal{Q} \subset \mathcal{E} \times \mathcal{R} \times \mathcal{E} \times \mathcal{T}$ be a set of all temporal tuples (h, r, t, τ), where $h, t \in \mathcal{E}$, $r \in \mathcal{R}$, $\tau \in \mathcal{T}$, that are facts. Let $\mathcal{Q}^c$ be the complement of $\mathcal{Q}$. A temporal knowledge graph $\mathcal{G}$ is a subset of $\mathcal{Q}$. As the same as [12], in this work, we only consider TKGs where each fact has a single time-stamp. Temporal knowledge graph completion (TKGC) is to infer $\mathcal{Q}$ from $\mathcal{G}$.

Knowledge graph embedding models are very useful for static knowledge graph completion. An entity embedding and a relation embedding are fundamental factors of knowledge graph embedding. An entity embedding and a relation embedding are defined as follows. An entity embedding, $EEMB : \mathcal{E} \rightarrow \psi$, and a relation embedding, $REMB : \mathcal{R} \rightarrow \psi$, is a function that maps each entity $h, t \in \mathcal{E}$ and $r \in \mathcal{R}$, respectively, to a hidden representation in ψ, where ψ is the class of non-empty tuples of vectors and/or matrices. Moreover, there are two things in a knowledge graph embedding model: 1 - the $EEMB$ and $REMB$ functions, 2 - a score function which takes $EEMB$ and $REMB$ as input and computes a score for a given tuple. The parameters of hidden representations are learned from the data.

Motivated by the above definition, the authors in [12] proposed an entity embedding function consisting of an entity and times as the input. They called such an embedding function a diachronic entity embedding, for simple a diachronic embedding (DE). The definition of a diachronic embedding is as follows. A diachronic entity embedding, denoted by $DEEMB : (\mathcal{E}, \mathcal{T}) \rightarrow \psi$, is a function that maps every pair (e, τ), where $e \in \mathcal{E}$ is an entity and $\tau \in \mathcal{T}$, to a hidden representation in ψ, where ψ is the class of non-empty tuples of vectors and/or matrices. Moreover, the authors in [12] indicated that it can take a static KG embedding score function and make it temporal by replacing its entity embedding with diachronic entity embedding.

3   Proposed Models

3.1   DE-SimplE Model

Before giving our proposed models, we describe the DE-SimplE model that was given by the authors in [12]. DE-SimplE is the combination of a diachronic entity embedding $DEEMB$ and SimplE model. They defined the $DEEMB$ function as follows. Let $z_e^\tau \in \mathbb{R}^d$ be a vector in $DEEMB(e, \tau)$. In [12], z_e^τ is defined as follows:

$$z_e^\tau[n] = \begin{cases} a_e[n]\sigma(w_e[n]\tau + b_e[n]), & \text{if } 1 \leq n \leq \gamma d \\ a_e[n], & \text{if } \gamma d < n \leq d \end{cases} \quad (1)$$

where $a_e \in \mathbb{R}^d$ and $w_e, b_e \in \mathbb{R}^d$ are (entity-specific) vectors with learnable parameters and σ is an activation function. Sometimes, entities may have some features that change over time and some other features that remain fixed. So they claimed that the first part of Eq. (1) contains γd elements of the vector that capture temporal features and the second part of Eq. (1) contains $(1 - \gamma)d$ remaining elements that capture static features. By learning w_es and b_es, the model learns how to turn entity features on and off at different points in time. a_es is used to control the importance of the features. Since the sine function can model several on and off states, they used sine as the activation function in Eq. 1. Note that $DEEMB(e, \tau) = (\ldots, z_e^\tau, \ldots)$.

SimplE Embedding Model: A static knowledge graph completion, was introduced by S. M. Kazemi and D. Poole [13]. SimplE is a simple enhancement of Canonical Polyadic (CP) to allow two embeddings of each entity to e learned dependently. For each entity e in CP, there is an information flow issue between two vectors $\overrightarrow{z_e}$ and $\overleftarrow{z_e}$. For each relation $r \in \mathcal{R}$, there exists the inverse of the relation r, denoted by r^{-1}. SimplE takes the inverse of the relation to address this issue. The relation embedding function of SimplE is defined as follows $REMB(r) = (\overrightarrow{z_r}, \overleftarrow{z_r})$, where $\overrightarrow{z_r}$ is in Canonical Polyadic and $\overleftarrow{z_r}$ is the embedding of r^{-1}. Let $\phi(h, r, t)$ be the score for a tuple (h, r, t). We denote the

n^{th} element of a vector z by $z[n]$. Next, let $\langle z_1, \ldots, z_k \rangle = \sum_{n=1}^{d}(z_1[n] * \ldots * z_2[n])$ be the sum of the element-wise product of the elements of the k vectors. In SimplE, $\phi(h, r, t)$ is the average of two CP scores:

$$\phi(h, r, t) = \frac{1}{2}\left(\langle \overrightarrow{z_h}, \overrightarrow{z_r}, \overleftarrow{z_t} \rangle + \langle \overrightarrow{z_t}, \overleftarrow{z_r}, \overleftarrow{z_h} \rangle\right)$$

where $\langle \overrightarrow{z_h}, \overrightarrow{z_r}, \overleftarrow{z_t} \rangle$ is the score for (h, r, t) and $\langle \overrightarrow{z_t}, \overleftarrow{z_r}, \overleftarrow{z_h} \rangle$ is the sore for (t, r^{-1}, h).

Learning: Liked the existent studies on the static (temporal) knowledge graph completion models, in [12] the authors split the facts in a KG into train, valid and test sets. A mini-batch, denoted by $\mathcal{B}$, is a subset of the train set. Model parameters are learned by using stochastic-gradient descent with mini-batches $\mathcal{B}$. Let $q = (h, r, t, \tau) \in \mathcal{B}$, they generate two queries as follows: the first one is $(h, r, ?, \tau)$ and the second one is $(?, r, t, \tau)$. For the first (or second) query, a candidate answer set $C(f, h)$ (or $C(f, t)$) is generated, where $C(f, h)$ (or $C(f, t)$) contains h (or t) and some other entities selected randomly from $\mathcal{E}$. The cross entropy loss $\mathcal{L}$ in [7,14], which has been used and shown good results for both static and temporal knowledge graph completion. The cross entropy loss is computed by the following formula:

$$\mathcal{L} = -\left(\sum_{f=(h,r,t,\tau)\in\mathcal{B}} \frac{\exp\left(\phi(h, r, t, \tau)\right)}{\sum_{t'\in C_{f,t}} \exp\left(\phi(h, r, t', \tau)\right)} + \frac{\exp\left(\phi(h, r, t, \tau)\right)}{\sum_{h'\in C_{f,h}} \exp\left(\phi(h', r, t, \tau)\right)}\right)$$

For all the models in [12], the authors minimize this cross entropy loss. Moreover, they also proved the following result.

Proposition 1 (Goel et al. [12]). *Symmetry, anti-symmetry, and inversion can be incorporated into DE-SimplE in the same way as SimplE.*

3.2   Our Proposed Model

When we study the RotatE model - a knowledge graph embedding by relational rotation in complex space, we realize that the RotatE model can infer all three types of relation patterns by the following result in [3].

Lemma 1 (Sun et al. [3]). *RotatE can infer the symmetry, anti-symmetry, inversion, composition pattern.*

Motivated by Proposition 1 and Lemma 1, we try to substitute the part of SimplE from the DE-SimplE model by the RotatE to get a new model, denoted by DE-RotatE model. The RotatE embedding model, a static knowledge graph completion, was introduced by Sun et al. in [3]. Given a triple (h, r, t), where h and t is the head entity and the tail entity, respectively, and r is the relation. By using the Euler's identity, they map h, t to the complex embedding, i.e. $z_h, z_t, z_r \in \mathbb{C}^k$, where $\mathbb{C}$ is the complex space. Next, they define the functional mapping induced by each relation r as an element-wise rotation from the head entity h to the tail entity t. Hence, $z_t = z_h \circ z_r$, where $|z_r[i]| = 1$, $z_r[i] \in \mathbb{C}$ and $\circ$ is the e Hadamard (or element-wise) product. Clearly, for each element in the embeddings $z_t[i] = z_h[i]z_r[i]$. By using Eurler's identity, $z_r[i]$ is of the form $e^{i\theta_{z_r[i]}}$. Moreover, it is affects the phases of the entity embedding in the complex vector space. They also shown the distance function of RotatE as follows:

$$d_r(z_h, z_t) = \| z_h \circ z_r - z_t \|$$

We construct our model by the following cases:

a. **The first case.** We substitute the SimplE model from the DE-SimplE model with the RotatE model and keep all the remaining parts of the DE-SimplE model. It is called the *DE-RotatE* model.
b. **The second case.** To improve the performance of models in [12], the authors claimed to change the σ activation function of Eq. 1. Besides that their results are best, when σ is the *sine* function. We propose to use the *sinc* function as the activation function of Eq. 1 in our DE-RotatE model. Thus, the new one is called the *DE-RotatE-sinc* model.

4   Experiment and Results

Datasets: In order to show the performance of our proposed models, we evaluate our model on three popular benchmarks for TKG completion, including ICEWS14 – corresponding to the facts in 2014, ICEWS05-15 – corresponding to the facts between 2005 to 2015, and GDELT. The first two datasets, introduced by Boschee et al. [4], are sparse. The last one, GDELT, is a dense KG introduced by Leetaru et al. [5]. Let $|\mathcal{E}|$, $|\mathcal{R}|$, $|\mathcal{T}|$, $|train|$, $|validation|$, $|test|$, $|G|$ be the number of entities, relation, time, facts in training sets, validation sets, test sets, and KG, respectively. The summary of all three datasets are shown in the Table 1.

Table 1. Statistics on ICEWS14, ICEWS05-15, and GDELT [12]

Datasets	$|\mathcal{E}|$	$|\mathcal{R}|$	$|\mathcal{T}|$	$|train|$	$|validation|$	$|test|$	$|\mathcal{G}|$
ICEWS14	7,128	230	365	72,826	8,941	8,963	90,730
ICEWS05-15	10,488	251	4,017	386,962	46,275	46,092	479,329
GDELT	500	20	366	2,735,685	341,961	341,961	3,419,607

Baseline: Since the performance of the DE-SimplE model is the best in [12], this model is the baseline for our experiments.

Metrics: We estimate our proposed models on the link prediction task. Like in [12], we replace the head and the tail entities with all the entities from $\mathcal{E}$ in turn for each quadruple in the test set. Then we compute the scores of all the corrupted quadruples $((h, r, ?, \tau)$ and $(?, r, t, \tau))$ and rank all the candidate entities according to the scores under the time-wise filtered settings. We report the following metrics for comparison: (a) Mean Rank (MR, the mean of all the predicted ranks); (b) Mean Reciprocal Rank (MRR, the mean of all the reciprocals of predicted ranks); (c) $Hits@n$ (the proportion of ranks not larger than n). So the lower MR, the larger MRR, and the larger $Hits@n$ indicate better performance.

Implementation: We implement our proposed models and the baseline in PyTorch and train in Google Colab+. The resources of Google Colab+ are as follows: System RAM 3.3/83.5 GB, GPU RAM 7.0/40.0 GB, Disk 25.5/166.8 GB. We follow a similar experimental setups as in [12] by using ADAM optimizer and learning rate = 0.001, batch size = 512, negative ratio = 500, embedding size = 100, and validating every 20 epochs that selecting the model giving the best validation MRR. For the dropout, we choose 0.2 and γ is selected from $\{16, 32, 64\}$.

Results: We run our proposed models with a different number of epochs depending on the datasets. Since the authors in [12] did not show the result of MR, we retrain the DE-SimplE model with the ICEWS dataset. For the ICEWS14 dataset, we retrain the DE-SimplE model and train our proposed models with 500 epochs (the default epoch). The results are shown in Table 2. We realize that our proposed models' results are better than the DE-SimplE for all measures. Moreover, the performance of the DE-RotatE-sinc is better than the performance of the DE-RotatE.

Table 2. Results on ICEWS14 with 500 epochs. Best results are **bold**. The second good results are *italic*

Model	$MR \downarrow$	$MRR \uparrow$	$Hit@1 \uparrow$	$Hit@3 \uparrow$	$Hit@10 \uparrow$
DE-SimplE [12]	–	0.526	41.8	59.2	72.5
DE-SimplE (retrain)	225	0.523	41.3	59.0	72.9
DE-RotatE	*184*	*0.538*	*42.7*	*60.5*	*74.3*
DE-RotatE-sinc	**181**	**0.555**	**44.8**	**62.5**	**75.5**

Since the number of facts of the two remaining datasets, ICEWS05-15 and GDELT, are large, we train (retrain) our proposed models (the DE-SimplE model) with a small number of epochs. For the ICWES05-15 dataset, the number of epochs is 100. The results are shown in Table 3. Since the DE-RotatE-sinc model's performance is state-of-art for the ICEWS14, and ICEWS05-15 datasets, we decide to train this model for the GDELT dataset. Moreover, the GDELT dataset is so large, and our resource is limited we train the DE-RotatE-sinc model on GDELT with 80 and 90 epochs, see Table 4.

Table 3. Results on ICEWS05-15 with small numbers of epochs. Best results are **bold**, and the second good results are *italic* for each dataset.

Model	$MR \downarrow$	$MRR \uparrow$	$Hit@1 \uparrow$	$Hit@3 \uparrow$	$Hit@10 \uparrow$
DE-SimplE [12]	–	*0.513*	*39.2*	*57.8*	*74.8*
DE-SimplE (retrain)	104	0.499	37.9	56.4	73.6
DE-RotatE	104	0.495	37.4	55.8	73.0
DE-RotatE-sinc	**92**	**0.528**	**40.5**	**59.9**	**76.6**

Table 4. Results on GDELT with small numbers of epochs. Best results are **bold**, and the second good results are *italic* for each dataset.

Model	$MR \downarrow$	$MRR \uparrow$	$Hit@1 \uparrow$	$Hit@3 \uparrow$	$Hit@10 \uparrow$
DE-SimplE [12]	–	**0.230**	14.1	**24.8**	**40.3**
DE-RotatE-sinc (80 epochs)	66.93	0.221	14.12	23.75	37.58
DE-RotatE-sinc (90 epochs)	**66.90**	*0.222*	**14.18**	*23.80*	*37.66*

5 Related Works

Most previous temporal KGs are constructed by extending from the static KG embedding models. R. Goel et al. [12] introduced a novel approach for temporal knowledge graph completion. According to the definition of an entity embedding function for SKGs and diachronic word embedding, R. Goel et al. defined diachronic entity embedding, denoted by DE. They also indicated that their diachronic entity embedding may combine with TransE, DistMult, SimplE, or other static models to obtain the temporal version by replacing the entity embedding function as diachronic embedding one. R. Goel et al. proved that DE-SimplE, a combination of DE and SimplE, is fully expressive. Moreover, DE-SimplE is symmetry, anti-symmetry, and inversion. By experiments with the ICEWS14, ICEWS05-15, GDETL datasets, R. Goel et al. showed that the performance of DE-SimplE is state-of-art [12].

Very recently, authors in [15] introduce a new TimeLine-Traced Knowledge Graph Embedding (TLT-KGE) model for TKGC. The authors construct this model by using the powerful ability of the complex vector space and hypercomplex vector space. In addition, in [16,17], the authors introduce a box embedding model for TKGC.

Inspired by Euler's identity $e^{i\theta} = \cos \theta + i \sin \theta$, Z. Sun et al. in [3] introduced the KGE by relational rotation in complex space, denoted by RotatE. This model takes relation as a rotation from a head entity to a tail entity in complex space. Moreover, Z. Sun et al. claimed that RotatE can capture inversion and composition patterns as well as symmetry and anti-symmetry. Motivated by DE's and RotatE's features, we propose new models, DE-RotatE and DE-RotatE-sinc models, by replacing the SimplE model from DE-SimplE model as the RotatE model.

6 Conclusion

This paper presents new models, denoted by DE-RotatE and DE-RotatE-sinc, for temporal knowledge graph completion. DE, which stands for diachronic entity embedding, is a function that maps every pair of entity and time-stamp to a hidden representation. Our models, extensions of the DE-SimplE model, combine the diachronic entity embedding, the RotatE model, and the sinc activation function. Since the performance of RotatE is very good [3, 18] and its characteristics are the same SimplE's ones by Proposition 1 and Lemma 1, we try to combine DE with RotatE. Moreover, we also change the sin function from the model in [12] by the sinc function. By experimental, we show that the results of our models are better than the results in [12] with all three datasets, say ICEWS14, ICEWS05-15, GDELT. All the results are shown in Tables 2, 3 and 4. By using the TLT-KGE, the authors in [15] show that the performance of their model is good ($MRR \sim 0.63$). On the other hand, the authors in [16,17] use a box embedding model for TKGC. These approaches are beneficial for us to continue our researches. In the future, we are going to improve our models' performance by combining the advantages of the TLT-KGE, the box embedding ... New results of KGC have been updated in [18].

References

1. Zhang, F., Yuan, N.J., Lian, D., Xie, X., Ma, W.Y.: Collaborative knowledge base embedding for recommender systems. In: Proceedings of the 22nd ACM SIGKDD International Conference on Knowledge Discovery and Data Mining, pp. 353–362 (2016)
2. Yang, B., Mitchell, T.: Leveraging knowledge bases in LSTMs for improving machine reading. In: Proceedings of the 55th Annual Meeting of the Association for Computational Linguistics, vol. 1, pp. 1436–1446 (2017)
3. Sun, Z., Deng, Z.H., Nie, J.Y., Tang, J.: RotatE: knowledge graph embedding by relational rotation in complex space. In: Proceedings of the ICLR, pp. 1–18 (2019)
4. Boschee, E., Lautenschlager, J.: ICEWS Coded Event Data, Harvard Dataverse (2015)
5. Leetaru, K., Schrodt, P.A.: GDELT: global data on events, location, and tone, 1979–2012. ISA Ann. Convention **2**, 1–49 (2013)
6. Vrandečić, D., Krötzsch, M.: Wikidata: a free collaborative. Commun. ACM **57**(10), 78–85 (2014)
7. García-Durán, A., Dumančić, S., Niepert, M.: Learning sequence encoders for temporal knowledge graph completion. In: Proceedings of the 2018 Conference on Empirical Methods in Natural Language Processing, pp. 4816–4821 (2018)
8. Bordes, A., Usunier, N., García-Durán, A., Weston, J., Yakhnenko, O.: Translating embeddings for modeling multi-relational data. In: Advances in Neural Information Processing Systems, pp. 2787–2795 (2013)

9. Yang, B., Yih, W., He, X., Gao, J., Deng, L.: Embedding entities and relations for learning and inference in knowledge bases. ICLR (2015)

10. Dasgupta, S.S., Ray, S.N., Talukdar, P.: HyTE: hyperplane-based temporally aware knowledge graph embedding. In: Proceeding of the 2018 Conference on Empirical Methods in Natural Language Processing, pp. 2001–2011 (2018)

11. Wang, Z., Zhang, J., Feng, J., Chen, Z.: Knowledge graph embedding by translating on hyperplanes. In: Proceedings of the AAAI Conference on Artificial Intelligence, vol. 28, no. 1 (2014)

12. Goel, G., Kazemi, S.M., Brubaker, M., Poupart, P.: Diachronic embedding for temporal knowledge graph completion. In: The Thirty-Fourth AAAI Conference on Artificial Intelligence (AAAI 2020), vol. 34, no. 4, pp. 3988–3995 (2020)

13. Kazemi, S.M., Poole, D.: SimplE embedding for link prediction in knowledge graphs. In: Proceedings of the 32nd International Conference on Neural Information Processing Systems (Montréal, Canada) (NIPS 2018), pp. 4289–4300. Curran Associates Inc., Red Hook (2018)

14. Kadlec, R., Bajgar, O., Kleindienst, J.: Knowledge base completion: baselines strike back. In: Proceedings of the 2nd Workshop on Representation Learning for NLP, pp. 69–74 (2017)

15. Zhang, F., Zhang, Z., Ao, X., Zhuang, F., Xu, Y., He, Q.: Along the time: timeline-traced embedding for temporal knowledge graph completion. In: Proceedings of the 31st ACM International Conference on Information & Knowledge Management (CIKM 2022), pp. 2529–2538 (2022)

16. Abboud, R., Ceylan, I.I., Lukasiewicz, T., Salvatori, T.: BoxE: a box embedding model for knowledge base completion. In: NeurIPS, vol. 33, pp. 9649–9661 (2020)

17. Messner, J., Abboud, R., Ceylan, I.I.: Temporal knowledge graph completion using box embeddings. In: Proceedings of the AAAI Conference on Artificial Intelligence, vol. 36, no. 7, pp. 7779–7787 (2022)

18. Cai, B., Xiang, Y., Gao, L., Zhang, H., Li, Y., Li, J.: Temporal Knowledge Graph Completion: A Survey. arXiv:2201.08236v1 [cs.AI] (2022)

19. Xu, C., Nayyeri, M., Alkhoury, F., Yazdi, H.S., Lehmann, J.: TeRo: a time-aware knowledge graph embedding via temporal rotation. In: Proceedings of the 28th International Conference on Computational Linguistics, pp. 1583–1593 (2020)

DCA-Based Weighted Bagging: A New Ensemble Learning Approach

Van Tuan Pham[1(✉)], Hoai An Le Thi[1,2] ⓘ, Hoang Phuc Hau Luu[1], and Pascal Damel[1]

[1] Université de Lorraine, LGIPM, Département IA, 57000 Metz, France
{van-tuan.pham,hoai-an.le-thi,hoang-phuc-hau.luu,
pascal.damel}@univ-lorraine.fr
[2] Institut Universitaire de France (IUF), Paris, France

Abstract. Ensemble learning is a highly efficient method that combines multiple machine learning models to improve the accuracy and robustness of predictions. Bagging is a popular approach of ensemble learning that involves training a group of independent classifiers on various subsets of the training data and combining their predictions using a voting or average technique. In this study, we present a new weighted bagging approach based on difference of convex functions algorithm (DCA), called BaggingDCA. The proposed algorithm combines multiple base models trained on different subsets of the training data and assigns a weight to each model based on its performance. The weights of the ensemble model are determined thanks to LS-DC, a unified approach for various loss functions in machine learning, both convex and non-convex. We evaluated our proposed algorithm on several benchmark datasets and compared its performance to existing bagging methods. We show that the proposed algorithm using both convex and nonconvex losses improves upon standard bagging, and also outperforms dynamic weighting bagging in terms of prediction accuracy.

Keywords: Bagging · DCA · DC programming · Ensemble learning · LS-DC

1 Introduction

Ensemble learning is defined as a method for combining predictions from multiple individual machine learning models to produce a final prediction [4]. This is commonly referred to as "the wisdom of the crowd", which indicates that the collective intelligence of a group of individuals is often more reliable and accurate than the intelligence of any individual member. Any form of machine learning algorithm may be used as an ensemble learner (e.g., Decision trees, K-nearest neighbor, SVM, Neural network, etc.). Ensemble learning has been proven as an effective method in machine learning and successfully applied in a variety of domains, including image recognition, natural language processing, and financial prediction [13, 22, 23].

© The Author(s), under exclusive license to Springer Nature Singapore Pte Ltd. 2023
N. T. Nguyen et al. (Eds.): ACIIDS 2023, LNAI 13996, pp. 121–132, 2023.
https://doi.org/10.1007/978-981-99-5837-5_11

Bagging [3], which stands for bootstrap aggregating, is one of the most popular techniques in ensemble learning. Bagging is a well-studied subject in machine learning and the practice of bagging has been demonstrated to be an effective method in a variety of fields, including classification, regression, and anomaly detection. It involves aggregating the predictions from a variety of models that were trained on various subsets of the training data in order to obtain the final outcome. For the past few years, bagging has been the topic of extensive study, and various modifications and additions have been proposed to further improve its effectiveness. Although bagging has demonstrated outstanding performance and is widely used in many applications, it appears that some modifications can be made on how the base learners are combined to improve the bagging predictions [2,15,16].

In machine learning, there are numerous loss functions because they are intended to optimize different objectives, provide different levels of robustness, interpretability, computing efficiency and are compatible with various regularization techniques. For instance, when the objective is to minimize the difference between the predicted output and the actual output, the least square loss is a valid choice. However, when the objective is to minimize the impact of noisy or mislabelled data points, the ramp loss is a better option. While having many loss functions to choose from can be advantageous in some cases, it is important to consider the potential downsides, such as added complexity in designing machine learning algorithms. Zhou and Zhou [24] defined a type of loss with a DC (Difference of Convex) decomposition, named LS-DC loss, and demonstrated that all commonly used losses, both convex and nonconvex loss functions, as well as for both classification and regression problems, are LS-DC loss or can be approximated by LS-DC loss. As numerous loss functions can now be utilized with the same implementation and in the same machine learning algorithm, this approach makes the algorithm powerfully applicable and has a great deal of potential. Zhou and Zhou applied the concept of LS-DC to the SVM algorithm for solving the large-scale problem. In this study, we bring this method to the bagging technique in a new way that enhances the accuracy and robustness of model prediction. Thus, the primary objectives of our study can be summarized as follows:

1. To propose a new weighted bagging algorithm based on DCA that provides an efficient training method for various kinds of base learners and loss functions.
2. Evaluate the performance of the proposed method through extensive experimentation that compares its performance to popular existing methods using widely accepted benchmark datasets.

2   Weighted Bagging Approach

2.1   Bagging

Bagging, introduced by L. Breiman in 1996 [3] is a machine learning ensemble meta-algorithm intended to increase the robustness and accuracy of machine learning algorithms. As an ensemble learning method, bagging can improve the performance of a given learning algorithm by combining the outputs of many classifiers via majority voting or averaging. Furthermore, by combining models trained on various bootstrap resample versions of the training set, bagging decreases variance and aids in avoiding overfitting. Although it is frequently used in conjunction with decision trees techniques, it may be applied to any type of machine learning algorithm.

Random forest [5], is a tree-based variant of the bagging method. Each tree in the random forest algorithm only examines a bootstrapped set of samples, and it also randomly picks subsets of features (aka. feature bootstrapping or random subspace method) used in each data sample. As the result, random forest is computationally efficient than the standard bagging approach thanks to the fact that it only needs to deal with a subset of features. In addition, feature bootstrapping also helps to improve variance by decreasing correlation between trees and the ability to rank the importance of features [1]. Brown and Mues demonstrated that random forest worked effectively when dealing with samples with a significant class imbalance in the context of credit scoring problem [6].

Bagging approaches provide a variety of advantages, such as the ability to enhance accuracy and avoid overfitting, scalability, the capability to execute parallel processing. However, the standard approach of bagging assumes that all models in the ensemble have equal prediction ability, which may not be true in practice. Equal weights strategies may not be as effective as weighted strategies for a variety of reasons, such as models may be better at predicting certain subsets of data or models in the ensemble may have different levels of accuracy.

2.2 Weighted Bagging

Suppose there is a bootstrap aggregating model from n individual base learner (aka. weak classifier or weak leaner), the overall output of the bagging models is a weighted combination of the individual classifier outputs. This can be represented by the following equation

$$f(x) = \sum_{j=1}^{n} w_j f_j(x) \tag{1}$$

where $f(x)$ is the aggregated model, $f_j(x)$ is the jth classifier predictor, w_j is the weight for combining the jth classifier, and x is an input vector of the ensemble model.

The weights must be determined properly in order for the model to perform effectively. Naturally, a straightforward way is to assign an equal weight to each classifier in a standard bagging scheme. Another approach is to assign each classifier a weight value according to its contribution to the aggregating model. There have been studies on weighted bagging in the literature. Breiman [3] proposed the bagging method by using a majority vote to combine multiple classifier outputs with equal weights. Wang et al. [23] proposed a dynamic weighting approach that assigns a weight to each weak learner based on their performance.

Leblanc and Tibshirani [12] suggested using weights for the bagging algorithm with a non negativity constraint.

Given a training dataset $\{(x_i, y_i)\}_{i=1}^m$ where m is the number of samples of the training dataset, y_i is the corresponding label (either 1 or -1). We denote the output of the aggregated model at data point i^{th} as $\hat{y}_i$. In order to find an optimal set of weights, we will minimize the following optimization problem

$$\min_{w \in R^n} K(w) = \frac{1}{m} \sum_{i=1}^m \ell(y_i, \hat{y}_i) = \frac{1}{m} \sum_{i=1}^m \ell\left(y_i, \sum_{j=1}^n w_j f_j(x_i)\right) \qquad (2)$$

$$\text{s.t. } \sum_{j=1}^n w_j = 1,$$

$$w_j \geq 0, \quad \forall j = 1, \ldots, n.$$

where ℓ, which is given arbitrarily by the user depending on the training purposes, is a loss to measure the discrepancy between the predicted values and the true targets. The idea of finding a good combination of base learners via optimizing this discrepancy is traced back to [17] where the quadratic loss was studied. Here we aim to investigate a more general class of losses (possibly nonconvex) that can be used flexibly in many cases.

3   Optimization Problem Based on DCA

3.1   An Overview of DC Programming and DCA

DC programming and DCA are well recognized as effective tools for nonconvex programming and global optimization (see, e.g., [19,20]). Standard form of a DC program is stated as:

$$\inf \{f(w) := G(w) - H(w) : w \in \mathbb{R}^p\}, (\text{P}_{dc})$$

where the functions G and $H \in \Gamma_0 (\mathbb{R}^p)$ are convex. Here $\Gamma_0 (\mathbb{R}^p)$ denotes the set of proper lower-semicontinuous convex functions from a set $\mathbb{R}^p$ to $\mathbb{R} \cup \{+\infty\}$. Such a function f is called a DC function, and $G - H$ is called a DC decomposition of f, while G and H are DC components of f. A convex constraint $w \in \mathcal{C}$ can be incorporated into the objective function of (P_{dc}) by using the indicator function χ_c of $\mathcal{C}$ (which is defined by $\chi_{\mathcal{C}}(w) := 0$ if $w \in \mathcal{C}, +\infty$ otherwise):

$$\inf\{f(w) := G(w) - H(w) : w \in \mathcal{C}\} = \inf \{[G(w) + \chi_{\mathcal{C}}(w)] - H(w) : w \in \mathbb{R}^p\}.$$

A point w^* is called a critical point of $G - H$ if it satisfies the generalized Kuhn-Tucker condition $\partial G(w^*) \cap \partial H(w^*) \neq \emptyset$, while it is called a strong DC critical point if $\emptyset \neq \partial H(w^*) \subset \partial G(w^*)$. The standard DCA scheme is presented in the algorithm below:

Algorithm. Standard DCA scheme

Initialization: Let $w^0 \in \mathbb{R}^p$ be a best guess. Set $k = 0$.
repeat
   1. Calculate $\overline{\beta}^k \in \partial h\left(w^k\right)$.
   2. Calculate $w^{k+1} \in \text{argmin}\left\{G(w) - \left\langle w, \overline{\beta}^k\right\rangle : w \in \mathbb{R}^p\right\}$.
   3. $k \leftarrow k + 1$.
until convergence of $\left\{w^k\right\}$.

The convergence properties and theoretical foundation of the DCA are addressed in [10,11]. Important convergence properties worth mentioning include the following:

- DCA is a descent method without line search (the sequence $\left\{G\left(w^k\right) - H\left(w^k\right)\right\}$ is decreasing) but with global convergence (i.e., it converges from an arbitrary starting point).
- If $G\left(w^{k+1}\right) - H\left(w^{k+1}\right) = G\left(w^k\right) - H\left(w^k\right)$, then w^k is a critical point of $G - H$. In this case, DCA terminates at k-th iteration.
- If the optimal value w of (P_{dc}) is finite, then every limit point of $\{w^k\}$ is a critical point of $G - H$.

In the last two decades, various DCA-based algorithms were proposed to efficiently address a variety of large-scale problems in a variety of application domains (see, for example, [8,9,18]).

3.2 Our Method Based on DCA

According to Zhou and Zhou [24], a loss ℓ is called an LS-DC loss if there exists a constant $A > 0$ such that the associated function ψ, i.e. $\ell(y, \hat{y}) = \psi(1 - y\hat{y})$, has the following DC decomposition

$$\psi(u) = Au^2 - \left(Au^2 - \psi(u)\right). \tag{3}$$

It is pointed out in [24] that almost all commonly used losses in machine learning are LS-DC or can be approximated by an LS-DC loss.

Let $\ell(y, t)$ be any LS - DC loss associated with ψ, the optimization problem (2) can be written as the following DC program

$$\min_{w \in \mathbb{R}^n} K(w) = \frac{1}{m}\sum_{i=1}^{m}\psi\left(1 - y_i\sum_{j=1}^{n}w_j f_j(x_i)\right) := G(w) - H(w) \tag{4}$$

$$\text{s.t. } \sum_{j=1}^{n}w_j = 1,$$

$$w_j \geq 0, \quad \forall j = 1, \ldots, n.$$

with the DC components being given by

$$G(w) := \frac{1}{m} A \sum_{i=1}^{m} \left(1 - y_i \sum_{j=1}^{n} w_j f_j(x_i)\right)^2, \tag{5}$$

$$H(w) := \frac{1}{m} \sum_{i=1}^{m} \left(A\left(1 - y_i \sum_{j=1}^{n} w_j f_j(x_i)\right)^2 - \psi\left(1 - y_i \sum_{j=1}^{n} w_j f_j(x_i)\right) \right). \tag{6}$$

At iteration k, we have the subgradient of H computed as

$$\nabla H(w^k) = \frac{1}{m} \sum_{i=1}^{m} 2A(1 - y_i \sum_{j=1}^{n} w_j^k f_j(x_i))(-y_i f_j(x_i))_{j=1,2,\ldots,n}$$

$$- \frac{1}{m} \sum_{i=1}^{m} \psi'\left(1 - y_i \sum_{j=1}^{n} w_j^k f_j(x_i)\right)(-y_i f_j(x_i))_{j=1,2,\ldots,n} \tag{7}$$

$$= \frac{1}{m} \sum_{i=1}^{m} \left(\psi'(1 - y_i \sum_{j=1}^{n} w_j^k f_j(x_i)) - 2A\left(1 - y_i \sum_{j=1}^{n} w_j^k f_j(x_i)\right) \right)(y_i f_j(x_i))_{j=1,\ldots,n}.$$

where $\psi'(u)$ denotes a subgradient of ψ at u. Thus, we solve the following convex subproblem

$$\min_{w \in R^n} G(w) - \langle \nabla H(w^k), w \rangle = \min_{w \in R^n} \frac{A}{m} \sum_{i=1}^{m} \left(1 - y_i w^T F(x_i)\right)^2 - \langle \nabla H(w^k), w \rangle$$

$$= \min_{w \in R^n} -\frac{2A}{m} \sum_{i=1}^{m} w^T y_i F(x_i) + \frac{A}{m} \sum_{i=1}^{m} \left(w^T F(x_i)\right)^2 - \langle \nabla H(w^k), w \rangle$$

$$= \min_{w \in R^n} \frac{A}{m} w^T \left(\sum_{i=1}^{m} F(x_i) F(x_i)^T \right) w - w^T \left(\frac{2A}{m} \sum_{i=1}^{m} y_i F(x_i) + \nabla H(w^k) \right) \tag{8}$$

$$\text{s.t. } \sum_{j=1}^{n} w_j = 1, w_j \geq 0, \quad \forall j = 1, \ldots, n.$$

The optimization problem (8) is a quadratic program. Because w has a relatively small dimension (e.g. commonly, number of base classifiers $n < 1000$), the solution can be efficiently solved using existing packages.

The strategy to solve (8) at each iteration and seek for the optimal weights of the problem (2) is a standard DC algorithm and is presented in Algorithm 1:

Algorithm 1. Weights calculation based on DCA

Input: Given a training set $\mathcal{D} = \{(\boldsymbol{x}_i, y_i)\}_{i=1}^m$ with $\boldsymbol{x}_i \in \mathbb{R}^d$ and $y_i \in \{-1, +1\}$;
Any LS-DC loss function $\psi(u)$ with parameter $A > 0$;
Set of n weak classifiers $F = \{f_1(x), f_2(x), \ldots, f_n(x)\}$;
Set $k = 0$, $w^0 = \mathbf{1} \in R^n$;
repeat
   1. Compute $\nabla H(w^k)$ in (7).
   2. Solve (8) to obtain w^{k+1}.
   3. Set $k \leftarrow k + 1$.
until Stopping criterion.
Output: w

The schema of the bagging method is as follows: Each base learner, using a specific technique such as decision trees, knn, or neural networks, is trained using a sample of instances taken from the original dataset by sampling with replacement. Each sample normally includes the same amount of instances as the original dataset in order to ensure a sufficient number of instances per base learner. Bootstrapping features can also be used when combining multiple base learner models for further reducing the variance of models and improving the overall performance, such as in random forest. After that, a base learner is sequentially trained on each random subset sample, or more quickly through parallel training. Finally, to determine the prediction of an unseen instance, majority vote of the base learners' predictions is performed. Thus, by integrating LS-DC to the bagging scheme, we have a new bagging algorithm referred to as BaggingDCA, and is presented in the Algorithm 2

Algorithm 2. BaggingDCA

Input: Given a training set $\mathcal{D} = \{(\boldsymbol{x}_i, y_i)\}_{i=1}^m$ with $\boldsymbol{x}_i \in \mathbb{R}^d$ and $y_i \in \{-1, +1\}$;
Base learning algorithm L;
Number of base classifiers n;
for $j = 1 \ldots n$ **do**
   1. Generate a bootstrap sample $\mathcal{D}_j$ from $\mathcal{D}$ by sampling with replacement (with feature bootstrapping).
   2. Train a base classifier $f_j(x) = L(\mathcal{D}_j)$.
end for
3. Calculate $w \in R^n$ using Algorithm 1.
4. Generate bagging model f(x)$=\sum_{j=1}^n w_j f_j(x)$.
Output: $f(x)$

---

4 Experiment Setting and Datasets

In this section, we analyze the performance of our proposed algorithm BaggingDCA with two algorithms: Standard Bagging (StdBagging) and dynamic weighting bagging [23] (Authors used Lasso-logistic regression as the base learner and referred to it as LLRE; for the convenience of our experimental purposes with several base learners, we refer to the method as DWBagging). In this study, we examine three comparison algorithms with four different base learners: Decision-trees, K-Nearest neighbor, LinearSVM and Neural network. Each algorithm will perform with feature bootstrapping, that mean each base model will be trained on a subset of the features. In the case of training tree-based estimators with subset of features, our bagging algorithm is the standard random forest algorithm of L. Breiman [5]. We utilize several datasets with different properties and characteristics in our experiment. Regarding the loss functions, we employ two LS-DC loss functions (one convex and one nonconvex) and their corresponding subdiferentials in our experiments:

1. Least squares: $\psi(u) = u^2$ is a convex loss function; $\nabla\psi(u) = 2u$.
2. Smoothed ramp loss (As ramp loss is not LS-DC, this is a smoothed approx-

   imation): $\psi(u) = \begin{cases} \frac{2}{a}u_+^2, & u \leq \frac{a}{2}, \\ a - \frac{2}{a}(a-u)_+^2, & u > \frac{a}{2}, \end{cases}$, with $a > 0$ is a nonconvex func-

   tion; $\partial\psi_a(u) = \begin{cases} \frac{4}{a}u_+, & u \leq \frac{a}{2} \\ \frac{4}{a}(a-u)_+, & u > \frac{a}{2} \end{cases}$.

Numerical experiments will be performed with the two loss functions: Least squares and smoothed ramp loss. Thus, we denote our algorithms as BaggingDCA1, BaggingDCA2 corresponding with BaggingDCA with least squares loss, and BaggingDCA with smoothed ramp loss.

4.1   Experimental Setup

The three comparison algorithms are implemented in MATLAB and executed on a PC with an Intel Core i7-8700 CPU @3.20GHz×6 16GB. The PC runs Windows 10 with MATLAB-2021a. For creating bootstrap samples with replacement to train each base learner, we use the function: *ransample(n_learner,n_learner,true)* in Matlab. We use the function: *ransample(m_feature,m_sub,false)* for choosing feature subsets, with the number of subset features to draw from m features is set: $m\_sub = sqrt(m)$. For DCA algorithm, the constant A is set as 1, the truncated parameter a of the non-convex loss is set as 2 and the initial weights are set as a vector of all ones. The stop condition of DCA is set as 10^{-5}. To solve the quadratic problem in (8), we use the function **quadprog**[1] in MATLAB. The number of base-classifiers: $n = 500$ is set for all algorithms. The base learners: Decision-trees, K-NN, LinearSVM and Neural network for the comparison algorithms can be obtained from corresponding functions: **fitctree**[2], **fitcknn**[3], **fitclinear**[4], **fitcnet**[5] in Matlab. All parameters of the four functions are set as default values.

---

[1] https://www.mathworks.com/help/optim/ug/quadprog.html.
[2] https://www.mathworks.com/help/stats/fitctree.html.
[3] https://fr.mathworks.com/help/stats/fitcknn.html.
[4] https://fr.mathworks.com/help/stats/fitclinear.html.
[5] https://fr.mathworks.com/help/stats/fitcnet.html.

4.2 Datasets

Benchmark Datasets from LibSVM: To assess the generalizability and scalability of the proposed algorithm, in the first experiment, we use nine well-known benchmark datasets from the LIBSVM website[6] and Mendeley Data[7]. Most datasets consist of binary data (Mnist-1–7 is a subset of Mnist with labels 1 and 7) and contain a variety of features and instances. Table 1 summarizes the information included in the datasets. This table illustrates that the selected datasets include datasets with varying levels of size, features, and class proportion. Additional information on these datasets is available on the LIBSVM website.

Table 1. Nine benchmark datasets from LibSVM.

Dataset	#of features	#of instances	#of label 1	#of label -1
Splice	60	3,175	1,648	1,527
Gisette	5,000	7,000	3,500	3,500
Mushroom	112	8,124	3,916	4,208
Australian	14	690	307	383
Phishing	49	10,000	5,000	5,000
Adult	123	48,840	11,687	37,153
W8a	300	49,761	1,491	48,270
Madelon	500	2,600	1,300	1,300
Mnist-1–7	784	15,170	7,877	7,293

In the experiments, for evaluation purposes, the binary datasets that have labels in $\{+1, -1\}$ will be split into two separate sets: Trainset and testset. To avoid overfitting and preserve the proportions of the classes, we use the function *StratifiedShuffleSplit* in Sklearn 1.1.1[8] to split the dataset with proportions: 80% and 20% for trainset and testset, respectively.

5 Numerical Experiment and Discussion

Experiment on Nine Benchmark Datasets: Due to the random nature of bagging algorithms, to prevent biased results caused by the random sampling of subsets of data and subsets of features, we execute each base classifier of each comparison method ten times, and then take the mean and standard deviation of the results for reporting. For comparison purpose, the accuracy on testsets of related algorithms are reported. The values that are highlighted in bold are the ones with the best results. The result of the experiment is presented in Table 2.

---

[6] https://www.csie.ntu.edu.tw/~cjlin/libsvm/.

[7] https://data.mendeley.com/datasets/h3cgnj8hft/1.

[8] https://scikit-learn.org/.../sklearn.model_selection.StratifiedShuffleSplit.html.

Table 2. Performance comparison (mean accuracy±std.) of related algorithms on seven benchmark datasets.

Dataset	Algorithm	Base learner			
		Decision tree	K-NN	Linear SVM	Neural net
Splice	StdBagging	95.47±0.56	93.93±0.71	80.95±0.79	93.83±0.74
	DWBagging	95.62±0.56	94.04±0.68	81.12±0.86	94.02±0.74
	BaggingDCA1	**96.07±0.56**	94.99±0.81	85.65±1.03	94.75±0.93
	BaggingDCA2	95.90±0.52	94.99±0.81	84.99±1.05	94.72±0.95
Gisette	StdBagging	95.86±0.16	93.39±0.31	93.31±0.24	91.26±0.14
	DWBagging	95.85±0.18	93.44±0.28	93.35±0.20	91.25±0.18
	BaggingDCA1	**96.10±0.12**	93.81±0.46	94.68±0.45	92.69±0.41
	BaggingDCA2	96.07±0.25	93.81±0.46	94.26±0.44	92.43±0.50
Mushroom	StdBagging	95.85±0.75	99.11±0.37	92.55±0.89	95.94±0.78
	DWBagging	95.91±0.74	99.15±0.38	92.60±0.91	96.02±0.79
	BaggingDCA1	**99.92±0.24**	99.71±0.32	99.33±0.41	99.86±0.22
	BaggingDCA2	**99.92±0.24**	99.63±0.37	99.21±0.41	99.85±0.27
Australian	StdBagging	87.74±0.83	79.27±0.62	83.80±0.90	83.43±0.98
	DWBagging	87.74±0.83	79.34±0.77	84.01±0.80	83.43±0.60
	BaggingDCA1	**88.33±0.61**	84.23±1.24	85.77±0.85	86.86±1.33
	BaggingDCA2	88.18±0.72	84.23±1.24	84.89±0.49	86.79±1.47
Phishing	StdBagging	98.91±0.12	99.36±0.15	96.60±0.63	97.97±0.45
	DWBagging	99.01±0.10	99.40±0.16	96.85±0.66	98.12±0.42
	BaggingDCA1	**100.00±0.00**	99.99±0.03	99.98±0.05	**100.00±0.00**
	BaggingDCA2	**100.00±0.00**	99.99±0.03	99.98±0.06	**100.00±0.00**
Adult	StdBagging	77.09±0.05	76.20±0.30	76.07±0.00	76.07±0.00
	DWBagging	77.23±0.10	76.19±0.27	76.07±0.00	76.27±0.06
	BaggingDCA1	**83.87±0.22**	83.63±0.23	83.44±0.24	83.30±0.37
	BaggingDCA2	83.76±0.22	82.91±0.41	82.96±0.35	83.14±0.38
W8a	StdBagging	97.06±0.01	97.07±0.03	97.04±0.02	97.06±0.01
	DWBagging	97.16±0.03	97.07±0.03	97.04±0.02	97.10±0.04
	BaggingDCA1	**98.07±0.13**	97.44±0.09	97.46±0.13	97.49±0.17
	BaggingDCA2	97.82±0.09	97.65±0.09	97.65±0.06	97.86±0.12
Madelon	StdBagging	65.19±1.41	63.04±2.01	61.00±0.85	59.38±0.86
	DWBagging	65.63±1.67	62.98±1.86	60.98±0.74	59.78±1.24
	BaggingDCA1	**71.25±1.34**	70.27±1.91	61.62±0.71	61.42±0.57
	BaggingDCA2	71.22±1.24	70.27±1.91	61.27±0.62	61.31±0.81
Mnist-1–7	StdBagging	99.60±0.05	99.51±0.01	98.95±0.01	99.55±0.04
	DWBagging	99.61±0.03	99.51±0.01	98.94±0.03	99.55±0.04
	BaggingDCA1	**99.74±0.05**	99.70±0.04	99.43±0.09	99.74±0.10
	BaggingDCA2	**99.74±0.05**	99.70±0.04	99.39±0.06	99.74±0.10

Comments: Table 2 compares the average accuracy of all comparison methods. Based on the experimental result, it is evident that our proposed algorithms give better results than StdBagging and DWBagging in most cases on nine benchmark datasets. There are cases where our algorithm is approximately 4% to 6% more accurate than the StdBagging and DWBagging techniques, such as in the Mushroom and Adult datasets, or Madelon in the case of Decision tree. When it comes to the base classifiers, Tree-based models typically produce superior outcomes to K-nearest neighbor, LinearSVM and Neural network. Considering the comparative results of the two loss functions, the method utilizing the least squares loss function performs slightly better than the method using the smoothed ramp loss function in the majority datasets.

6   Conclusion and Future Work

In this study, we propose a new bagging algorithm based on the LS-DC scheme, which is applicable to the majority of existing loss functions for both classification and regression problems. The proposed algorithm inherits the benefits of standard bagging, such as decreasing variance, the ability to set up parallel computations and helping to avoid overfitting, additionally, it improves the accuracy of the training model. Both convex (Least squares) and nonconvex (Ramp) losses employed in the experiments improve upon the classical combination using equal weights and also the dynamic weighting bagging algorithm. Moreover, it is known that truncated losses, of which ramp loss is an example, are more resistant to outliers than convex losses. One of our future objectives is to confirm the robustness of the proposed bagging scheme using truncated losses.

In our research, we focused primarily on dealing with the bagging method using four base classifiers. The experiments were designed to thoroughly test the robustness and generalizability of the proposed method on various datasets. The experimental results were examined to assess the significance of the improvements over the existing approaches and to identify any potential areas for future improvement. There are some other studies that employ alternative ensemble approaches with different base classifiers, such as pasting, boosting, and stacking with various types of base learner: Logistic regression, Neural network, or a mixture of learners [4,7,14,21]. In the future, we will intensively study whether our proposed algorithm may improve these types of ensemble learning problems.

References

1. Altman, N., Krzywinski, M.: Ensemble methods: bagging and random forests. Nat. Methods **14**(10), 933–935 (2017)
2. Błaszczyński, J., Stefanowski, J.: Neighbourhood sampling in bagging for imbalanced data. Neurocomputing **150**, 529–542 (2015)
3. Breiman, L.: Bagging predictors. Mach. Learn. **24**(2), 123–140 (1996)
4. Breiman, L.: Stacked regressions. Mach. Learn. **24**(1), 49–64 (1996)
5. Breiman, L.: Random forests. Mach. Learn. **45**(1), 5–32 (2001)
6. Brown, I., Mues, C.: An experimental comparison of classification algorithms for imbalanced credit scoring data sets. Expert Syst. Appl. **39**(3), 3446–3453 (2012)

7. Jovanović, R.Ž, Sretenović, A.A., Živković, B.D.: Ensemble of various neural networks for prediction of heating energy consumption. Energy Build. **94**, 189–199 (2015)

8. Le Thi, H.A., Le, H.M., Pham Dinh, T.: Feature selection in machine learning: an exact penalty approach using a difference of convex function algorithm. Mach. Learn. **101**(1), 163–186 (2015)

9. Le Thi, H.A., Nguyen, M.C.: DCA based algorithms for feature selection in multiclass support vector machine. Ann. Oper. Res. **249**(1–2), 273–300 (2017)

10. Le Thi, H.A., Pham Dinh, T.: The dc (difference of convex functions) programming and DCA revisited with dc models of real world nonconvex optimization problems. Ann. Oper. Res. **133**(1), 23–46 (2005)

11. Le Thi, H.A., Pham Dinh, T.: Dc programming and DCA: thirty years of developments. Math. Program. **169**(1), 5–68 (2018)

12. LeBlanc, M., Tibshirani, R.: Combining estimates in regression and classification. J. Am. Stat. Assoc. **91**(436), 1641–1650 (1996)

13. McIntosh, T., Curran, J.R.: Reducing semantic drift with bagging and distributional similarity. In: Proceedings of the Joint Conference of the 47th Annual Meeting of the ACL and the 4th International Joint Conference on Natural Language Processing of the AFNLP, pp. 396–404 (2009)

14. Mi, X., Zou, F., Zhu, R.: Bagging and deep learning in optimal individualized treatment rules. Biometrics **75**(2), 674–684 (2019)

15. Mordelet, F., Vert, J.P.: A bagging SVM to learn from positive and unlabeled examples. Pattern Recogn. Lett. **37**, 201–209 (2014)

16. Moretti, F., Pizzuti, S., Panzieri, S., Annunziato, M.: Urban traffic flow forecasting through statistical and neural network bagging ensemble hybrid modeling. Neurocomputing **167**, 3–7 (2015)

17. Perrone, M.P., Cooper, L.N.: When networks disagree: ensemble methods for hybrid neural networks. In: How We Learn; How We Remember: Toward An Understanding of Brain and Neural Systems: Selected Papers of Leon N Cooper, pp. 342–358. World Scientific (1995)

18. Pham, V.T., Luu, H.P.H., Le Thi, H.A.: A block coordinate DCA approach for large-scale kernel SVM. In: Nguyen, N.T., Manolopoulos, Y., Chbeir, R., Kozierkiewicz, A., Trawinski, B. (eds.) Computational Collective Intelligence. ICCCI 2022. Lecture Notes in Computer Science(), vol. 13501, pp. 334–347. Springer, Cham (2022). https://doi.org/10.1007/978-3-031-16014-1_27

19. Pham Dinh, T., Le Thi, H.A.: Convex analysis approach to dc programming: theory, algorithms and applications. Acta Math. Vietnam **22**(1), 289–355 (1997)

20. Pham Dinh, T., Le Thi, H.A.: Recent advances in DC programming and DCA. In: Nguyen, N.-T., Le-Thi, H.A. (eds.) Transactions on Computational Intelligence XIII. LNCS, vol. 8342, pp. 1–37. Springer, Heidelberg (2014). https://doi.org/10.1007/978-3-642-54455-2_1

21. Rasti, R., Teshnehlab, M., Phung, S.L.: Breast cancer diagnosis in DCE-MRI using mixture ensemble of convolutional neural networks. Pattern Recogn. **72**, 381–390 (2017)

22. Saffari, A., Leistner, C., Santner, J., Godec, M., Bischof, H.: On-line random forests. In: 2009 IEEE 12th International Conference on Computer Vision Workshops, ICCV Workshops, pp. 1393–1400. IEEE (2009)

23. Wang, H., Xu, Q., Zhou, L.: Large unbalanced credit scoring using lasso-logistic regression ensemble. PLoS ONE **10**(2), e0117844 (2015)

24. Zhou, S., Zhou, W.: Unified SVM algorithm based on LS-DC loss. Mach. Learn., 1–28 (2021)

Deep-Learning- and GCN-Based Aspect-Level Sentiment Analysis Methods on Balanced and Unbalanced Datasets

Huyen Trang Phan[1,2], Ngoc Thanh Nguyen[3], Yeong-Seok Seo[1(✉)], and Dosam Hwang[1]

[1] Department of Computer Engineering, Yeungnam University, Gyeongsan, South Korea
ysseo@yu.ac.kr

[2] Faculty of Information Technology, Nguyen Tat Thanh University, Ho Chi Minh City, Vietnam

[3] Department of Applied Informatics, Wroclaw University of Science and Technology, Wrocław, Poland
ngoc-thanh.nguyen@pwr.edu.pl

Abstract. With the growth of social networks, an increasing number of publicly available opinions are posted on them. Sentiment analysis, especially aspect-level sentiment analysis (ALSA), of these opinions has emerged and is a concern for many researchers. ALSA aims to gather, evaluate, and aggregate sentiments regarding the aspects of a topic of concern. Previous research has demonstrated that deep learning and graph convolutional network (GCN) methods can effectively improve the performance of ALSA methods. However, further investigation is required, especially when comparing the performance of deep learning-based and GCN-based ALSA methods on balanced and unbalanced datasets. In this study, we aimed to investigate two hypotheses: (i) the effectiveness of ALSA methods can be improved by deep learning and GCN techniques on balanced and unbalanced datasets, especially GCNs, over BERT representations, and (ii) the balanced data slightly affect the accuracy and F_1 score of the deep learning and GCN-based ALSA methods. To implement this study, we first constructed balanced Laptop, Restaurant, and MAMS datasets based on their original unbalanced datasets; then, we experimented with 17 prepared methods on six unbalanced and balanced datasets; finally, we evaluated, discussed, and concluded the two hypotheses.

Keywords: Aspect-level sentiment analysis · Graph Convolutional Network · Deep learning · Balanced dataset · Unbalanced dataset

1 Introduction

We live in a digital society surrounded by systems, websites, and social networks with limitless connectivity and information sharing [18]. This has resulted in

N. T. Nguyen et al. (Eds.): ACIIDS 2023, LNAI 13996, pp. 133–144, 2023.
https://doi.org/10.1007/978-981-99-5837-5_12

an ever-increasing volume of generated data, especially textual data regarding people's opinions and judgments. These data have become a significant source for various applications such as recommendation systems, decision-making systems [15], and sentiment analysis systems [14].

Sentiment analysis (SA) of a text is the process of using natural language processing techniques to analyze and identify the emotional polarizations or attitudes of speakers expressed in opinions about a particular reality entity, which is any distinct or identifiable object; the term "reality entity" typically refers to individuals, events, organizations, systems, and products. The SA degree, level, and target are the main characteristics of SA. Sentiment degrees may be divided into two levels (positive and negative), three levels (negative, positive, and neutral), or multi-level (strongly negative, negative, strongly positive, positive, and neutral). The SA target is the object to which the affection is directed. The SA level can be divided into the aspect, sentence, and document levels. Document-level SA determines the general sentiments about a topic expressed throughout the document and can become sentence-level SA when the text consists of only one sentence. Sentence-level SA determines the sentiment polarity of a particular topic expressed in a sentence. Aspect-level SA (ALSA) is used to identify feelings regarding a part or detail of an entity.

Several state-of-the-art methods have been proposed to improve the performance of ALSA methods, and methods based on deep learning and graph convolutional networks (GCNs) are proven to be the most effective. Many studies have compared and evaluated the performance of ALSA methods. Komang et al. [20] compared ALSA-based deep learning methods. Qaiser et al. [17] compared machine learning-based SA methods. Alantari et al. [1] empirically compared machine learning-based SA methods using online consumer reviews. Do et al. [3] compared deep-learning-based ALSA methods. Wang et al. [21] further reviewed the recent achievements of different deep-learning-based ALSA models. As we can see, the previous studies comprehensively compared deep learning-based and machine learning-based ALSA methods regarding their accuracy, completeness, and detail. However, few studies have focused on comparing the performances of two different groups of ALSA methods, and fewer have compared the performances of ALSA methods on balanced and unbalanced datasets. This study compares the performance of two state-of-the-art ALSA method groups, deep learning and GCNs, on two different types of datasets: balanced and unbalanced. The main objectives of this study are to (i) investigate whether deep learning and GCN-based techniques can be effectively leveraged for ALSA on balanced and unbalanced datasets and (ii) determine how avoiding balanced datasets can influence the accuracy and F_1 score of deep learning and GCN-based ALSA. Our main contributions are summarized as follows. (i) We reconstruct three unbalanced datasets to create three balanced datasets. (ii) We assess the performances of 17 state-of-the-art deep learning and GCN-based ALSA models on three unbalanced and three balanced datasets. (iii) We provide evaluations and discussions regarding the performance of deep-learning-based and GCN-based ALSA methods on two types of dataset groups.

The remainder of this paper is organized as follows: Sect. 2 introduces the background material. Section 3 presents the research problem. Section 4 describes the experimental design of the ALSA method and the datasets. Section 5 presents the experimental results of the ALSA method on the prepared datasets. Finally, Sect. 6 concludes the study and presents future research directions.

2 Backgrounds Materials

This section introduces the primary materials used to construct the deep learning and GCNs models for ALSA. Most deep-learning- and GCN-based ALSA methods surveyed in this study used at least one of the following techniques for their models.

2.1 Word Embeddings

Many deep learning and GCN models have been improved for use in ALSA. The text forms most of the primary input data for the deep-learning- and GCN-based ALSA methods. Therefore, for these methods to capture critical features more precisely, it is essential to use word-embedding models to convert the text into numerical vectors. The state-of-the-art word embeddings can be categorized into three categories: basic embeddings (one-hot encoding, term frequency-inverse document frequency, bags of words), distributed embeddings (Word2vec[1], Glove[2], Fasttext[3]), and universal language embeddings (BERT[4] and ELMo[5]).

2.2 Long Short-Term Memory

Long Short-Term Memory (LSTM) includes a forward LSTM ($\overrightarrow{lstm}$) to read the text from left to right and a backward LSTM ($\overleftarrow{lstm}$) to read the text from right to left. For the i-th word in the text, the forward and backward LSTM $\overrightarrow{lstm}$ and $\overleftarrow{lstm}$ are performed as follows:

For $\overrightarrow{lstm}(x_i)$:

$$\overrightarrow{G_i} = \begin{bmatrix} \overrightarrow{h}_{i-1} \\ x_i \end{bmatrix} \tag{1}$$

$$f_i = sigmoid(W_f.\overrightarrow{G_i} + b_f) \tag{2}$$

$$in_i = sigmoid(W_{in}.\overrightarrow{G_i} + b_{in}) \tag{3}$$

$$o_i = sigmoid(W_o.\overrightarrow{G_i} + b_o) \tag{4}$$

---

[1] https://code.google.com/archive/p/word2vec/.
[2] https://nlp.stanford.edu/projects/glove/.
[3] https://fasttext.cc/docs/en/crawl-vectors.html.
[4] https://mccormickml.com/2019/05/14/BERT-word-embeddings-tutorial/.
[5] https://github.com/yuanxiaosc/ELMo.

$$c_i = f_i \odot c_{i-1} + in_i \odot tanh(W_c.\overrightarrow{G_i} + b_c) \tag{5}$$

$$\overrightarrow{h_i} = o_i \odot tanh(c_i) \tag{6}$$

where h_{i-1} is the previous hidden state of h_i and $h_0 = 0$.

For $\overleftarrow{lstm}(x_i)$:

$$\overleftarrow{G_i} = \begin{bmatrix} \overleftarrow{h}_{i+1} \\ x_i \end{bmatrix} \tag{7}$$

$$f_i = sigmoid(W_f.\overleftarrow{G_i} + b_f) \tag{8}$$

$$in_i = sigmoid(W_{in}.\overleftarrow{G_i} + b_{in}) \tag{9}$$

$$o_i = sigmoid(W_o.\overleftarrow{G_i} + b_o) \tag{10}$$

$$c_i = f_i \odot c_{i-1} + in_i \odot tanh(W_c.\overleftarrow{G_i} + b_c) \tag{11}$$

$$\overleftarrow{h_i} = o_i \odot tanh(c_i) \tag{12}$$

where x_i is the word embeddings. h_{i+1} is the next hidden state of h_i and $h_{m+1} = 0$ (m is the number of words in the text). c, f, oin are the operations. $\odot$ is element-wise production. $W_c, W_f, W_o, W_{in} \in R^{h \times (h+d)}$, $b^f, b^{in}, b^o, b^c \in R^h$ are LSTM parameters. d and h are the dimension of the word embeddings and the hidden vectors, respectively.

2.3   Graph Convolutional Network

GCNs can be considered an extended version of conventional deep-learning algorithms over a graph structure to encode the local information which contains unstructured data. The original GCN model proposed by Kipf et al. [8] is defined as follows:

$$H^1 = ReLU(M \cdot H^0 \cdot W^1 + b^1) \tag{13}$$

Hence,

$$H^2 = ReLU(M \cdot H^1 \cdot W^2 + b^2) \tag{14}$$

That means:

$$X = H^2 = ReLU(ReLU(M \cdot H^0 \cdot W^1 + b^1) \cdot W^2 + b^2)) \tag{15}$$

where $X \in R^{m \times d}$; H^0 is the word embeddings matrix. $ReLU$ is an activation function. $W^1 \in R^{d_h \times m}$ and $W^2 \in R^{m \times h}$ are the weight matrices created for the i-th layer. b^i are the biases of two layers, respectively. m is the number of words in the input. d and h are the dimension of the word embeddings and the hidden vectors, respectively.

$$M = D^{-0.5}AD^{-0.5} \tag{16}$$

D and M are the matrices of degree and normalized symmetric adjacency of matrix A, respectively, where:

$$D_{ii} = \sum_j A_{ij} \tag{17}$$

3   Research Problems

For a balanced and an unbalanced dataset, both comprising opinion data in the form of text posted on social media, as well as a set of deep-learning- and GCN-based ALSA models, the objective of this study is to investigate two hypotheses:

(i) The effectiveness of the ALSA methods can be improved by deep learning and GCN techniques on balanced and unbalanced datasets, particularly GCNs, over BERT representations.
(ii) Balanced data slightly affects the F_1 score and accuracy of the deep learning and GCN-based ALSA methods.

Based on these objectives, two research questions must be answered:

(i) Can deep learning and GCNs techniques be effectively used for ALSA on both balanced and unbalanced data?
(ii) Can avoiding balanced data influence the F_1 score and accuracy of deep learning and GCN-based ALSA methods?

4   Experiment Design

As mentioned previously, deep learning and GCN techniques are widely used for ALSA. However, further investigation into these techniques is required, particularly a comparison of their performance on balanced and unbalanced datasets.

4.1   Source Dataset

We used the Laptop and Restaurant datasets in SemEval[6] [16] and MAMS[7] [6] dataset to retrain the ALSA methods. For each dataset, we first retained its original form and considered it an unbalanced dataset; afterwards, we reconstructed it by adjusting the number of samples to be equal in terms of aspect sentiment labels and considered it a balanced dataset. A balanced dataset was created using randomly selected samples from the original data. The reason is no benchmark dataset separately according to the vast amounts. The statistics for the data in the unbalanced and balanced datasets are presented in Table 1.

The MAMS dataset includes sets of opinions related to laptops and restaurants, each of which has at least two aspects with different corresponding sentiment polarities. The laptop dataset includes sets of opinions related to laptops, each containing one aspect and one corresponding sentiment polarity. The restaurant dataset includes sets of opinions related to restaurants, each of which also has one aspect and one corresponding sentiment polarity.

The effectiveness of the model in correctly classifying sentiment polarity was tested against a held-out test set containing aspects with assigned labels. We randomly split these datasets into training and test datasets for both the unbalanced and balanced datasets. We then randomly selected a developing set from the training set.

---

[6] https://alt.qcri.org/semeval2014/task4/index.php?id=data-and-tools.
[7] https://github.com/siat-nlp/MAMS-for-ABSA.

Table 1. Main characteristics of baseline and MEMO datasets

Dataset	Training set				Testing set			
	Positive	Neutral	Negative	Total	Positive	Neutral	Negative	Total
Unbalanced dataset								
Laptop	604	260	516	1380	119	165	100	384
Restaurant	1170	327	414	1911	412	74	102	588
MAMS	2526	3745	2021	8292	293	455	239	987
Balanced dataset								
Laptop	460	460	460	1380	128	128	128	384
Restaurant	637	637	637	1911	196	196	196	588
MAMS	2764	2764	2764	8292	329	329	329	987

4.2 Model Training and Tuning

Given the complexity of various deep-learning- and GCN-based ALSA algorithms, selecting an optimal algorithm using only theoretical arguments can be challenging. Hence, in this study, we evaluated 17 common deep-learning- and GCN-based ALSA algorithms frequently used in computer science research. These classification algorithms fall into the following categories:

Deep-learning-based: This category focuses on capturing the interactions between aspects and contexts in sentences using various approaches. LSTM [22] uses the standard LSTM for the ALSA task. TD-LSTM [19] uses two LSTM networks that follow both directions, left to right and the inverse. Subsequently, an advanced version of TD-LSTM called TC-LSTM [19] was proposed, which focuses on representing the interaction between a target and contexts by adding a target-aware connection component. ATAE-LSTM [22] focuses attention on the critical parts of the sentence by computing specific aspect-aware attention scores. AOA [5] uses an attention-over-attention model to focus on essential parts related to aspects and contexts. Additionally, IAN [11] creates an interactive attention network. MGAN [4] explores attention mechanisms such as fine- and coarse-grained mechanisms. CABASC [10] introduces a combination of sentence-level content and context attention mechanisms into a memory network to solve the semantic mismatch problem. LCF_BERT [23] uses a combination of a local context focus mechanism and multihead self-attention to capture long-term internal dependencies between contexts, such as local and global dependencies.

GCN-based: This category focuses on constructing graphs to extract various types of information such as syntactic, contextual, and sentiment knowledge. SDGCN [25] uses a position-encoder-based bidirectional attention mechanism to represent the aspect-specific information in the GCN layers to extract the sentiment dependence of different aspects. ASGCN [24] constructs a directional graph for each sentence using a dependency tree to capture syntactic information. ASCNN [24] is an ablation version of ASGCN that replaces GCN layers with a CNN. ASTCN [24] is a new and enhanced version of ASGCN that constructs directed dependency trees for use in a bidirectional GCN. AFFGCN [9] constructs a unidirectional sentence graph using affective words extracted from SenticNet. SenticGCN [9] constructs a dependency-tree-based sentence graph and leverages the affective information of SenticNet to enhance the dependencies between contexts and aspects. SenticGCN_BERT [9] is a version of SenticGCN that replaces the combination of GloVe embedding and LSTM layers with pretrained BERT.

Hyperparameters Setting: For a fair comparison, we reused the experimental settings of the study by Bin Liang et al. [9] for all datasets. For models using BERT embedding, the pretrained uncased BERT [2] with 768-dimension vectors was used. For models that did not use BERT, GloVe embeddings [13] with a dimension of 300 were used. The learning rate was 0.00002. The batch size was 16. The number of GCN layers was two. The size of the hidden vectors in the neural-network layers was 300. An Adam optimizer [7], with a learning rate of 0.001 was used to optimize and update the model parameters. Weight matrices and biases with a uniform distribution for all layers were initialized randomly.

Evaluation Metrics: The experimental results were evaluated using the F_1 score and accuracy based on scripts from Scikit-learn[8] [12]. We also conducted a paired test on the F_1 score and *accuracy* and accuracy to verify whether the improvements achieved by our models over the baselines were significant. We executed all selected SA methods on all datasets 10 times, and the mean and standard deviation statistics were extracted.

5   Experimental Results

The obtained F_1 score and accuracy used to evaluate the hypotheses in this study are statistically discussed in this section. We investigated the performance of the ALSA methods by tuning the hyperparameters and testing these models to answer the research questions.

5.1   Experimental Results on Unbalanced Dataset

This section focuses on answering the first research question, *"Can deep learning and GCN techniques be effectively used for ALSA on both balanced and unbalanced data?"*

Table 2 reports the F_1 score and accuracy for deep-learning- and GCN-based ALSA methods. The ALSA methods have been evaluated using parts of the original Laptop, Restaurant, and MAMS datasets. The accuracy values are sorted in ascending order.

All the methods achieved promising performance, with a few exceptional cases. The accuracy and F_1 score values of all methods were satisfied on average 67.47% and 61.32% respectively. SenticGCN_BERT, a GCN category method,

---

[8]   https://scikit-learn.org/stable/modules/classes.html#module-sklearn.metrics.

Table 2. Performance of ALSA methods over unbalanced datasets (%)

Method	Laptop		Restaurant		MAMS	
	Accuracy	F_1	Accuracy	F_1	Accuracy	F_1
CABASC	39.06	34.06	70.58	30.74	51.27	44.99
ATAE_LSTM	40.63	38.10	78.74	59.46	65.96	62.99
TD_LSTM	43.75	39.92	79.93	63.41	67.07	64.86
LSTM	44.27	43.49	82.31	71.16	73.15	71.34
TC_LSTM	46.09	45.60	80.10	61.61	66.57	63.54
IAN	46.88	46.75	82.48	71.15	73.46	72.20
AOA	48.17	47.89	83.33	70.05	74.37	72.75
MGAN	51.04	49.00	83.16	70.76	74.16	72.64
ASCNN	49.48	49.46	83.33	71.11	73.96	72.35
ASTCN	49.22	47.03	83.67	71.37	75.38	74.13
ASGCN	51.04	48.31	84.01	71.90	74.57	72.98
SDGCN	48.18	45.93	83.16	71.71	74.37	72.97
ATTSenticGCN	48.96	48.62	81.97	69.80	73.96	72.67
AFFGCN	51.04	48.31	84.01	71.90	74.57	72.98
SenticGCN	48.70	47.30	82.99	70.30	74.47	73.01
LCF_BERT	57.29	55.55	89.29	81.95	80.75	80.55
SenticGCN_BERT	61.72	61.85	89.97	79.45	81.76	81.30

achieved the highest F_1 score and accuracy, whereas CABASC, a deep-learning category method, achieved the lowest accuracy and F_1 score. Most methods achieved relatively good results on the Restaurant and MAMS datasets but did not achieve good results on the Laptop dataset. More than 94% (16 of 17) of the models achieved an accuracy greater than 65% on the Restaurant and MAMS datasets. Significantly, the GCN-based methods obtained the best results with accuracy values greater than 70%, which was interpreted as a good performance. The highest accuracy of 89.97% was achieved on the Restaurant dataset by the SenticGCN-BERT method, whereas the lowest accuracy was 48.18% achieved on the Laptop dataset by SDGCN. In addition, more than 88% of the models (15 of 17) achieved an F_1 score greater than 63%; more than 76% of this set (13 of 17) had F_1 scores greater than 71%. LCF_BERT obtained the highest value (81.95%) on the Restaurant dataset, while CABASC demonstrated the lowest value (30.74%) on the Restaurant dataset. For the Laptop dataset, only 5 out of 17 methods demonstrated an accuracy greater than 50% and only 2 out of 17 demonstrated an F_1 score greater than 50%. In addition, it is worth noting that there was a significant difference between the accuracy value and the F_1 score value for unbalanced datasets.

Based on these observations, most deep learning and GCN classifiers have achieved high accuracy and F_1 score values, such as LCF_BERT, SenticGCN_BERT, SenticGCN, and AFFGCN, because they represent a realistic context for classifying the sentiment polarities of aspects. Therefore, we conclude that we can improve the effectiveness of ALSA methods based on deep learning and GCN algorithms. The null hypothesis of "We cannot improve the effectiveness of ALSA methods using deep learning and GCN techniques on balanced or unbalanced datasets." is rejected. Moreover, for unbalanced datasets, the F_1 score more accurately reflected the effectiveness of the methods than the accuracy value. Thus, when evaluating the efficacy of SA methods at the aspect level, an additional measure of the F_1 score should be used.

5.2 Experimental Results on Balanced Dataset

This section focuses on answering the second research question: *"Can avoiding balanced data influence the F_1 score and accuracy of deep-learning- and GCN-based ALSA methods?"*

In this experiment, all methods were developed using balanced data from the Laptop, Restaurant, and MAMS datasets mentioned in Sect. 5.1. Table 3 reports the F_1 scores and accuracies of the methods. The performances are arranged in ascending order of accuracy values.

Table 3. Performance of ALSA methods over balanced datasets (%)

Method	Laptop		Restaurant		MAMS	
	Accuracy	F_1	Accuracy	F_1	Accuracy	F_1
LSTM	52.34	55.66	60.54	60.25	69.30	69.30
ATAE_LSTM	58.59	58.32	55.27	54.07	55.12	55.06
TC_LSTM	60.42	60.42	61.90	60.90	66.46	66.50
TD_LSTM	60.16	60.22	63.95	63.74	65.15	65.16
AOA	64.58	64.30	65.48	63.83	72.14	72.15
IAN	63.54	63.66	69.90	69.49	71.83	71.81
MGAN	64.84	64.69	70.41	70.04	72.75	72.75
CABASC	56.77	56.98	57.65	57.40	54.20	53.92
ASCNN	63.02	62.96	65.14	64.95	70.82	70.74
ASTCN	65.10	64.99	69.39	69.11	73.35	73.23
ASGCN	64.84	64.47	69.58	68.97	73.15	73.11
SDGCN	65.10	64.54	66.33	65.87	72.54	72.47
ATTSenticGCN	61.72	61.38	66.33	66.33	70.31	70.30
AFFGCN	64.84	64.47	69.56	68.97	73.15	73.11
SenticGCN	64.84	64.89	69.39	69.29	72.85	72.69
LCF_BERT	71.09	71.20	78.23	78.48	81.05	81.05
SenticGCN_BERT	70.31	70.79	80.27	80.27	81.05	81.03

Although all methods achieved promising results, their performance was lower than that for the unbalanced datasets, with accuracy and F_1 score values averaging 66.80% and 66.67%, respectively. LCF BERT, a deep learning method, and SenticGCN BERT, a GCN method, achieved the highest, whereas LSTM, ATAE LSTM, and CABASC, all deep learning methods, reached the lowest. Most methods achieved good results on the Restaurant and MAMS datasets and demonstrated improved performance on the Laptop dataset. More than 64% (11 of 17) of the models achieved an accuracy of greater than (65%) on the Restaurant and MAMS datasets. Significantly, two deep learning-based (MGAN and LCF_BERT) and one GCN-based (SenticGCN_BERT) method obtained the best results with values greater than 70%, which were interpreted as good performance. The highest accuracy of 81.05% was achieved on the Restaurant dataset using SenticGCN_BERT and LCF_BERT, whereas the lowest accuracy was 52.34% on the Laptop dataset using LSTM. In addition, on both the Restaurant and MAMS datasets, more than 88% of the models (15 of 17) achieved an F_1 score greater than 60%; however, only more than 17% of this set (3 of 17) demonstrated F_1 scores greater than 70%. SenticGCN_BERT obtained the highest F_1 scores of 80.27% and 81.03% on the Restaurant and MAMS datasets, respectively, whereas LSTM demonstrated the lowest F_1 score (60.25%) on the Restaurant dataset. For the Laptop dataset, 17 out of 17 methods had an accuracy greater than 50%, and 17 out of 17 had an F_1 score greater than 50%. In addition, it is worth noting that there is only a slight difference between the accuracy value and the F_1 score value for balanced datasets. All the methods obtained different ranks based on their accuracy and F_1 scores. For example, in the experiment with unbalanced datasets, CABASC ranked first in terms of accuracy and F_1 score; in the experiment with balanced datasets, it ranked seventh. Thus, we can conclude that CABASC demonstrated the best accuracy and F_1 score when using unbalanced datasets.

Tables 2 and 3 show that most models performed slightly better when the training set was unbalanced on the Restaurant and MAMS datasets. However, most of the models achieved better performance when the training set was balanced on the Laptop dataset. In general, we can say that the ALSA models have similar accuracy and F_1 scores with balanced data, but these values differ with unbalanced data. The experimental comparison of the accuracy and F_1 score of the deep-learning- and GCN-based methods on balanced and unbalanced datasets separately was used to test the null hypothesis "Avoiding balanced data will not influence the F_1 score and accuracy of deep-learning- and GCN-based ALSA methods. "The test provided an accuracy value of 1.54%, 14.79%, and 15.27% and an F_1 score value of 0.28%, 0.8%, and 16.57% on the MAMS, Laptop, and Restaurant datasets, respectively. This means that the null hypothesis is rejected because the differences in the F_1 score and accuracy between the ALSA methods on the balanced and unbalanced datasets are significant; these differences are also found to be statistically significant. Therefore, we conclude that balanced data slightly affect the accuracy and F_1 score of the deep-learning- and GCN-based ALSA methods.

6   Conclusion

This study ascertained whether deep learning and GCN models can be used effectively for ALSA. In addition, it compared the performance of deep-learning- and GCN-based ALSA methods for an imbalanced and a balanced dataset. For this purpose, we designed experiments to compare 17 supervised deep-learning- and GCN-based ALSA methods. To implement this study, we first constructed balanced variants of the Laptop, Restaurant, and MAMS datasets based on the original unbalanced datasets. We then experimented with the 17 prepared methods on six unbalanced and balanced datasets. Finally, we evaluated, discussed, and concluded our findings by answering the initially posed research questions. The conclusions are summarized as follows: (i) the effectiveness of ALSA methods can be improved by deep learning and GCN techniques, especially GCNs over BERT representations; and (ii) the balanced data slightly affect the accuracy and F_1 score of the deep-learning- and GCN-based ALSA methods.

Acknowledgment. This research was supported by the 2020 Yeungnam University Research Grant. This work was supported by the National Research Foundation of Korea (NRF) grant funded by the Korea government (MSIT) (No. NRF-2023R1A2C1008134).

References

1. Alantari, H.J., Currim, I.S., Deng, Y., Singh, S.: An empirical comparison of machine learning methods for text-based sentiment analysis of online consumer reviews. Int. J. Res. Mark. **39**(1), 1–19 (2022)
2. Devlin, J., Chang, M.W., Lee, K., Toutanova, K.: BERT: pre-training of deep bidirectional transformers for language understanding. arXiv preprint: arXiv:1810.04805 (2018)
3. Do, H.H., Prasad, P.W., Maag, A., Alsadoon, A.: Deep learning for aspect-based sentiment analysis: a comparative review. Expert Syst. Appl. **118**, 272–299 (2019)
4. Fan, F., Feng, Y., Zhao, D.: Multi-grained attention network for aspect-level sentiment classification. In: Proceedings of the 2018 Conference on Empirical Methods in Natural Language Processing, pp. 3433–3442 (2018)
5. Huang, B., Ou, Y., Carley, K.M.: Aspect level sentiment classification with attention-over-attention neural networks. In: Thomson, R., Dancy, C., Hyder, A., Bisgin, H. (eds.) SBP-BRiMS 2018. LNCS, vol. 10899, pp. 197–206. Springer, Cham (2018). https://doi.org/10.1007/978-3-319-93372-6_22
6. Jiang, Q., Chen, L., Xu, R., Ao, X., Yang, M.: A challenge dataset and effective models for aspect-based sentiment analysis. In: Proceedings of the 2019 Conference on Empirical Methods in Natural Language Processing and the 9th International Joint Conference on Natural Language Processing (EMNLP-IJCNLP), pp. 6280–6285 (2019)
7. Kingma, D.P., Ba, J.: Adam: a method for stochastic optimization. arXiv preprint: arXiv:1412.6980 (2014)
8. Kipf, T.N., Welling, M.: Semi-supervised classification with graph convolutional networks. arXiv preprint: arXiv:1609.02907 (2016)

9. Liang, B., Su, H., Gui, L., Cambria, E., Xu, R.: Aspect-based sentiment analysis via affective knowledge enhanced graph convolutional networks. Knowl.-Based Syst. **235**, 107643 (2022)

10. Liu, Q., Zhang, H., Zeng, Y., Huang, Z., Wu, Z.: Content attention model for aspect based sentiment analysis. In: Proceedings of the 2018 World Wide Web Conference, pp. 1023–1032 (2018)

11. Ma, D., Li, S., Zhang, X., Wang, H.: Interactive attention networks for aspect-level sentiment classification. arXiv preprint: arXiv:1709.00893 (2017)

12. Pedregosa, F., et al.: Scikitlearn: machine learning in Python (2011). Accessed 28 Mar 2022

13. Pennington, J., Socher, R., Manning, C.D.: Glove: global vectors for word representation. In: Proceedings of the 2014 Conference on Empirical Methods in Natural Language Processing (EMNLP), pp. 1532–1543 (2014)

14. Phan, H.T., Nguyen, N.T., Hwang, D.: Convolutional attention neural network over graph structures for improving the performance of aspect-level sentiment analysis. Inf. Sci. **589**, 416–439 (2022)

15. Phan, H.T., Nguyen, N.T., Tran, V.C., Hwang, D.: An approach for a decision-making support system based on measuring the user satisfaction level on twitter. Inf. Sci. **561**, 243–273 (2021)

16. Pontiki, M., Galanis, D., Pavlopoulos, J., Papageorgiou, H., Androutsopoulos, I., Manandhar, S.: Semeval-2014 task 4: Aspect based sentiment analysis. In: Proceedings of the 8th International Workshop on Semantic Evaluation (SemEval 2014), pp. 27–35. Association for Computational Linguistics, Dublin (2014). https://doi. org/10.3115/v1/S14-2004, https://aclanthology.org/S14-2004

17. Qaiser, S., Yusoff, N., Remli, M., Adli, H.K.: A comparison of machine learning techniques for sentiment analysis. Turk. J. Comput. Math. Educ. (2021)

18. Schouten, K., Frasincar, F.: Survey on aspect-level sentiment analysis. IEEE Trans. Knowl. Data Eng. **28**(3), 813–830 (2015)

19. Tang, D., Qin, B., Feng, X., Liu, T.: Effective LSTMS for target-dependent sentiment classification. arXiv preprint: arXiv:1512.01100 (2015)

20. Trisna, K.W., Jie, H.J.: Deep learning approach for aspect-based sentiment classification: a comparative review. Appl. Artif. Intell. **36**(1), 2014186 (2022)

21. Wang, J., Xu, B., Zu, Y.: Deep learning for aspect-based sentiment analysis. In: 2021 International Conference on Machine Learning and Intelligent Systems Engineering (MLISE), pp. 267–271. IEEE (2021)

22. Wang, Y., Huang, M., Zhu, X., Zhao, L.: Attention-based LSTM for aspect-level sentiment classification. In: Proceedings of the 2016 Conference on Empirical Methods in Natural Language Processing, pp. 606–615 (2016)

23. Zeng, B., Yang, H., Xu, R., Zhou, W., Han, X.: LCF: a local context focus mechanism for aspect-based sentiment classification. Appl. Sci. **9**(16), 3389 (2019)

24. Zhang, C., Li, Q., Song, D.: Aspect-based sentiment classification with aspect-specific graph convolutional networks. arXiv preprint: arXiv:1909.03477 (2019)

25. Zhao, P., Hou, L., Wu, O.: Modeling sentiment dependencies with graph convolutional networks for aspect-level sentiment classification. Knowl.-Based Syst. **193**, 105443 (2020)

Applying the Malmquist Productivity Index Model to Assess the Performance of the Coffee Enterprise in DakLak Vietnam

Phan Van-Thanh[1,2]([✉]), Zbigniew Malara[3], Duong Thi Ai Nhi[4], and Nguyen Ngoc-Thang[4]([✉])

[1] Quang Binh University, Dong Hoi, Vietnam
thanhkem2710@gmail.com, pvthanh.tg@vku.udn.vn
[2] The University of Danang, Vietnam-Korea University of Information and Communication Technology, Danang, Vietnam
[3] Wroclaw University of Science and Technology, Wrocław, Poland
zbigniew.malara@pwr.edu.pl
[4] Tay Nguyen University, Buon Ma Thuot, Vietnam
{dtanhi,nnthang}@ttn.edu.vn

Abstract. To assess the performance of the coffee industry in DakLak, this study focused on three aspects in terms of efficiency change, productivity growth, and technological change based on the concept of the DEA - Malmquist model. A total of seven companies are chosen with realistic data collected from their company's website over the period 2020–2022. The empirical results showed that there are 4 companies with an efficiency change greater than 1 and 3 companies with an efficiency change of less than 1. However, the decomposition of total factor productivity change (increasing by 6.12%) showed that technical change has decreased by 4.83% while technical efficiency increased by 14.59% during the period, arising mainly from technical efficiency. This study provides empirical research on the efficiency of the Vietnamese coffee industry to get a brief view of how efficient the coffee industry is, and what needs to change to improve the performance of this sector. Furthermore, this paper is very valuable for both academic study and real value.

Keywords: Coffee industry · Data envelopment analysis (DEA) · Malmquist index · DakLak province · Vietnam

1 Introduction

Vietnam is the world's second-largest coffee producer, behind Brazil, according to the United States Department of Agriculture, with a total production of about 1.78 million tons in 2022. (USDA). The Vietnamese economy depends heavily on the coffee industry, with the export of coffee making up a sizeable share of the nation's agricultural exports (Nguyen et al. 2019). Robusta coffee is the most popular kind in Vietnam, which supports the livelihoods of over 2.6 million households and contributes to over 3% of the nation's GDP (World Bank 2019).

© The Author(s), under exclusive license to Springer Nature Singapore Pte Ltd. 2023
N. T. Nguyen et al. (Eds.): ACIIDS 2023, LNAI 13996, pp. 145–156, 2023.
https://doi.org/10.1007/978-981-99-5837-5_13

In 2021, Vietnam's coffee production increased again, as the country's weather conditions have been favorable. According to the USDA, Vietnam's coffee production in 2021 reached 1.74 million tons (USDA 2021). The Vietnamese government has also introduced policies to encourage the adoption of sustainable coffee farming practices, which are expected to increase yields and improve the quality of coffee produced in the country (Vietnam Investment Review 2020). In 2022, with continued investment in the sector, favorable weather conditions, and the adoption of sustainable farming practices, Vietnam's coffee production remained strong. According to USDA, Vietnam's coffee production in 2022 reached nearly 1.9 million tons and it is forecast to continue its upward trend in the coming years.

The Covid-19 outbreak has disrupted the coffee supply chain, decreased consumer demand for coffee in several regions, and impacted coffee producers' earnings (Duc et al. 2021). Despite these difficulties, Vietnamese coffee businesses have been taking steps to increase the effectiveness of their manufacturing methods. Using contemporary technologies, such as precision agriculture and innovative irrigation systems, has been one of the important methods to boost agricultural yields and decrease water usage (Do et al. 2021). Vietnamese coffee businesses have been making investments in the growth of specialty coffee, which has greater value and better market prospects. Specialty coffee manufacturing demands more care and attention during cultivation and processing, leading to greater quality and premium prices (Pham et al. 2021).

Vietnamese coffee businesses have also been concentrating on sustainable practices, such as the use of organic farming techniques, supporting biodiversity, and minimizing waste in the production process, in order to increase the efficiency and sustainability of coffee production (Nguyen et al. 2021).

In order to improve the efficiency in the coffee sector, we need to have a comprehensive investigation and assessment to find out the problems in operation. That reason why of this study is to investigate the changes in efficiency productivity of 07 companies over three years (2020–2022) by the Malmquist index and its catch-up, frontier-shift components in order to see how efficient the coffee industry is, and how it change during the above period. Through this study, we not only reflect the actual situation of Vietnames coffee industry, but also propose a new procedure for helping the top manager to evaluate the productivity, efficiency change of coffee industry based on DEA- Malmquist model.

2 Related Works

Malmquist Productivity Index (MPI) is a very useful benchmarking tool for calculating the productivity change of a Decision Making Units (DMU) to examine the relative company progress among competitors. In the past two decades, more and more publications have adopted DEA and/or MPI as evaluation techniques to assess company efficiency.

Berg et al. (1992) presented the first application of the Malmquist index to measure productivity growth in the Norwegian banking system in the pre and post-deregulation era (1980–1989), using the value-added approach. Their analysis showed that the productivity exhibited a lackluster performance in the pre-deregulation era. However, the productivity increased remarkably in the post-deregulation era suggesting that deregulation led to more competitive environment, especially for larger banks. Likewise, using a generalized Malmquist productivity index, Grifell-Tatjé and Lovell (1997) analyzed the sources of productivity change in Spanish banking over the period 1986–1993 and found that the commercial banks had a lower rate of productivity growth compared to saving banks, but a higher rate of potential productivity growth.

Raphael (2013) investigated the nature of efficiency and productivity change by using the Malmquist Productivity Index (MPI). He measures the productivity change of Tanzanian commercial banks for seven years. His results showed that most commercial banks recorded an improvement in efficiency change by 67%, a technical change improvement by 83%, pure technical change improvement by 67 and scale efficiency change by 50%. Generally, the mean efficiency change of Large Domestic Banks (LDB) is higher compared to the rest of the groups; hence manages to push the frontier of possibility outwards concerning other groups, followed by small banks with a mean efficiency change of 10.3% while the Large Foreign Banks (LFB) recorded an efficiency change of 1.8%, similarly the mean total factor of productivity of small banks were higher compared with the rest of the groups, by recording productivity improvement of 57.9% exceeding LDB and LFB with 51.4% and 54%, respectively. Generally, both groups of commercial banks experienced technological progress. However, the efficiency gains during the period of the study were due to improvements in technical efficiency rather than scale efficiency.

Recently, Thayaparan and Pratheepan (2014) applied data envelopment analysis (DEA) total factor productivity and its components to measure in terms of efficiency change, technical efficiency change, pure efficiency change and scale change. They found that all six banks operate on average at 87.2% of overall efficiency and it reveals the less performance of the banks. This less performance was achieved due to the less progress in technical change than efficiency change and the finding highlights that technical change has been the main constraint to achieving a high level of total factor productivity of commercial banks in Sri Lanka. Among the private banks, Seylan Bank has the highest efficiency of 1.033 than the other banks and among the state banks, Peoples' Bank have the values of 0.773 than Bank of Ceylon. The overall results concluded that comparatively selected private banks are more efficient than state banks in the study period in Sri Lanka.

3 Methodology

3.1 Malmquist Index Introduction

The Malmquist index evaluates the productivity change of a DMU between two time periods. It is defined as the product of "Catch up" and "Frontier shift" terms. The catch up (or recovery) term relates to the degree that a DMU attains for improving its efficiency, while the frontier-shift (or innovation) term reflects the change in the efficient frontiers surrounding the DMU between the two time periods. It deal with a set of n DMUs (x_j, y_j) $(j = 1, \ldots .n)$ each having m inputs denoted by a vector $x_j \in R^m$ and q outputs denoted by a vector $y_j \in R^q$ over the periods 1 and 2. We assume $x_j > 0$ and $y_j > 0 (\forall j)$. The notations $(x_0, y_0)^1$ and $(x_0, y_0)^2$ are employed for designating DMU_0 $(0 = 1, \ldots, n)$ in the periods 1 and 2 respectively.

The production possibility set $(X, Y)^t$ ($t = 1$ and 2) spanned by $\{(x_j, y_j)^t\}$ ($j = 1,$ ..., n) is defined by:

$$(X, Y)^t = \{(x, y) | x \geq \sum_{j=1}^{n} \lambda_j x_j^t, 0 \leq y \leq \sum_{j=1}^{n} \lambda_j y_j^t, L \leq e\lambda \leq U, \lambda \geq 0\}$$

where e is the row vector with all elements being equal to one and $\lambda \in R^n$ is the intensity vector, and L and U are the lower and upper bounds of the sum. $(L, U) = (0, \infty), (1, 1), (1, \infty)$ and $(0, 1)$ correspond to the CCR (Charnes et al. 1978), BCC (Banker et al. 1984), IRS (Increasing Returns to Scale) and DRS (Decreasing Returns to Scale) models respectively. The production possibility set $(X, Y)^t$ is characterized by frontiers that are composed of $(x, y) \in (X, Y)^t$ such that it is not possible toimprove any element of the input x or any element of the output y without worsening some other input or output. See Cooper et al. (2007) for further discussion on this and related subjects. We call this frontier set the frontier technology at the period t. In the Malmquist index analysis, the efficiencies of DMUs $(x_0, y_0)^1$ and $(x_0, y_0)^2$ are evaluated by the frontier technologies 1 and 2 in several ways.

3.1.1 Catch-Up Effect

The catch up effect is measured by the following formula.

$$\text{Catch up} = \frac{\text{efficiency of} (x_0, y_0)^2 \text{ with respect to the period 2 frontier}}{\text{efficiency of} (x_0, y_0)^1 \text{ with respect to the period 1 frontier}} \quad (1)$$

We evaluate each element (efficiency) of the above formula by the non-parametric DEA models as described later. A simple example in the case of a single input and output technology is illustrated in Fig. 1.

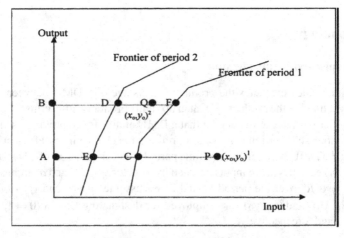

Fig. 1. Catch-up (Source: [4])

The catch-up effect (in input-orientation) can be computed as:

$$\textbf{Catch up} = \frac{\frac{BD}{BQ}}{\frac{AC}{AP}} \tag{2}$$

(Catch up) > 1 indicates progress in relative efficiency from period 1 to 2, while (Catch up = 1) and (Catch up) < 1 indicate respectively no change and regress in efficiency.

3.1.2 Frontier Shift Effect

In addition to the catch-up term, we must take account of the frontier-shift effect in order to evaluate totally the efficiency change of the DMU, since the catch-up is determined by the efficiencies as measured by the distances from the respective frontiers. In the Figure 1 case, The reference point C of (x_0^1, y_0^1) moved to E on the frontier of period 2. Thus, the frontier-shift effect at (x_0^1, y_0^1) is evaluated by

$$\varphi_1 = \frac{AC}{AE} \tag{3}$$

This is equivalent to

$$\varphi_1 = \frac{\frac{AC}{AP}}{\frac{AE}{AP}} = \frac{efficiency\ of\ (x_0,y_0)^1 with\ respect\ to\ the\ period\ 1\ frontier}{efficiency\ of (x_0,y_0)^1 with\ respect\ to\ the\ period\ 2\ frontier} \tag{4}$$

The numerator of the Eq. (4) right is already obtained in (1). The denominator is measured as the distance from the period 2 production possibility set to (x_0^1, y_0^1). Likewise, the frontier-shift effect at (x_0^2, y_0^2) is expressed by

$$\varphi_1 = \frac{\frac{BF}{BQ}}{\frac{BD}{BQ}} = \frac{efficiency\ of\ (x_0,y_0)^2\ with\ respect\ to\ the\ period\ 1\ frontier}{efficiency\ of (x_0,y_0)^2\ with\ respect\ to\ the\ period\ 2\ frontier} \tag{5}$$

We can evaluate the numerator of the above by means of the DEA models. Using $\varphi 1$ and $\varphi 2$, we define "Frontier shift" effect by their geometric mean as:

$$\textbf{Frontier shift} = \varphi = \sqrt{\varphi_1 \varphi_2} \tag{6}$$

(Frontier shift) > 1 indicates progress in the frontier technology around the DMU_0 from period 1 to 2, while (Frontier shift) = 1 and (Frontier shift) < 1 indicate respectively the status quo and regress in the frontier technology.

3.1.3 Malmquist Index

The Malmquist index (MI) is computed as the product of (Catch up) and (Frontier shift), i.e.,

$$\textbf{Malmquist index} = (\text{Catch up}) \times (\text{Frontier shift}). \tag{7}$$

It is an index representing Total Factor Productivity (TFP) of the DMU, in that it reflects progress or regress in efficiency of the DMU along with progress or regress of the frontier technology.

We now employ the following notation for the efficiency score of DMU $(x_0, y_0)^{t_1}$ measured by the frontier technology t_2.

$$\delta^{t_2}((x_0, y_0)^{t_1}) \ (t_1 = 1, 2 \text{ and } t_2 = 1, 2). \tag{8}$$

Using this notation, the catch up effect (C) in (F_1) can be expressed as

$$C = \frac{\delta^2((x_0, y_0)^2)}{\delta^1((x_0, y_0)^1)} \tag{9}$$

The frontier-shift effect is described as

$$F = \left[\frac{\delta^1((x_0, y_0)^1)}{\delta^2((x_0, y_0)^1)} \times \frac{\delta^1((x_0, y_0)^2)}{\delta^2((x_0, y_0)^2)} \right]^{\frac{1}{2}} \tag{10}$$

As the product of C and F, we obtain the following formula for the computation of MI.

$$MI = \left[\frac{\delta^1((x_0, y_0)^2)}{\delta^1((x_0, y_0)^1)} \times \frac{\delta^2((x_0, y_0)^2)}{\delta^2((x_0, y_0)^1)} \right]^{\frac{1}{2}} \tag{11}$$

This last expression gives an another interpretation of MI, i.e., the geometric means of the two efficiency ratios: the one being the efficiency change measured by the period 1 technology and the other the efficiency change measured by the period 2 technology.

As can be seen from these formulas, the MI consists of four terms:
$\delta^1((x_0, y_0)^1)$, $\delta^2((x_0, y_0)^2)$, $\delta^1((x_0, y_0)^2)$ and $\delta^2((x_0, y_0)^1)$. The first two are related with the measurements *within* the same time period, while the last two are for *intertemporal* comparison.

MI > 1 indicates progress in the total factor productivity of the DMU_0 from period 1 to 2, while MI = 1 and MI < 1 indicate respectively the status quo and decay in the total factor productivity.

3.2 Research Procedure

The working procedure in this study is primarily collecting the information of proceeding coffee industry data and also collecting all related documents as this study draft action plan as referenced. Then after confirming the subject and proceeding industrial analysis, the procedure of this study is shown in Figure as below (Fig. 2):

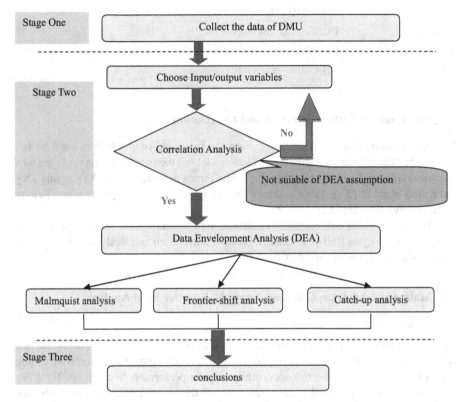

Fig. 2. Flow chart of coffee enterprise performance evaluation (Source: Authors' own study)

4 Empirical Results

4.1 Collect the Data

After survey and collected all enterprises related to business coffee in ĐakLak province, a total of 13 companies are operating in the market. However, some companies have not published financial statements or just established only one or two years does not meet the requirements of this study. So only 07 companies with full data are selected to be our DMUs to assess the performance. To convenient for presentation in the manuscript, the name of all companies is coded in Table 1.

Table 1. Code of companies (Source: Code by authors after the entities surveyed)

DMU_s	Name of company
DMU_1	Buon Ho Coffee Company
DMU_2	Cu Pul Coffee two member limited liability company
DMU_3	Thang Loi Coffee Joint Stock Company
DMU_4	Robanme Coffee limited company
DMU_5	Duc Lap Coffee company
DMU_6	Dak Man VietNam limited
DMU_7	721 Coffee one member company

4.2 Input/output Factors Selection and Correlation Analysis

The DEA requires that the relationship between input and output factors must be iso-tonicity. Hence, after using Pearson correlation analysis the outputs in this study are total revenue $Y_{(1)}$ and the profit after tax (Y_2). The inputs are fixed asset (X_1), equity (X_2) and a staff number (X_3). Table 2 shows the results of Pearson correlation coefficients obtained from the year 2020 to 2022.

The result in Table 2 shows that the correlation between the input and output variables satisfies isotonicity in Bao et al. (2010), positive correlation and well complies with the prerequisite condition of the DEA model.

4.3 Malmquist Analysis, Catch up Analysis, Frontier Shift Analysis

4.3.1 Malmquist Productivity Index

Malmquist productivity indexes provide us with the opportunity of comparing the productivity change within the coffee industry as well as to compare the productivity change within groups, hence giving the opportunity of poor performers to catch up. Total factor productivity as the word implies refers to all factors pertaining to the production of companies, more specifically the change in total factor productivity entails changes in efficiency and changes in technology. When interpreting the Malmquist total factor productivity we consider all of its components greater than one indicates improvement or progression. On the other hand, the value less than one refers to the deterioration of regression. The values equal to one referred to as no improvement has been observed (Table 3).

Table 4 reveals the terms of the Malmquist productivity index during the periods of 2020–2022 as follows. Four companies have obtained a score more than one are DMU_2, DMU_3, DMU_5 and DMU_7. DMU_3 is the top performer in the system with highest efficiency of 1.46511 indicating that it has increased at 46.5% followed by DMU_5 and DMU_7 with 30.8% and 23.5%, respectively. DMU_2 standing on fourth with MPI equal 1.177 equivalents increased at 17.7%. Other companies have less than one indicated that they have negative growth in total factor productivity. More specifically, we saw that DMU_3 have the top performer in seven companies. However, in the year of 2021–2022,

Table 2. Pearson correlation coefficients inputs and outputs in 2020–2022 (Source: Authors' own analysis).

2020	X_1	X_2	X_3	Y_1	Y_2
X_1	1	0.935105	0.916639	0.685013	0.419510
X_2	0.935105	1	0.838620	0.695738	0.326586
X_3	0.916639	0.838620	1	0.548458	0.673741
Y_1	0.685013	0.695738	0.548458	1	0.126076
Y_2	0.419510	0.3265864	0.673741	0.126076	1
2021	X_1	X_2	X_3	Y_1	Y_2
X_1	1	0.842251	0.985304	0.621757	0.682861
X_2	0.842251	1	0.822172	0.798965	0.541510
X_3	0.985304	0.822172	1	0.581854	0.726212
Y_1	0.621757	0.798965	0.581854	1	0.137436
Y_2	0.682861	0.541510	0.726212	0.137436	1
2022	X_1	X_2	X_3	Y_1	Y_2
X_1	1	0.719567	0.958436	0.556471	0.676066
X_2	0.719567	1	0.811176	0.812817	0.433101
X_3	0.958436	0.811176	1	0.591018	0.740682
Y_1	0.556471	0.812817	0.591018	1	0.290429
Y_2	0.676066	0.433101	0.740682	0.290429	1

Table 3. Malmquist index summary of annual means (Source: Authors' own analysis).

Malmquist	2020 => 2021	2021 => 2022	Average
DMU_1	0.3806681	0.896442	0.6385551
DMU_2	1.3536717	1.0014631	1.1775674
DMU_3	2.4245598	0.5056602	1.46511
DMU_4	0.6322184	0.5964072	0.6143128
DMU_5	1.2887277	1.3272821	1.3080049
DMU_6	0.9605355	1.0200749	0.9903052
DMU_7	1.1478469	1.3215787	1.2347128
Average	1.1697469	0.9527012	1.061224
Max	2.4245598	1.3272821	1.46511
Min	0.3806681	0.5056602	0.6143128
SD	0.6555369	0.3197016	0.3294103

the performance of these companies dramatically decreased with the MPI index equal 0.505. While two companies (DMU_5 and DMU_7) increasing. In addition, the average Malmquist productivity index of seven companies show that it was increasing in during the period 2020–2022 by 6.12% due to the technological change than efficiency change.

4.3.2 Components of the Malmquist Productivity Index: (1) Catch up

"Catch up" effect is so-called "Technical Efficiency Change". The change values for technical efficiency are bigger than 1, which indicates that the DMU improves technical efficiency. Table 4 shows the average technical efficiency change from 2020 to 2022.

Table 4. Average Technical efficiency change of companies from 2020 to 2022 (Source: Authors' own analysis).

Catch-up	2020 => 2021	2021 => 2022	Average
DMU_1	0.521796067	1.160592058	0.8411941
DMU_2	1.013078434	1.000190352	1.0066344
DMU_3	2.419865947	0.560012755	1.4899394
DMU_4	0.468949027	0.724409445	0.5966792
DMU_5	1.37549544	1.210629132	1.2930623
DMU_6	1	1	1
DMU_7	1.947407874	1.640268979	1.7938384
Average	1.249513256	1.042300389	1.1459068
Max	2.419865947	1.640268979	1.7938384
Min	0.468949027	0.560012755	0.5966792
SD	0.722226297	0.350502618	0.4072294

Table 4 shows that the technical efficiency change of DMU_4 increased in the period time 2020 to 2022 but the change in values for technical efficiency of this company is less than 1, While DMU_3 decreased in the year 2021–2022 falling at 0.560. The highest growth achievers were DMU_7, DMU_5, and DMU_1 with double-digit growth. There are two companies (DMU_2 and DMU_6) have a technical efficiency equal to 1. Overall, the mean growth rate of the industry is around 14.59% for the year 2020–2022.

4.3.3 Components of the Malmquist Productivity Index: (2) Frontier Shift

Technical change or the so-called "innovation" or "frontier shift" effect is the second component of the Malmquist productivity change index. This component captures the effect of the shift in the frontier of the productivity change for an exposition of the effect of technical change on productivity change using production functions. The average technique change from 2020 to 2022 was shown as follows:

Table 5 above is showing the average technical change for a haft of companies over the three years to be less than one except DMU_2 which has obtained a score of 1.16873

Table 5. Average Technique change of commercial bank from 2020 to 2022 (Source: Authors' own analysis).

Frontier	2020 => 2021	2021 => 2022	Average
DMU_1	0.7295343	0.7724006	0.7509674
DMU_2	1.3361963	1.0012725	1.1687344
DMU_3	1.0019397	0.9029441	0.9524419
DMU_4	1.3481601	0.8233013	1.0857307
DMU_5	0.9369189	1.0963573	1.0166381
DMU_6	0.9605355	1.0200749	0.9903052
DMU_7	0.589423	0.8057085	0.6975657
Average	0.9861011	0.917437	0.9517691
Max	1.3481601	1.0963573	1.1687344
Min	0.589423	0.7724006	0.6975657
SD	0.2828889	0.1239394	0.1711012

indicating that it has increased at 16.8%. So, that the reason why the average annual technological progress declined 4.83%.

5 Conclusions

The coffee industry is one of the industries that play an important role in the development of the country. Therefore, in order to improve the performance of business enterprises in this field. we need to have a comprehensive investigation and assessment to find out the problems in operation. In this study, the total factor productivity change, technical change and technical efficiency are identified with total factor productivity change (increasing 6.12%) showed that technical change has decreased by 4.83% while technical efficiency increased by 14.59%. Furthermore, the result also find out the main impact to MPI comes from technical efficiency. Through this study, we not only reflect on the actual situation of the Vietnamese coffee industry but also propose a new procedure for helping the top manager to evaluate the performance of the coffee industry. And this also helps top managers, policy makers to bring more technological advancement and enhancement to increase the performance of the Vietnamese coffee industry, innovative and adequate methods of responding to the picture resulting from the assessment and conducive to increasing the potential of the entities examined. In future research, more companies and more variables in this industry can be assessed with the intention of confirming the obtained results of the study and the usefulness of this approach.

References

Banker, R.D., Charnes, A., Cooper, W.W.: Some models for estimating technical and scale inefficiencies in data envelopment analysis. Manag. Sci. **30**(9), 1078–1092 (1984)

Berg, S.A., Førsund, F.R., Jansen, E.S.: Malmquist indices of productivity growth during the deregulation of Norwegian banking, 1980–89. Scand. J. Econ., S211–S228 (1992)

Charnes, A., Cooper, W.W., Rhodes, E.: Measuring the efficiency of decision making units. Eur. J. Oper. Res. **2**(6), 429–444 (1978)

Cooper, W.W., Seiford, L.M., Tone, K.: Data Envelopment Analysis: A Comprehensive Text with Models, Applications, References and DEA-Solver Software, 2nd edn., p. 490. Springer, New York (2007). ISBN 387452818

Duc, N.T., Nguyen, L.V., Do, M.H. et al.: The impact of COVID-19 pandemic on coffee exportation and vietnamese farmers' livelihood. Agriculture **11**, 79 (2021)

Do, N.T.T., Nguyen, H.N., Do, T.T. et al.: Farmers' willingness to adopt smart farming technologies in Vietnam. Agronomy **11**, 670 (2021)

Grifell-Tatjé, E., Lovell, C.K.: The sources of productivity change in Spanish banking. Eur. J. Oper. Res. **98**(2), 364–380 (1997)

Nguyen, T.T., Tran, D.V., Vuong, T.T., et al.: The economic efficiency of coffee production in the central highlands, Vietnam. J. Asian Agric. Dev. **16**(1), 35–53 (2019)

Nguyen, T.T., Nguyen, N.T., Vuong, T.T., et al.: The influence of climate change on the sustainable development of the coffee industry in Vietnam. Environ. Sci. Pollut. Res. **28**, 49564–49577 (2021)

Pham, T.T., Nguyen, T.T.T., Nguyen, T.H.T., et al.: The sustainability of the coffee value chain in Vietnam: an empirical analysis. J. Clean. Prod. **279**, 123733 (2021)

Raphael, G.: X-efficiency in Tanzanian commercial banks: an empirical investigation. Res. J. Finance Account. **4**(3), 12–22 (2013)

Thayaparan, A., Pratheepan, T.: Evaluating total factor productivity growth of commercial banks in Sri Lanka: an application of Malmquist index. J. Manag. Res. **6**(3), 58–68 (2014)

World Bank. Vietnam Country Overview (2019). https://www.worldbank.org/en/country/vietnam/overview

A Cluster-Constrained Graph Convolutional Network for Protein-Protein Association Networks

Nguyen Bao Phuoc, Duong Thuy Trang, and Phan Duy Hung[✉]

FPT University, Hanoi, Vietnam
{phuocnbhe153036,trangdthe150573}@fpt.edu.vn, hungpd2@fe.edu.vn

Abstract. Cluster-GCN is one of the effective methods for studying the scalability of Graph Neural Networks. The idea of this approach is to use METIS community detection algorithm to split the graph into several sub-graphs, or communities that are small enough to fit into a common Graphics Processing Unit. However, METIS algorithm still has some limitations. Therefore, this research aims to improve the performance of cluster-GCN by changing the community detection algorithm, specifically by using Leiden algorithm as an alternative. Originally, Leiden algorithm is claimed to be powerful in identifying communities in networks. Nevertheless, the common feature of community detection algorithms makes nodes in the same community tend to be similar. For that reason, this research also proposes to add constraints such as minimum/maximum community size and overlapping communities to increase community diversity. By combining the above approaches with a careful analysis of protein data, our experiments on a single 8 GB GPU show that the performance of Cluster-GCN on the ogbn-proteins dataset could be improved by a 0.98% ROC-AUC score.

Keywords: Graph Convolution Network · graph clustering · Leiden algorithm · graph mining · Constraint clustering

1 Introduction

In recent years, Graph Convolution Networks (GCNs) have become a popular research topic because of its outstanding performance in various complex network learning tasks. Furthermore, the advancement in data mining technologies leads to exponential growth in the amount of data that needs to be processed. Since this is a primary concern for scientists, many strategies have been proposed to scale up the GCNs. One of the key goals is to make the models run faster with less memory consumption.

Inspired by the ever-growing demand for analytics and downstream application with available data, this work aims to improve the performance of GCNs for large datasets in science, particularly deep graph learning in protein functions annotation tasks - a biological application that is essential to contribute to biomedicine and pharmaceuticals. The protein-protein association networks are as follows: each node in the graph represents a specific protein and its edge signifies a type, or many types of relationship with other

© The Author(s), under exclusive license to Springer Nature Singapore Pte Ltd. 2023
N. T. Nguyen et al. (Eds.): ACIIDS 2023, LNAI 13996, pp. 157–169, 2023.
https://doi.org/10.1007/978-981-99-5837-5_14

proteins such as physical interactions, co-expression, or homology [1]. Previously, many adaptations of GCNs in this domain have demonstrated notable results [2]. Nevertheless, as the size of the graphs continues to increase, handling large graphs with advanced GCNs architecture is challenging since a new node embedding is interrelated to other nodes' information [3–6].

To reduce this burden, several GCNs architectures that utilize node-wise and layer-wise sampling have been proposed [4–6]. These methods, however, are still affected by the neighborhood expansion problem and high memory usage when neural networks go deeper. Notably, Chiang et al. propose Cluster-GCN [3] to tackle those issues with another approach: subgraph sampling. To be more specific, a clustering algorithm (i.e., METIS [7]) is used in the preprocessing phase to split a graph into several clusters, then, several clusters are chosen randomly to form a subgraph before fitting into the Graphics Processing Unit (GPU) for training. By using this strategy, the amount of information used to train in each iteration is much smaller than the entire graph, which not only leads to a significant reduction in runtime and memory but also achieves outstanding performance with several large datasets [4].

Originally, METIS algorithm [7] is used in the sampling phase of Cluster-GCN architecture. It is one of the most popular community detection algorithms, known for being efficient for large datasets. The authors also state that the advantage of the METIS over random partition is less information lost during training because it retains more edge. However, METIS algorithm still has some drawbacks. Firstly, as a cut-based k-way multilevel graph partition method, METIS needs a predefined number of clusters. Secondly, the clustering results are not guaranteed to be locally optimal [8]. To overcome the above weaknesses, modularity was suggested as a quality measure of graph clustering [9]. Especially, modularity-based community detection (e.g., Louvain [10], Leiden [11]) has been shown to overcome the limitations of cut-based algorithms and achieve better performance [8, 12, 13].

Cluster-GCN [4] architecture has laid the foundation for many advanced methods of Graph Learning [14–17]. This motivates us to deep dive into Cluster-GCN with the hypothesis that improve subgraph sampling can further boost the performance of the model. To verify the hypothesis, this research proposes to change METIS algorithm with Leiden algorithm. Moreover, community detection in general leads to nodes within communities having similar characteristics. When being applied to Cluster-GCN, it leads to skewed node distribution. Therefore, this research proposes to use additional constraints such as minimum/maximum community size and overlapping communities to balance these distributions and increase the performance of the model.

2 Data Preparation

2.1 Dataset Introduction

In this study, we evaluate our proposed improvement for the Cluster-GCN with ogbn-proteins dataset [19], which is an undirected, weighted, and typed (according to species) graph. Especially, nodes represent proteins, and edges indicate different types of biologically meaningful associations between proteins. All edges come with 8-dimensional features, where each dimension represents the approximate confidence of a single association type and takes on values between 0 and 1 (the closer the value to 1, the more confident about the association).

One of the interesting features that this dataset exhibits is the highly diverse graph statistics, as shown in Table 1 below:

Table 1. Statistics of currently-available ogbn-proteins dataset.

Dataset	#Graphs	#Node	#Edge	#Labels	#Features
ogbn-proteins	1	132534	39561252	112	8

The challenge is to predict the probability of each function for a node in a multi-label binary classification setup, where there are 112 labels to predict in total.

2.2 Pre-processing Data

Aggregate Edge Features to Node Features. As introduced in [19], all edges contain valuable information besides speciesID. Those features symbolize the various relationships between nodes; thus, they are important and should be considered during training to boost the prediction capacity of the model. However, Cluster-GCN [3] architecture does not have the function to handle edge features but node features so that before sampling takes place, instead of using the speciesID, the model aggregates edge features of each node's entire neighbors to construct a set of node features. This step can be formulated as follow:

$$x_i = aggr_{j \in \mathcal{N}(i)}(e_{ij}) \tag{1}$$

where x_i is the feature of node i, $aggr(.)$ is the differentiable and permutation invariant functions such as add, mean or max, $\mathcal{N}(i)$ is the set of i neighbors and e_{ij} is the edge feature between node i and node j. By doing this, each node feature can contain some information about the whole complete graph.

3 Methodology

To bring the most accurate comparison results, this work uses a baseline model, which is the network architecture of the Cluster-GCN [3]. Specifically, the original graph is divided into a sequence of communities using the community detection algorithm. Then, Stochastic Multiple Partition is used to merge several communities into a subgraph, which is then fitted into GPU for mini-batch training. The overall architecture is described in Fig. 1.

As mentioned in Sect. 1, Leiden algorithm [11] has addressed the limitations of cut-based community detection (e.g., METIS [7]). On the one hand, because this algorithm is agglomerative hierarchical clustering, it creates communities of various sizes (Fig. 2).

Consequently, there may be a community that is still too large to fit into GPU. On the other hand, the resolution limit of modularity [8] leads to very small communities or even singleton graphs, which contain diminutive information to be able to train. To tackle those problems, this work proposes maximum and minimum community size to balance the number of nodes between communities.

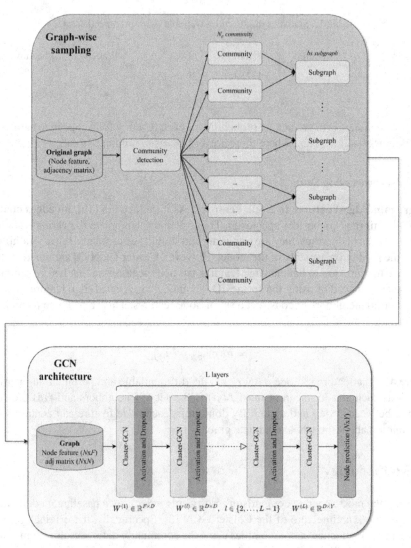

Fig. 1. Architecture illustration. The baseline model consists of two phases: Graph-wise sampling and mini-batch training.

3.1 Minimum and Maximum Community Size Constraint

Maximum Constraint. This work proposes a parameter named ct_{max} to control the maximum size of the communities extracted by the Leiden algorithm. Specifically, all communities whose sizes are larger than ct_{max} are applied Leiden algorithm recursively until all communities are smaller than ct_{max} in size.

Minimum Constraint. Similar to maximum constraint, this work proposes a parameter name ct_{min} to control the minimum size of the communities extracted by the Leiden algorithm. Intuitively, the minimum constraint can be implemented by merging communities whose sizes are smaller than ct_{min} using the maximum modularity gain like the local moving nodes in the Leiden algorithm. Nevertheless, the modularity optimization fails to identify communities smaller than a certain scale [8]. Therefore, instead of modularity, this work proposes to use Edge Ratio (ER):

$$ER\big(C_i, C_j\big) = \frac{E\big(C_i, C_j\big)}{\big|C_i + C_j\big|} + eps \tag{2}$$

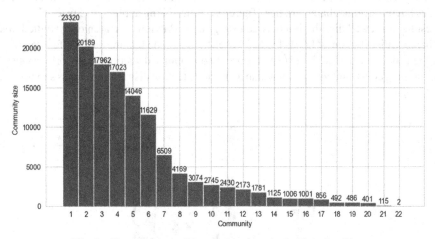

Fig. 2. Size of communities extracted by the Leiden algorithm

Where $ER\big(C_i, C_j\big)$ is the edge ratio between community C_i and community C_j, $E\big(C_i, C_j\big)$ is the number of edges between community C_i and community C_j, $\big|C_i + C_j\big|$ is the total number of edges of those two communities and eps is the small number used to prevent zero edge ratio. If only the number of edges is considered as a criterion to merge communities, small communities tend to merge with the largest communities due to the fact that the greater number of nodes, the more likely there will be links between them. Therefore, the denominator in Eq. (2) acts as the normalization term to get rid of the influence of the number of nodes on the merging process.

With edge ratio criterion defined in Eq. (2), all communities whose size is smaller than ct_{min} is considered to compute the edge ratio with all other communities. To increase generality for the merging process and exploit the diversity in the community of networks, instead of maximum edge ratio, community C_j is randomly merged with community C_i with probability:

$$Pr\left(C_i = C_j\right) \cong \frac{ER\left(C_i, C_j\right)}{\sum_{C_k \in (C - C_i)} ER(C_i, C_k)} \tag{3}$$

Combine ct_{max} and ct_{min}. Those two constraints can be used together. Hence, all communities are guaranteed to have size within $[ct_{min}, ct_{max}]$. In order to do that, Eq. (3) need to modified to avoid creating communities larger than ct_{max}:

$$Pr\left(C_i = C_j\right) \cong \begin{cases} \frac{ER(C_i, C_j)}{\sum_{C_k \in (C - C_i)} ER(C_i, C_k)} & if \left|C_i + C_j\right| \leq ct_{max} \\ 0 & otherwise \end{cases} \tag{4}$$

3.2 Overlapping Community Constraint

As mentioned in Sect. 1, community detection in general leads to communities where nodes having similar characteristics. Therefore, when being applied to Cluster-GCN [3], it results in skewed distribution between subgraphs. Previous research only studied this imbalance distribution of labels [3, 18], nevertheless, node features also suffer from this problem (Fig. 3). To tackle this limitation of community detection, this work proposes overlapping constraint: $ct_{overlap}$ to balance the node feature distribution across subgraphs.

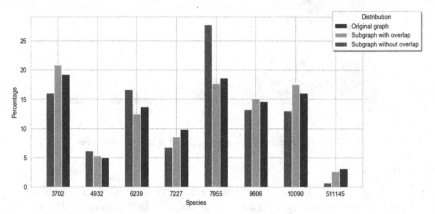

Fig. 3. Node species distribution of subgraph generated by Leiden algorithm (with and without $ct_{overlap}$) and the original graph in ogbn-proteins dataset. With $ct_{overlap}$, the species distribution is much more similar to the original graph

This work experiments on ogbn-proteins dataset [19] where each node has a speciesID to denote its origin, therefore, balancing node feature distribution becomes balancing the species across subgraphs. Assume f_i^C is the frequency of appearance of species i in the community C, this community is assigned to species i^*, which has the most appearance within the community:

$$S(C) = \{i^*\} = \left\{ \underset{i \in S(\mathbf{G})}{\operatorname{argmax}} f_i^C \right\} \tag{5}$$

where $S(C)$ is the set of species appear in community C and $S(\mathbf{G})$ is the set of all species in the ogbn-proteins dataset. Then, each community C is added to:

$$SP_i = \{C | i \in S(C)\} \tag{6}$$

where SP_i is the set of communities whose nodes mostly come from species i. From here, the Stochastic Multiple Partition from Cluster-GCN [3] is applied. However, the randomness is restricted within species. Specifically, assume the number of batches used for training mini-batch gradient descent is bs, a subgraph is generated by merging $\frac{|SP_i|}{bs}$ random communities of each species $i \in S(\mathbf{G})$.

Nevertheless, assigning a species for a community based on the species of mostly nodes within the community will ignore the case when there are two or more species whose frequency of appearance is approximately the same (Fig. 4). To solve the above problem, this work proposed the overlapping constraint: $0 \leq ct_{overlap} \leq 1$. Instead of assigning only one species for each community, Eq. (5) is modified so that community C can reach two or more species based on the frequency of appearance of species within community:

$$S(C) = \left\{ i | f_i^C \geq ct_{overlap} \times (f_{i^*}^C) \right\} \tag{7}$$

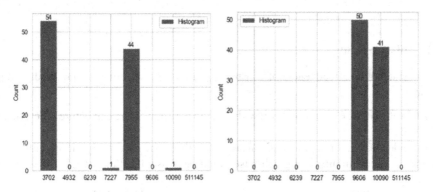

Fig. 4. Communities with multiple species.

if $ct_{overlap}$ close to 1, each community tends to belong to only one species. In contrast, if $ct_{overlap}$ close to 0, each community tends to belong to all species, which reduces the effectiveness of dividing communities based on species.

4 Experiment

4.1 Evaluation Metrics and Hyperparameter Setting

To evaluate the effectiveness of constraints, dataset is chosen and preprocessed as described in Sect. 2 of this paper. The authors of this dataset average of ROC-AUC scores across the 112 tasks as standardized evaluator to ensure fairness. Furthermore, this work uses the random partition as a benchmark for comparison due to its superiority over the METIS algorithm [18]. Common hyperparameters are list in Table 2. Inspired by [18], to avoid the situation that lost edges are impossible to be retrieved during training, the community is generated at each epoch instead of once before training. The implementation is on Pytorch Geometric [20]. Experiments are run on a single NVIDIA GeForce RTX 3070 8 GB.

Table 2. Common hyperparameters

Hyperparameter	Value
Number of layers	{3, 4, 5, 6, 7}
Number of subgraphs	20
Edge feature aggregation type	Mean
Number of epochs	1000
Learning rate	0.005
Propagation dropout rate	0.5
Hidden channels	256

4.2 Ablation Study

To understand the effectiveness of constraints, this work experiments on three different models:

- r-Cluster-GCN: Cluster-GCN [9] with random partition. The architecture is described in Fig. 1.
- b-LeidenGCN (ours): Bounded Cluster-GCN using the Leiden algorithm with ct_{min} and ct_{max} constraints. The graph-wise sampling phase of b-LeidenGCN is described in Fig. 5.
- ob-LeidenGCN (ours): Overlapping and Bounded Cluster-GCN architecture, b-LeidenGCN with adding $ct_{overlap}$ constraint. The graph-wise sampling phase of ob-LeidenGCN is described in Fig. 6.

Effect of Maximum/Minimum Constraint. With ct_{min} and ct_{max} constraints, the number of nodes within each community is guaranteed to be in range $[ct_{min}, ct_{max}]$. Therefore, each batch is considered approximately equal in terms of size (Fig. 7) and b-LeidenGCN is able to train with GPU. The results are shown in Table 3.

Table 3. Performance comparison. Result of r-Cluster-GCN is taken from paper [18] with mean and standard deviation of 10 runs, other models run 5 times. The constraints of b-LeidenGCN are set to $ct_{min} = 80$ and $ct_{max} = 100$. The number of layers is 3.

Algorithm	ROC-AUC
r-Cluster-GCN	**0,7771 ± 0,0025**
b-LeidenGCN (ours)	0.7630 ± 0.0030
ob-LeidenGCN (ours)	0.7746 ± 0.0007

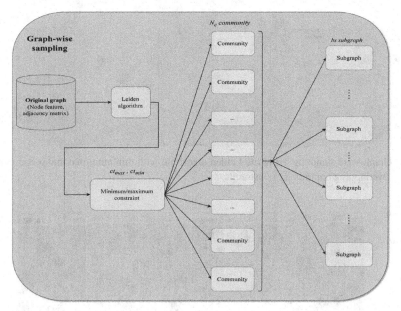

Fig. 5. Graph-wise sampling using the Leiden algorithm with minimum/maximum community size constraints.

Effect of Overlapping Constraint. By introducing $ct_{overlap}$ constraints, the node feature distribution across batches is adjusted to be similar (Fig. 3). From Table 3, the performance of ob-LeidenGCN is improved 1.16% compared to b-LeidenGCN, which shows the effect of balancing the node feature distribution across batches. To see more clearly the effect of $ct_{overlap}$ constraint, ob-LeidenGCN is tested with different $ct_{overlap}$ values, detailed results are listed in Table 4.

Deeper Models. As discussed in Sect. 2.2, node features are generated by the aggregation of neighbor's edge features. Thus, node features alone already contain a lot of information. Consequently, r-Cluster-GCN, with very small number of edges retained [18], become the best method in Table 3. Moreover, the capability of community detection algorithms to retain a large number of edges is not exposed when the number of layers is set to be small. To clarify this, r-Cluster-GCN and ob-LeidenGCN are trained with a

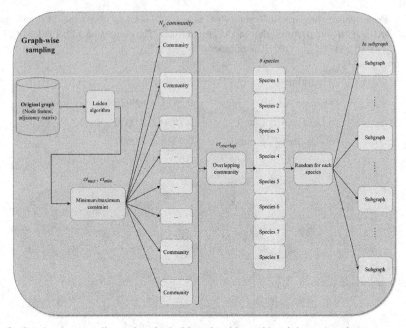

Fig. 6. Graph-wise sampling using the Leiden algorithm with minimum/maximum community size and overlapping community constraints.

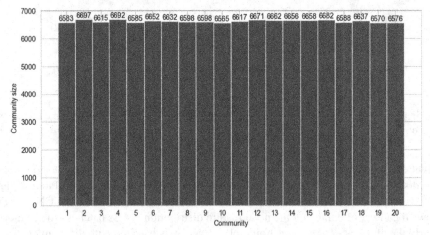

Fig. 7. Number of nodes across batches after using maximum/minimum community size constraint. The number of batches is set to 20

larger number of layers, the results are listed in Table 5. Except for the case when the number of layers is 3, ob-LeidenGCN has the best performance among mentioned models. The highest score is achieved when the number of layers equals 5, ob-LeidenGCN archives 79.4%, which is almost 1% better than r-Cluster-GCN. For both models, when

Table 4. The effect of $ct_{overlap}$ constraints. Constraint minimum/maximum community size are fixed to: $ct_{min} = 80$ and $ct_{max} = 100$. The number of layers is 5

Algorithm	ROC-AUC
ob-LeidenGCN with $ct_{overlap} = 0.1$	0.7839313367
ob-LeidenGCN with $ct_{overlap} = 0.2$	0.7867266871
ob-LeidenGCN with $ct_{overlap} = 0.3$	0.7872546432
ob-LeidenGCN with $ct_{overlap} = 0.4$	0.788294836
ob-LeidenGCN with $ct_{overlap} = 0.5$	**0.7943720784**
ob-LeidenGCN with $ct_{overlap} = 0.6$	0.7875076298
ob-LeidenGCN with $ct_{overlap} = 0.7$	0.7894462016
ob-LeidenGCN with $ct_{overlap} = 0.8$	0.7910601422
ob-LeidenGCN with $ct_{overlap} = 0.9$	0.7913429585

the number of layers is larger than 5, it fails to converge and results in a loss of ROC-AUC. As stated in [18], the optimization for deeper GCNs becomes more difficult and the message passing needs to be refined so that the information from the first few layers is not impeded.

Table 5. Performance comparison of deeper models. The constraints are fixed to $ct_{min} = 80$ and $ct_{max} = 100$ and $ct_{overlap} = 0.5$. The results are the mean of 5 independent runs. The best results are printed in bolded.

Algorithm	ROC-AUC				
	3 layers	4 layers	5 layers	6 layers	7 layers
r-Cluster-GCN	**0.7771**	0.7833	0.7842	0.7643	0.7597
ob-LeidenGCN (ours)	0.7746	**0.7848**	**0.7940**	**0.7864**	**0.7780**

5   Conclusion and Future Work

This work proposes using Leiden algorithm with additional constraints to leverage the performance of Cluster-GCN. The results show that our approach achieves better classification outcome than random partition, which previously has been shown to outperform METIS. This demonstrated that fine tuning the graph-wise sampling step can improve model performance to some extent. We believe that this work suggests a direction for building more complex and deeper architectures. The work is also a good reference for image pattern recognition problems [21–24].

In the future, we can expand the scope to a variety of community detection algorithms to analyze the effect of each algorithm on model performance. Furthermore, we aim to experiment with more datasets from different domains to demonstrate the effect of adding constraints. In order to do that, a generalization for the overlapping constraint needs to be studied.

References

1. Szklarczyk, D., et al.: STRING v11: protein–protein association networks with increased coverage, supporting functional discovery in genome-wide experimental datasets. Nucleic Acids Res. **47**(D1), D607–D613 (2019)
2. Kipf, T.N., Welling, M.: Semi-supervised classification with graph convolutional networks. arXiv:1609.02907 (2016)
3. Chiang, W.-L., Liu, X., Si, S., Li, Y., Bengio, S., Hsieh, C.-J.: Cluster-GCN: an efficient algorithm for training deep and large graph convolutional networks. In: Proceedings of the 25th ACM SIGKDD International Conference on Knowledge Discovery & Data Mining, pp. 257–266 (2019)
4. Hamilton, W., Ying, Z., Leskovec, J.: Inductive representation learning on large graphs. In: Advances in Neural Information Processing Systems, 30 (2017)
5. Chen, J., Ma, T., Xiao, C.: FastGCN: fast learning with graph convolutional networks via importance sampling. arXiv:1801.10247 (2018)
6. Chen, J., Zhu, J., Song, L.: Stochastic training of graph convolutional networks with variance reduction. arXiv:1710.10568 (2017)
7. Karypis, G., Kumar, V.: A fast and high quality multilevel scheme for partitioning irregular graphs. SIAM J. Sci. Comput. **20**(1), 359–392 (1998)
8. Shiokawa, H., Onizuka, M.: Scalable graph clustering and its applications. In: Alhajj, R., Rokne, J. (eds.) Encyclopedia of Social Network Analysis and Mining, pp. 2290–2299. Springer, New York (2018). https://doi.org/10.1007/978-1-4939-7131-2_110185
9. Newman, M.E.J., Girvan, M.: Finding and evaluating community structure in networks. Phys. Rev. E **69**(2), 026113 (2004)
10. Blondel, V.D., Guillaume, J.-L., Lambiotte, R., Lefebvre, E.: Fast unfolding of communities in large networks. J. Stat. Mech. Theory Exp. **2008**(10), P10008 (2008)
11. Traag, V.A., Waltman, L., Van Eck, N.J.: From Louvain to Leiden: guaranteeing well-connected communities. Sci. Rep. **9**(1), 5233 (2019)
12. Liu, Y., Shah, N., Koutra, D.: An empirical comparison of the summarization power of graph clustering methods. arXiv:1511.06820 (2015)
13. Xu, H., Lou, D., Carin, L.: Scalable gromov-wasserstein learning for graph partitioning and matching. In: Advances in Neural Information Processing Systems, 32 (2019)
14. Li, G., Xiong, C., Thabet, A., Ghanem, B.: Deepergcn: All you need to train deeper gcns. arXiv:2006.07739 (2020)
15. Liu, W., Tang, Z., Wang, L., Li, M.: DCBGCN: an algorithm with high memory and computational efficiency for training deep graph convolutional network. In: Proceedings of the 3rd International Conference on Advanced Electronic Materials, Computers and Software Engineering (AEMCSE), pp. 16–21. IEEE (2020)
16. Luo, M., et al.: A novel high-order cluster-GCN-based approach for service recommendation. In: Xu, C., Xia, Y., Zhang, Y., Zhang, LJ. (eds.) ICWS 2021. LNCS, vol. 12994, pp. 32–45. Springer, Cham (2022). https://doi.org/10.1007/978-3-030-96140-4_3
17. Li, G., Muller, M., Ghanem, B., Koltun, V.: Training graph neural networks with 1000 layers. In: International Conference on Machine Learning, pp. 6437–6449. PMLR (2021)

18. Xion, C.: Deep GCNs with random partition and generalized aggregator. Ph.D thesis. https://repository.kaust.edu.sa/bitstream/handle/10754/666216/ChenxinXiong_Masters_Thesis.pdf. Accessed 10 Feb 2023
19. Hu, W., et al.: Open graph benchmark: datasets for machine learning on graphs. Adv. Neural. Inf. Process. Syst. **33**, 22118–22133 (2020)
20. Fey, M., Lenssen, J.E.: Fast graph representation learning with PyTorch geometric. arXiv: 1903.02428 (2019)
21. Hung, P.D., Kien, N.N.: SSD-Mobilenet implementation for classifying fish species. In: Vasant, P., Zelinka, I., Weber, GW. (eds.) ICO 2019. AISC, vol. 1072, pp. 399–408. Springer, Cham. https://doi.org/10.1007/978-3-030-33585-4_40
22. Hung, P.D., Su, N.T., Diep, V.T.: Surface classification of damaged concrete using deep convolutional neural network. Pattern Recognit. Image Anal. **29**, 676–687 (2019)
23. Hung, P.D., Su, N.T.: Unsafe construction behavior classification using deep convolutional neural network. Pattern Recognit. Image Anal. **31**, 271–284 (2021)
24. Duy, L.D., Hung, P.D.: Adaptive graph attention network in person re-identification. Pattern Recognit. Image Anal. **32**, 384–392 (2022)

Fuzzy Logic Framework for Ontology Concepts Alignment

Adrianna Kozierkiewicz[ID], Marcin Pietranik[✉][ID], and Wojciech Jankowiak

Faculty of Information and Communication Technology, Wroclaw University of
Science and Technology, Wybrzeze Wyspianskiego 27, 50-370 Wroclaw, Poland
{adrianna.kozierkiewicz,marcin.pietranik}@pwr.edu.pl,
242432@student.pwr.edu.pl

Abstract. The problem of ontology alignment appears when interoperability of independently created ontologies is expected. The task can be described as collecting a set of pairs of elements taken from such ontologies that relate to the same objects from the universe of discourse. In our previous research, we introduced incorporating fuzzy logic in the considered task. It was used to combine several different similarity measures calculated between elements taken from different ontologies to eventually provide an unequivocal decision on whether or not a pair of such elements can be treated as mappable. Up until now, we focused solely on the level of instances and relations. Therefore, in this paper, we propose our novel approach to designating ontology mappings on the level of concepts. The developed methods were experimentally verified, yielding very promising results. We used the widely accepted benchmarks provided by the Ontology Alignment Evaluation Initiative, which are considered the state-of-the-art datasets used to evaluate solutions to any ontology-related problem.

Keywords: ontology alignment · fuzzy logic · knowledge management

1 Introduction

Ontology alignment is a well-known, broadly researched topic addressing the problem of establishing a set of mappings between independently created and maintained ontologies. These mappings connect elements from ontologies that somehow relate to the same objects taken from the real world. In other words - ontology alignment is the task of designating pairs of elements taken from ontologies that are as similar as possible. The topic has recently re-gained researchers' attention due to the emerging topic of knowledge graphs and entity alignments. It puts focus more on the aspect of matching instances, representing real-world objects, rather than on matching schemas describing those objects.

© The Author(s), under exclusive license to Springer Nature Singapore Pte Ltd. 2023
N. T. Nguyen et al. (Eds.): ACIIDS 2023, LNAI 13996, pp. 170–182, 2023.
https://doi.org/10.1007/978-981-99-5837-5_15

Even though the described task is very simple to describe and understand, the solution is not easy to achieve. First of all, ontologies are very complex knowledge representation methods. Using them as a backbone of the knowledge base entails the decomposition of some domains into elementary classes (which form the level of concept), defining how they can interact with each other (forming the level of relations), and finally defining instances of concepts. In consequence, independent ontologies representing the same universe of discourse may significantly differ.

In the literature, a plethora of different solutions to this task can be found. A majority of methods can be stripped down to calculating similarities between two concepts taken from two ontologies.

These similarities can be based on the analysis of different aspects that describe the content of ontologies. For example, they can involve comparing concept names, how their hierarchies are constructed, which instances they include, what are their attributes, etc.

Such methods when used separately, cannot be expected to yield satisfying outcomes. Therefore, there must be a method of combining several different values into one, interpretable output. The most obvious approach would involve calculating an average similarity value from all partial functions. If such similarity is higher than some accepted threshold, then such a pair of concepts are added to the final result.

In our previous publications, we have proposed a different approach to combining several similarity values to decide whether or not some elements taken from independent ontologies are matchable. We developed a set of fuzzy inference rules that can be used to aggregate the aforementioned similarities. In other words - we proposed including another layer of experts' knowledge about ontology alignment in the process.

Since the level concepts is the most covered in literature in [8] we focused on the level of relations, and in [9] on the level of instances. Therefore, the level of concepts remained to be addressed with our fuzzy approach. Thus, the main research tasks solved in the paper involve:

1. To develop a set of functions for calculating similarities between concepts based on their different features.
2. To develop a set of fuzzy inference rules that could be used to reason about how close two concepts taken from different ontologies are.

Along with previously developed tools (which target levels of instances and relations) presented in our earlier publications, the described methodology forms a fuzzy logic-based framework for ontology alignment. It is the main contribution of this article, which is organized as follows. Section 2 contains an overview of similar research along with its critical analysis. Section 3 provides mathematical foundations used throughout the paper. Section 4 forms the main contribution of the article. It describes partial similarity measures calculated between concepts of two ontologies. and how they can be combined into a single method for concept alignment employing fuzzy logic. The proposed solution was experimentally verified, and the collected results can be found in Sect. 5. Section 6 provides a summary and brief overview of our upcoming research plans.

2   Related Works

Since their introduction ontologies have become a popular knowledge representation format. They provide a flexible structure that allows to model, store, and process knowledge concerning some assumed universe of discourse ([19]). However, in many real cases, the knowledge can be distributed among multiple independent ontologies ([7]). Informally speaking, for a better understanding of the modelled part of the universe of discourse we need a whole picture of knowledge provided by the aforementioned distributed ontologies, thus a "bridge" between such ontologies is required ([12]).

As aforementioned in the previous section, a vast majority of ontology alignment methods come down to calculating similarities between two concepts taken from two ontologies. This approach spread across the initial research in the field ([3]) and more contemporary publications ([17]).

These similarities can be based on the analysis of different aspects that describe the content of ontologies. For example, they can involve comparing concept names ([2]) or how their hierarchies are constructed ([1]), etc.

To compare different approaches to ontology alignment an experimental verification is needed, which requires solid input data. In the context of ontology alignment, widely-accepted benchmark datasets are provided by the Ontology Alignment Evaluation Initiative ([16]), which contains a set of carefully curated ontologies along with alignments between them, that are treated as correct. Having such a dataset it is easy to compare results obtained from some ontology alignment tool with such arbitrarily given mappings and calculate measures like Precision and Recall.

According to [15] one of the most prominent ontology alignment systems are AML (also referred to as AgreementMaker), and LogMap. The description of AML can be found in ([4]). The solution is built on top of a set of matchers, each calculating different similarities among elements extracted from processed ontologies. The matchers are used in a cascade, each narrowing the results created by previous matchers to form the final result. The main focus is put on computational efficiency - AML has been designed to handle large ontologies while providing good quality outcomes.

[11] is solely devoted to LogMap. It is based on indexing lexical data extracted from ontologies and enriching them using WordNet. Then it uses interval labelling schema to create extended concepts' hierarchies which are then used as a base for final ontology alignments. Further steps of mapping repair involve finding internally inconsistent alignments which can be excluded from the final outcome.

A machine learning approach was proposed in [13] where authors develop semantic similarity measures to use them as the background knowledge which in consequence can provide constraints that improve machine learning models.

To the best of our knowledge, very little research attempt to utilize fuzzy logic in the task of ontology alignment. Only a few publications ([5]) describes useful research on the given subject. Therefore, having in mind good results yielded from our previous publications, in this paper, we incorporate fuzzy logic for the task of ontology alignment on the level of concepts.

3   Basic Notions

A pair (A, V), in which A denotes a set of attributes and V denotes a set of valuations of these attributes ($V = \bigcup_{a \in A} V_a$, where V_a is a domain of an attribute a), represents the real world. A single (A, V)-based ontology taken from the set $\tilde{O}$ (which contains all (A, V)-based ontologies) is defined as a tuple:

$$O = (C, H, R^C, I, R^I) \tag{1}$$

where:

- C is a finite set of concepts,
- H is a concepts' hierarchy,
- R^C is a finite set of binary relations between concepts $R^C = \{r_1^C, r_2^C, ..., r_n^C\}$, $n \in N$, such that $r_i^C \in R^C$ ($i \in [1, n]$) is a subset of $r_i^C \subset C \times C$,
- I is a finite set of instance identifiers,
- $R^I = \{r_1^I, r_2^I, ..., r_n^I\}$ is a finite set of binary relations between instances.

As easily seen the proposed ontology definition distinguishes four levels of abstraction: concepts, relations between concepts, instances, and relations. In this paper, we will mainly focus on the level of concepts. Each concept c from the set C is defined as:

$$c = (id^c, A^c, V^c, I^c) \tag{2}$$

where:

- id^c is an identifier (name) of the concept c,
- A^c represents a set of concept's c attributes,
- V^c represents a set of domains attributes from A^c, $V^c = \bigcup_{a \in A^c} V_a$,
- I^c represents a set of concepts' c instances.

In order to "translate" a content of some source ontology to the content of some other target ontology one must provide a set of correspondences between their elements. This set is called an alignment, and between two (A, V)-based ontologies $O_1 = (C_1, H_1, R^{C_1}, I_1, R^{I_1})$ and $O_2 = (C_2, H_2, R^{C_2}, I_2, R^{I_2})$ it can defined in the following way:

$$Align(O_1, O_2) = \{Align_C(O_1, O_2), Align_I(O_1, O_2), Align_R(O_1, O_2)\} \tag{3}$$

It includes three sets each containing correspondences between elements taken from the level of concepts, instances, and relations. Due to the limited space and the fact that this paper is solely devoted to the level of concepts, we will provide a detailed definition only for the level of concepts:

$$Align_C(O_1, O_2) = \{(c_1, c_2)|c_1 \in C_1 \wedge c_2 \in C_2\} \tag{4}$$

where c_1, c_2 are concepts from O_1 and O_2 respectively. The set $Align_C(O_1, O_2)$ contains only concepts pairs that have been processed and eventually marked as equivalent by fuzzy-based alignment algorithm described in the next section.

4   Fuzzy Based Approach to Concept Alignment

The main aim of our work is to determine the mappings between two ontologies on the concept level. In the proposed fuzzy method, we distinguish four input variables and one output. The input elements are measures that examine the lexical, semantic, and structural degrees of similarity of the two concepts. We incorporate four similarity functions. For the two given concepts $c_1 \in C_1$ and $c_2 \in C_2$ taken from two ontologies O_1 and O_2 they include:

1. the value of similarity between sets of concepts' attributes
   $attributesSim(c_1, c_2) = \frac{|A^{c_1} \cap A^{c_2}|}{|A^{c_1} \cup A^{c_2}|}$;
2. the value $jaroWinklerSim(c_1, c_2)$ of Jaro-Winkler similarity ([6]) between identifiers id^{c_1} and id^{c_2};
3. the value $levenshteinSim(c_1, c_2)$ of Levenshtein similarity ([6]) between identifiers id^{c_1} and id^{c_2};
4. the value $wordNetSim(c_1, c_2)$ of Wu-Palmer ([14]) similarity between identifiers id^{c_1} and id^{c_2} calculated by incorporating WordNet as external knowledge source.

Function 1 is based on a simple Jaccard index. However, attributes are taken from the global set A of all possible attributes therefore, it is possible to designate structural concept similarity this way. Functions 2–3 in order to provide a similarity value compare identifiers of concepts that may look counterintuitive from a formal model point of view. However, in practical applications concepts are usually identified by their names which express their nature. Moreover, in OWL (which is the most commonly used ontology representation format [10]) the given names are enforced to be unique. Therefore it is very straightforward to use them to calculate a similarity between concepts. In our method, we initially perform basic preprocessing of those names which involve lemmatization and stop word removal, which is omitted to increase its clarity.

Eventually, all input variables are associated with the following set of linguistic terms: *low, medium, high*, which allows us to define them through triangular-shaped membership functions presented in Fig. 1.

The output of the proposed system (referred to as *connection*) can obtain one of three values: *{equivalent, related, independent}*. Based on these assumptions,

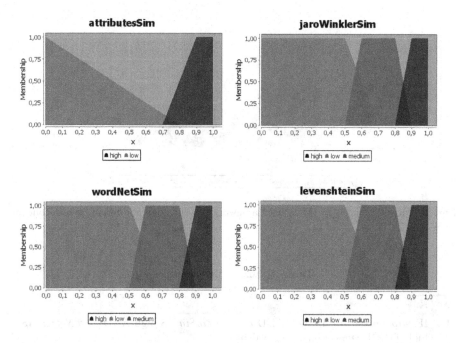

Fig. 1. Input variables of the fuzzy framework for ontology alignment on the concept level

in our approach, we decided that two concepts can be mapped if the obtained output of the fuzzy-processing of similarities calculated between them identifies the connection as *equivalent*. The described framework uses the minimum rule for the conjunction, the maximum rule for fuzzy aggregation, and the Mamdani type rule inference. The eventual defuzzification is performed based on the centroid method based on the values presented in Fig. 2. The inference rules were prepared by a group of experts and are presented in Table 1.

The fuzzy based algorithm for creating ontology alignment on the concept level is presented as Algorithm 1. The procedure accepts two ontologies O_1 and O_2 as its input and initially creates an empty set $Align_C(O_1, O_2)$ for the found mappings (Line 1). The it produces a cartesian product of two sets of concepts C_1 and C_2 (Line 2), creating a set of all concepts pairs from aligned ontologies. It iterates over its content (Lines 3–7). In each iteration the algorithm calculates four similarities $attributesSim(c_1, c_2)$, $jaroWinklerSim(c_1, c_2)$, $levenshteinSim(c_1, c_2)$ and $wordNetSim(c_1, c_2)$ (Line 3). Then the collected values are fuzzified (Line 4) and a potential mapping of two concepts is created with fuzzy inference rules taken from Table 1. This step yields final result. If the found mapping is identified as "*equivalent*" then it is added to the set $Align_C(O_1, O_2)$ (Lines 5–7).

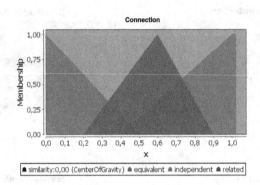

Fig. 2. Output variable of the fuzzy framework for ontology alignment on the concept level

Table 1. Fuzzy inference rules

No	Rule
1	IF *jaroWinklerSim* IS high AND *levenshteinSim* IS high AND *wordNetSim* IS high THEN *connection* IS equivalent
2	IF *jaroWinklerSim* IS high AND *levenshteinSim* IS high AND *wordNetSim* IS high AND *attributesSim* IS high THEN *connection* IS equivalent
3	IF *jaroWinklerSim* IS high AND *levenshteinSim* IS high THEN *connection* IS equivalent
4	IF *jaroWinklerSim* IS high AND *wordNetSim* IS high THEN *connection* IS equivalent
5	IF *levenshteinSim* IS high AND *wordNetSim* IS high THEN *connection* IS equivalent
6	IF *jaroWinklerSim* IS medium AND levenshteinSim IS medium AND *wordNetSim* IS medium THEN *connection* IS related
7	IF (*jaroWinklerSim* IS medium AND *levenshteinSim* IS medium) OR (*jaroWinklerSim* IS medium AND *wordNetSim* IS medium) THEN *connection* IS related
8	IF *jaroWinklerSim* IS low AND *levenshteinSim* IS low AND *wordNetSim* IS low THEN *connection* IS independent
9	IF (*jaroWinklerSim* IS low AND *levenshteinSim* IS medium) OR (*jaroWinklerSim* IS medium AND *levenshteinSim* IS low) AND *wordNetSim* IS low THEN *connection* IS independent
10	IF *jaroWinklerSim* IS low AND *levenshteinSim* IS low AND *wordNetSim* IS low AND *attributesSim* IS low THEN *connection* IS independent
11	IF *attributesSim* IS high THEN *connection* IS equivalent
12	IF *attributesSim* IS high AND (*jaroWinklerSim* IS low OR *jaroWinklerSim* is medium) THEN *connection* IS independent

---

Algorithm 1. Fuzzy based approach for relations alignment

Require: O_1, O_2
Ensure: $Align_C(O_1, O_2)$
1: $Align_C(O_1, O_2) = \{\}$
2: **for all** $(c_1, c_2) \in C_1 \times C_2$ **do**
3:    Calculate: $attributesSim(c_1, c_2)$, $jaroWinklerSim(c_1, c_2)$,
            $levenshteinSim(c_1, c_2)$, $wordNetSim(c_1, c_2)$;
4:    calculate $connection$ based on $attributesSim(c_1, c_2)$, $jaroWinklerSim(c_1, c_2)$,
         $levenshteinSim(c_1, c_2)$, $wordNetSim(c_1, c_2)$ and fuzzy inference rules;
5:    **if** $connection == equivalent$ **then**
6:        $Align_C(O_1, O_2) = Align_C(O_1, O_2) \cup \{(c_1, c_2)\}$;
7:    **end if**
8: **end for**
9: **return** $Align_C(O_1, O_2)$

---

5   Experimental Verification

5.1   Evaluation Procedure and Statistical Analysis

The proposed fuzzy framework for ontology alignment on the concept, described in Sect. 4, has been implemented and verified against a benchmark dataset provided by the Ontology Alignment Evaluation Initiative (OAEI) ([16]). OAEI is a non-profit organisation which since 2004 organises annual evaluation campaigns aimed at evaluating ontology mapping solutions. Organisers provide benchmarks that include a set of pairs of ontologies with their corresponding reference alignments, which should be treated as correct. These pairs of ontologies are then grouped in tracks and each track allows testing alignment systems for different features i.e. the Interactive matching track offers the possibility to compare different interactive matching tools which require user interaction and the Link Discovery track verifies how alignment systems deal with link discovery for spatial data where spatial data are represented as trajectories. Having the aforementioned reference alignments at your disposal makes it very easy to confront them with alignments generated be the evaluated solution, eventually calculating the common measures of Precision, Recall and F-measure.

This section presents the results of an experimental verification of the developed solution. We wanted to verify a hypothesis that the quality of mappings created by our framework is better or at least not worse than the other alignment system described in Sect. 2. The experiment was based on the *Conference*[1] track from OAEI dataset from the campaign conducted in 2019 ([18]). The accepted track includes seven ontologies *cmt, conference, confOf, edas, ekaw, iasted* and *sifkdd*. The experiment began with designating an alignment for each of 21 pairs of ontologies from this set using the implemented fuzzy framework. Then, thank to the provided reference alignments we were able to calculate values of Precision, Recall and F-measure. The obtained results can be found in Table 2.

---

[1] http://oaei.ontologymatching.org/2019/conference/index.html.

Table 2. Results of fuzzy-based concept alignment framework

Source ontology	Target ontology	Precision	Recall	F-measure
cmt	conference	0,71	0,45	0,55
cmt	confOf	0,80	0,40	0,53
cmt	edas	1,00	1,00	1,00
cmt	ekaw	1,00	0,63	0,77
cmt	iasted	0,80	1,00	0,89
cmt	sigkdd	1,00	0,80	0,89
conference	confOf	0,88	0,64	0,74
conference	edas	0,88	0,50	0,64
conference	ekaw	0,69	0,39	0,50
conference	iasted	0,80	0,31	0,44
conference	sigkdd	0,89	0,67	0,76
confOf	edas	0,90	0,64	0,75
confOf	ekaw	0,90	0,45	0,60
confOf	iasted	1,00	0,44	0,61
confOf	sigkdd	1,00	0,67	0,80
edas	ekaw	0,62	0,50	0,55
edas	iasted	0,78	0,39	0,52
edas	sigkdd	1,00	0,64	0,78
ekaw	iasted	0,83	0,56	0,67
ekaw	sigkdd	0,86	0,60	0,71
iasted	sigkdd	0,92	0,73	0,81

Data from Table 2 has been subjected to statistical analysis. All tests have been conducted for the significance level $\alpha = 0.05$. Before selecting a proper statistical test, we checked the distribution of obtained samples of: Precision, Recall and F-measure using a Shapiro-Wilk test. The p-value calculated for the sample of Precision equals 0.07538, the p-value for the Recall sample equals 0.113911 and the p-value for the F-measure sample equals 0.768263. The p-values of all samples were greater than the assumed α therefore we were allowed to claim that all of them come from the normal distribution. Thus, we have chosen the t-Student test for further analysis of experimental results.

The first null hypothesis we checked claims that the mean value of the Precision measure equals 0.82. The calculated p-value of the t-Student test equals 0.0282232 and the value of statistic equals 2.025. We can reject this hypothesis in favour of claiming that the mean value of the Precision measure is greater than 0.82.

The second verified null hypothesis claims that the mean value of the Recall measure equals 0.52. The obtained p-value equals 0.048257 and the value of statistic equals 1.744. It allows us to reject such null hypothesis and claim that the mean value of the Recall measure is greater than 0.52.

The final verified null hypothesis claims that the mean value of the F-measure equals 0.63. The t-Student test resulted with 1.887 with a p-value equal to 0.036876. As previously, it is possible to accept the alternative hypothesis which claims that the mean value of the F-measure is greater than 0.63.

5.2 Results Interpretation

Unfortunately, the OAEI organization did not provide partial results obtained by individual tools on specific pairs of ontologies, but only a summary of mean values for the entire dataset. Therefore, in order to compare the developed method with others, we decide to calculate the average values of Precision, Recall and F-measure obtained for individual ontology pairs from the dataset. The collected values of other tools and the proposed framework are presented in Table 3.

As easily seen in Table 3, the proposed tool for determining mappings between ontologies obtained a very good value of the Precision equal to 0.87, which is only 0.01 worse than the best result of 0.88 obtained by the *enda* and *StringEquiv* tools. On the other hand, the average value of the Recall measure obtained on the entire dataset was 0.59, which may not be a very good result, but is still satisfactory considering the results obtained by other tools. Most of the solutions achieved an average value of the Recall ranging from 0.54 to 0.64, with the best value of the Recall equal to 0.76 for the *SANOM*, and the worst for *ONTMAT1* equal to 0.49.

In the case of the most reliable indicator of the solution assessment, the F-measure which takes into account both Precision and Recall, the proposed tool was at the forefront of the tested solutions with a score of 0.69. The best result of the F-measure was obtained by the *SANOM* system - 0.77, while the worst equal 0.61 for Lily.

The conducted experiment proved that the developed tool obtains better values of measures for assessing the quality of solutions than most other tools tested in the Conference track. The dataset used in the experiment contained many reference mappings, the validity of which, from the subjective user point of view, seems questionable. For example, the concepts *Country* and *State* may seem similar, but the Cambridge Dictionary defines the former as "*an area of land that has its own government, army, etc.*", while the latter as "*one of the parts that some countries such as the US are divided into*".

Due to the nature of the benchmark dataset provided by OAEI, which included many difficult or unintuitive mappings, it can be assumed that the developed method is capable of achieving even better results under real-world conditions on real ontologies, and not on a synthetic dataset.

Table 3. Average results of ontology alignment tools

Alignment tool	Average Precision	Average Recall	Average F-measure
SANOM	0,78	0,76	0,77
AML	0,83	0,7	0,76
LogMap	0,84	0,64	0,73
Proposed solution	**0,87**	**0,59**	**0,69**
Wiktionary	0,80	0,58	0,68
DOME	0,87	0,54	0,67
edna	0,88	0,54	0,67
LogMapLt	0,84	0,54	0,66
ALIN	0,87	0,52	0,65
StringEquiv	0,88	0,5	0,64
ONTMAT1	0,82	0,49	0,61
Lily	0,59	0,63	0,61

6   Future Works and Summary

In recent years ontologies have become more and more popular because they provide a flexible structure that allows one to model, store, and process knowledge concerning some assumed universe of discourse. In many real cases, there is a need for integrating ontologies into a single unified knowledge representation and the initial step to achieving such a goal is designating a "bridge" between two ontologies. In the literature, this issue is referred to as ontology alignment. Formally, it can be treated as a task for providing a set of correspondences (mappings) of elements taken from two aligned ontologies. Even though the described task is very simple to describe and understand, the solution is not easy to achieve.

The practical application of the tackled issue appears when communication of two independently created information systems is required. Informally speaking, some kind of a bridge between them is expected. Utilizing ontologies in such systems is not uncommon. For example, many medical systems operating different therapeutic devices (e.g. CT or linear accelerator) incorporate ontologies as a backbone healthcare vocabularies (e.g. SNOMED-CT or ICD10) used to describe patients' treatment courses. Providing the interoperability of such systems requires their terminologies to be initially matched.

The article is the final element of our fuzzy logic ontology alignment framework, which addresses the level of concepts. We have taken inspiration from our previous publications ([8,9]) in which we focused on mapping ontologies on the level of relations and the level of instances. We proposed to aggregate several similarity values calculated between concepts taken from independently created ontologies by introducing another level of expert knowledge in the form of fuzzy inference rules. Such an approach makes it possible to decide whether or not some elements taken from independent ontologies are matchable.

The developed framework was experimentally verified and compared with other solutions known from the literature (*SANOM, AML, LogMap, Wiktionary, DOME, edna, LogMapLt, ALIN, StringEquiv, ONTMAT1* and *Lily*). This comparison was performed using a commonly accepted dataset provided by Ontology Alignment Evaluation Initiative. The collected results allow us to claim that our approach yield very good alignments.

In the upcoming future, we plan to focus on the aspect of scalability of the developed fuzzy framework. To do so we plan to conduct more extensive experiments using different datasets created by the Ontology Alignment Evaluation Initiative. Additionally, we plan the developed other similarity functions for the level of concepts to increase the flexibility of the proposed framework.

References

1. Ardjani, F., Bouchiha, D., Malki, M.: Ontology-alignment techniques: survey and analysis. Int. J. Modern Educ. Comput. Sci. **7**(11), 67 (2015)
2. Cheatham, M., Hitzler, P.: String similarity metrics for ontology alignment. In: Alani, H., et al. (eds.) ISWC 2013. LNCS, vol. 8219, pp. 294–309. Springer, Heidelberg (2013). https://doi.org/10.1007/978-3-642-41338-4_19
3. Euzenat, J.: An API for ontology alignment. In: McIlraith, S.A., Plexousakis, D., van Harmelen, F. (eds.) ISWC 2004. LNCS, vol. 3298, pp. 698–712. Springer, Heidelberg (2004). https://doi.org/10.1007/978-3-540-30475-3_48
4. Faria, D., Pesquita, C., Santos, E., Palmonari, M., Cruz, I.F., Couto, F.M.: The AgreementMakerLight ontology matching system. In: Meersman, R., et al. (eds.) OTM 2013. LNCS, vol. 8185, pp. 527–541. Springer, Heidelberg (2013). https://doi.org/10.1007/978-3-642-41030-7_38
5. Fernández, S., Velasco, J., Marsa-Maestre, I., Lopez-Carmona, M.A.: A fuzzy method for ontology alignment. In: Proceedings of the International Conference on Knowledge Engineering and Ontology Development (2012)
6. Gomaa, W.H., Fahmy, A.A.: A survey of text similarity approaches. Int. J. Comput. Appl. **68**(13), 13–18 (2013)
7. Harrow, I., et al.: Matching disease and phenotype ontologies in the ontology alignment evaluation initiative. J. Biomed. Semant. **8**(1), 1–13 (2017)
8. Hnatkowska, B., Kozierkiewicz, A., Pietranik, M.: Fuzzy based approach to ontology relations alignment. In: 2021 IEEE International Conference on Fuzzy Systems (FUZZ-IEEE), pp. 1–7. IEEE (2021)
9. Hnatkowska, B., Kozierkiewicz, A., Pietranik, M.: Fuzzy logic framework for ontology instance alignment. In: Groen, D., de Mulatier, C., Paszynski, M., Krzhizhanovskaya, V.V., Dongarra, J.J., Sloot, P.M.A. (eds.) Computational Science—ICCS 2022. ICCS 2022. Lecture Notes in Computer Science, vol. 13351, pp. 653–666. Springer, Cham (2022). https://doi.org/10.1007/978-3-031-08754-7_68
10. Hogan, A.: Web ontology language. In: The Web of Data, pp. 185–322. Springer, Cham (2020). https://doi.org/10.1007/978-3-030-51580-5_5
11. Jiménez-Ruiz, E.: LogMap family participation in the OAEI 2020. In: Proceedings of the 15th International Workshop on Ontology Matching (OM 2020), vol. 2788, pp. 201–203. CEUR-WS (2020)

12. Kolyvakis, P., Kalousis, A., Smith, B., Kiritsis, D.: Biomedical ontology alignment: an approach based on representation learning. J. Biomed. Semant. **9**(1), 1–20 (2018)
13. Kulmanov, M., Smaili, F.Z., Gao, X., Hoehndorf, R.: Semantic similarity and machine learning with ontologies. Briefings Bioinform. **22**(4), bbaa199 (2021)
14. Meng, L., Huang, R., Gu, J.: A review of semantic similarity measures in wordnet. Int. J. Hybrid Inf. Technol. **6**(1), 1–12 (2013)
15. Mohammadi, M., Rezaei, J.: Evaluating and comparing ontology alignment systems: an MCDM approach. J. Web Semant. **64**, 100592 (2020)
16. Ontology Alignment Evaluation Initiative. http://oaei.ontologymatching.org. Accessed 03 Sep 2020
17. Patel, A., Jain, S.: A novel approach to discover ontology alignment. Recent Adv. Comput. Sci. Commun. (Formerly: Recent Patents on Computer Science) **14**(1), 273–281 (2021)
18. Pour, M., et al.: Results of the ontology alignment evaluation initiative 2021. In: CEUR Workshop Proceedings 2021, vol. 3063, pp. 62–108. CEUR (2021)
19. Sampath Kumar, V., et al.: Ontologies for industry 4.0. Knowl. Eng. Rev. **34**, E17 (2019). https://doi.org/10.1017/S0269888919000109

Can Ensemble Calibrated Learning Enhance Link Prediction? A Study on Commonsense Knowledge

Teeradaj Racharak[1]([✉])[iD], Watanee Jearanaiwongkul[2][iD],
and Khine Myat Thwe[1]

[1] School of Information Science,
Japan Advanced Institute of Science and Technology, Nomi, Japan
racharak@jaist.ac.jp, khine.myat.thwe.14@gmail.com
[2] Asian Institute of Technology, Khlong Luang, Thailand
watanee.j@gmail.com

Abstract. Numerous prior works have shown how we can use knowledge graph embedding (KGE) models for ranking unseen facts that are likely to be true. Though these KGE models have been shown to make good performance on the ranking task with standard benchmark datasets, in practice only a subset of the top-k ranked list is indeed correct. This is due to the fact that most knowledge graphs are built under the open world assumption, while state-of-the-art KGE exploit the closed world assumption to build negative samples for training. In this paper, we show to address this problem by ensembling calibrated learning, following the principle that multiple calibrated models can make a stronger one. In experiments on the ConceptNet of commonsense knowledge base, we show significant improvement over each individual baseline model. This suggests that ensembling calibrated learning is a promising technique to improve link prediction with KGEs.

Keywords: Link prediction · Knowledge graph embedding · Ensemble learning · Commonsense knowledge base · ConceptNet

1 Introduction

Knowledge graphs (KGs) [18] have emerged as the de-facto standard to share large amounts of factual knowledge on the web. Numerous KGs, such as Word-Net [14], FreeBase [3], YAGO [24], DBpedia [10], and ConceptNet [23], have been published with different utilization purposes. These KGs have become a significant resource for many AI applications, such as question and answering systems and recommendation systems [29], academic search [32], social relationship recognition [31], and drug discovery [13]. This emphasizes an importance of investigating machine learning approaches for structured data.

A fundamental problem considered in KGs is *link prediction*, i.e., the task of predicting potential missing entities in a KG. Recently, numerous works such as [15,29] show that knowledge graph embedding (KGE) models can be used

N. T. Nguyen et al. (Eds.): ACIIDS 2023, LNAI 13996, pp. 183–194, 2023.
https://doi.org/10.1007/978-981-99-5837-5_16

to identify the top k potential prediction of missing entities (e.g. predicting an entity ? in $\langle$ *Tokyo, capitalOf, ?* $\rangle$). The main evaluation metrics of KGE models are typically based on ranking, e.g., mean rank (MR), mean reciprocal rank (MRR), and Hits@k. MR calculates the mean of the rank of each predicted entity, MRR calculates the mean of reciprocal of the predicted ranks, and Hits@k calculates the proportion of correct entities ranked in the top k. However, we preliminarily study the performance on the de-facto KGE models, e.g., TransE [4], RESCAL [16], ComplEx [27], DisMult [33], ConvE [6] (cf. Appendix A), and our preliminary results reveal that the models may not work well in practice because not all correct completion of the triples appear in the top ranked positions. In other words, a small k would affect recall while a large k would affect precision.

We suspect that this issue occurs because the training scheme of KGE models follows the closed world assumption (CWA)[1] [19]. That is, assume a (positive) KG $\mathcal{G} := \{\langle h_i, r_i, t_i \rangle \mid i \in \mathbb{N}\}$, where h, r, t represent head, relation, and tail entities in $\mathcal{G}$, respectively. A corresponding negative training set is generated by replacing the head entity h_i using all other entities and replacing the tail entity t_i using all other entities [4]. However, most KGs are built under the open world assumption (OWA) [2]. Hence, the generated negative triples could be incorrect. One could address this problem by manually annotating the negative samples, but this step is time-consuming and requires human experts.

Shi and Weninger [21] introduced a sophisticated technique that could deal with the open world assumption in KGs. On the one hand, Shi and Weninger's method work quite well in their experiment settings. On the other hand, this paper argues that this problem can be remedied by *ensemble learning*. Note that ensemble learning is a technique that has been successfully applied in numerous fields [35]. The idea behind ensemble learning is rather simple, i.e., instead of focusing on a strong model, we can use multiple weaker ones together instead.

A challenge behind the adoption of ensemble learning with KGEs is that each KGE method in the literature (cf. Sect. 2) does not have probabilistic interpretation; they use a distance function as a score so that facts in KGS can have higher scores than others. To deal with this challenge properly, we propose to calibrate each KGE model to match the expected distribution of probabilities of each triple before aggregating the final prediction (cf. Sect. 3). While our solution is conceptually simple, our extensive experiments with commonsense knowledge base (cf. Sect. 2) confirm that doing probability calibration for ensemble learning with KGEs help improve the overall performance. Section 4 shows the results of our experiments and comparison between the ensemble models and the corresponding KGE models. Indeed, our ensemble with ConvE and TransE is shown to increase up to 138% when compared with individual ConvE at Hits@1. Finally, we compare our proposal to the state-of-the-art in Sect. 5 and then conclude in Sect. 6.

---

[1] https://pykeen.readthedocs.io/en/stable/reference/negative_sampling.html.

2   Background

2.1   Commonsense Knowledge Bases

According to [34], commonsense knowledge should satisfy important characteristics, i.e., possessed and shared by everyone, fundamental and widely understood, implicit, massive in terms of volume and diversity, covered all domains of daily life, as well as it should contain default assumptions of the world, such as, birds can fly, fish cannot speak, etc. These characteristics lead to an adoption of open world assumption to the commonsense knowledge, i.e., absent statements are considered to have unknown truth rather than being invalid.

To make commonsense knowledge explicit and understandable by computers, commonsense knowledge bases (CKBs) are developed to represent commonsense knowledge in formal languages and often in form of a triple $\langle h, r, t \rangle$. There are a number of well-known CKBs that have been built, e.g., Cyc [11], WordNet [14], WebChild [26], Open Mind Common Sense (OMCS) [22], SenticNet [5], and ConceptNet [23]. This work uses ConceptNet (Fig. 1) as our experimental dataset. It contains knowledge from various sources having 21 million edges and 8 million nodes and covers English and other 83 languages.

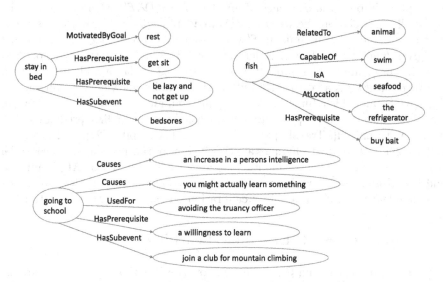

Fig. 1. A visualization of sampled triples in ConceptNet

2.2   Link Prediction with Knowledge Graph Embedding

A knowledge graph (KG) $\mathcal{G}$ can be formalized as a pair $(\mathcal{V}, \mathcal{E})$ where $\mathcal{V}$ denotes an entity set and $\mathcal{E} \subseteq \mathcal{V} \times \mathcal{V}$ represents a semantic relation between two entities.

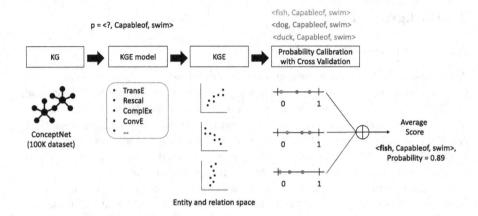

Fig. 2. The proposed calibrated KGE ensemble learning method

Given $h, t \in \mathcal{V}$ and $(h, t) \in \mathcal{E}$, we write $\langle h, r, t \rangle$ to indicate a link (edge) from h to t with label r, e.g., $\langle Tokyo, capitalOf, Japan \rangle$.

Typically, $\mathcal{G}$ is often incomplete in the sense that some links are missing. This assumption implies the existence of another $\mathcal{G}' := (\mathcal{V}, \mathcal{E}')$ where $\mathcal{E}' \supset \mathcal{E}$ is the set of all true relations between entities in $\mathcal{V}$. Link prediction is the task that predicts all and only the links in $\mathcal{E}' \setminus \mathcal{E}$. Specifically, predicting h given r and t or t given h and r. The former is denoted by $\langle ?, r, t \rangle$ and the latter is denoted by $\langle h, r, ? \rangle$. For example, $\langle ?, capitalOf, Japan \rangle$ is to predict the capital city of Japan.

An *embedding* is a vector in $\mathbb{R}^d$ where $d > 0$ and a knowledge graph embedding (KGE) model is a function $\mathcal{M}$ mapping from any h, r, t to a vector. Several techniques have been proposed to train a KGE model for link prediction. For instance, TransE [4], TransH [30], RESCAL [16], ComplEx [27], and DistMult [33]. In [29], embedding techniques are classified into translational distance model (e.g., TransE, TransH) and semantic matching model (e.g., RESCAL, ComplEx, and DistMult). In this work, we consider various embedding methods to apply ensemble learning with probability calibration on each KGE model's outputs.

3   Calibrated KGE Ensemble Learning

Let $\mathcal{G}$ be an input KG and $\mathcal{G}'$ be the (unknown) KG with all the valid relations. Our proposed approach, called *calibrated KGE ensemble learning*, is designed as pipeline to predict all the relations in $\mathcal{G}'$ given a broken triple p of the form $\langle ?, r, t \rangle$ or $\langle h, r, ? \rangle$. Formally, our approach has two additional inputs, i.e., p and a set of KGE models M that learn the representation of $\mathcal{G}$. The output is the set of all valid link prediction for p observable in $\mathcal{G}'$.

Recall that KGE techniques [29] generally have three steps: (1) representing entities and relations in a vector space, (2) defining a scoring function f that calculates the plausibility of triples in $\mathcal{G}'$, and (3) learning the entity and relation representations by maximizing the total plausibility of the triples. The scoring function of each KGE model is often different as it is defined from distinct perspectives. For example, in TransE, the scoring function is

$$f_{te}(\langle h, r, t \rangle) := ||\mathbf{h} + \mathbf{r} - \mathbf{t}|| \tag{1}$$

where $|| \cdot ||$ is the L1 norm, $\mathbf{h}$, $\mathbf{r}$, and $\mathbf{t}$ are the vectors associated to the h, r, and t, respectively. With ComplEx, it is

$$f_{ce} := Re(\langle \mathbf{h}, \mathbf{r}, \bar{\mathbf{t}} \rangle) \tag{2}$$

where $\langle \cdot \rangle$ computes the dot product, $Re(\cdot)$ is real component, and $\bar{\cdot}$ is the conjugate for complex vector. We refer the details of the scoring function f_M of each KGE model M in [29]. It is not difficult to observe that different f_M gives outputs in a different continuous vector space. Hence, ensembling multiple models M for $\mathcal{G}$ cannot be done immediately. It is necessary to match the expected distribution of probabilities for triples in $\mathcal{G}'$ (i.e. probability calibration). We detail the steps (cf. Fig. 2 as an illustration of our proposal) as follows:

1. Given a (fixed) set M of KGE models, each KGE model $M^* \in M$ is optimized based on its scoring function f_{M^*},
2. Each model M^* is calibrated to output the probability of all triples. This is to find a new function $\mathbf{Pr}_{M^*}$ that maps a score to a probability value,
3. Lastly, the pipeline aggregates all the probability outputs. In this work, we consider the average predicted probabilities of all predicted link p.

To ensure unbiased calibration, Step 2 can employ cross-validation (as shown in Fig. 2) by splitting the data into k folds. In this case, the output of $\mathbf{Pr}_{M^*}$ corresponds to the average of the predicted probabilities of all k estimators. We formally describe the probability calibration of Step 2 and ensemble learning of Step 3 in the following subsections.

3.1   Probability Calibration

Definition 1 (Calibration). *Let $\langle h, r, t \rangle$ be a triple in $\mathcal{G}$ and $f_X(\langle h, r, t \rangle)$ be a score corresponding to model X. Then, the calibration is defined as a probability function $\mathbf{Pr}(\langle h, r, t \rangle = \text{``TRUE''} \mid f_X(\langle h, r, t \rangle))$, where $\langle h, r, t \rangle = \text{``TRUE''}$ means a valid triple in $\mathcal{G}'$.*

Definition 1 spells out the calibration as a general function that can match the expected probability distribute, providing an advantage on the implementation with various calibration options[2]. In this work, we employ the logistic sigmoid as a concrete choice of the probability calibration. Hence, our calibration step is formally defined as follows:

$$\mathbf{Pr}(\langle h, r, t \rangle = \text{``TRUE''} \mid f_X(\langle h, r, t \rangle)) := \frac{1}{1 + \exp(\alpha f_X(\langle h, r, t \rangle)) + \beta} \tag{3}$$

---

[2] https://scikit-learn.org/stable/modules/calibration.html#calibration explains possible concrete implementation of the calibration.

Table 1. Statistic of ConceptNet-100K

	ConceptNet
Train	100,000
Validation	1,200
Test	2,400

where parameters $\alpha, \beta \in \mathbb{R}$ are to be determined when fitting the logistic regressor via maximum likelihood. Other concrete calibration choices can be found in [9,17]. We plan to investigate these choices as a future task.

3.2 Ensemble Learning

Given multiple KGE models M_i ($i \in \{1, 2, 3, \ldots, e\}$), our ensemble considers the average of all probabilities determined by each corresponding calibrator $\mathbf{Pr}_{M_i}$ as the final output of the pipeline for any input $\langle ?, r, t \rangle$ or $\langle h, r, ? \rangle$. Hence, the prediction for ? is determined by entities having the highest average probability.

$$predicted\_prob := \frac{1}{e} \sum_{i=1}^{e} \mathbf{Pr}_{M_i}(\langle h, r, t \rangle = \text{``TRUE''} \mid f_{M_i}(\langle h, r, t \rangle)) \quad (4)$$

Due to the fact that KGEs are defined under different assumptions, we suspect that not all the models can be ensembled. Therefore, we plan to investigate a subset of KGE models that can stay together. We will also detail our experiment to address this question in the next section.

4 Experimental Setup

4.1 Dataset

ConceptNet [23] is a knowledge graph that contains common sense knowledge from Open Mind Common Sense. In this work, we use ConceptNet-100K provided by [12][3] as the gold dataset. Table 1 shows its dataset statistic.

4.2 Training and Evaluation Protocols

Given a positive triples in ConceptNet-100K denoted by $\mathcal{G}_{CN} := \{\langle h_i, r_i, t_i \rangle\}_{i=1}^{u}$, where u represents the size of ConceptNet-100K (100,000), we generated a corresponding negative set by uniformly replacing each head entity using all other entities and replacing each tail entity using all other entities. We adopted the KL divergence to measure the loss on learning the facts according to each corresponding scoring function while training the representation of $\mathcal{G}_{CN}$.

We trained each KGE model and its corresponding ensemble version using Adagrad via LibKGE[4] based on the training and validation sets of ConceptNet-

---

[3] https://home.ttic.edu/~kgimpel/commonsense.html.
[4] https://github.com/uma-pi1/kge.

100K. The default setting was used for training and all the training options were configured via LibKGE. The maximum number of epochs used for training is set to 20 and the batch size is set to 100.

For a test triple in the test set. The head entity h was replaced by each of the entities of the vocabulary in turn to form all possible triples. Triple scores were calculated by the scoring function of each corresponding KGE model and then sorted in ascending order. After sorting, the ranking of the testing triple rk_i was recorded. This whole procedure was repeated while removing t instead of h. We adopted Hits@k to measure the performance of the proposed approach:

$$\text{Hits@}k = \sum_{i=1}^{u^{(\text{test})}} \mathbb{I}[\text{rk}_i \leq k] \tag{5}$$

where $u^{(\text{test})}$ denotes the number of the test set, $\mathbb{I}[expr]$ is the indicator function which outputs 1 if $expr$ is true and 0 otherwise.

4.3   Experiment

Prior to this writing, we tested the performance of numerous KGE models using Pykeen [1] (cf. Appendix A). We used Pykeen under the default setting of each model. The performance was evaluated according to mean rank (MR), mean reciprocal rank (MRR), and Hits@k. We selected Pykeen instead of LibKGE in this stage because it provides more comprehensive implementation of KGEs. As mentioned in the Introduction, the preliminary results show that, with the experimental dataset, it was easy to get high recall and low precision. These results led us to develop our calibrated KGE ensemble learning.

To test the performance of our proposed method, we selected top five performance KGE models from the preliminary experiment: TransE, RESCAL, ComplEx, DisMult, and ConvE, and trained each model using LibKGE as explained in the above subsection. Unsurprisingly, the performances reported from Pykeen and LibKGE are coincided. Note that we changed our implementation from Pykeen to LibKGE due to our familiarity on the framework when implementing our proposed ensemble method, which will describe next.

We extended the class KgeModel of LibKGE to implement the proposed calibrated KGE ensemble learning. In our extension, we used CalibratedClassifierCV of scikit-learn under the default values, meaning that we could successfully implemented the logistic sigmoid as our calibrator (cf. Eq. 3) and fitted under 5-fold cross validation to ensure unbiased probability calibration. Lastly, we implemented Eq. 4 in the same extended class to output the final prediction as illustrated in Fig. 2. Using this implementation, we can justify our question "whether or not all the KGE models can stay together" (cf. Subsect. 3.2). For this, we tried several combinations out of the selected KGE models and reported the results on Table 2. There are conclusions that likely to be remarked here from the table. (1) Ensembling many models (e.g. ComplEx, ConvE, TransE) does not guarantee to reach the best performance. (2) Surprisingly, involving the best performance

Table 2. Performance of the individual and ensemble models

Model	Hits@1	Hits@10	Hits@100	Hits@1000
TransE	0.088	**0.202**	**0.303**	**0.400**
RESCAL	0.076	0.138	0.198	0.281
ComplEx	**0.109**	0.177	0.240	0.313
DistMult	0.003	0.006	0.015	0.038
ConvE	0.080	0.195	0.205	0.381
RESCAL + TransE	0.092	0.142	0.247	0.314
ComplEx + TransE	0.138	0.251	**0.332**	0.428
ConvE + TransE	**0.190**	**0.281**	0.310	**0.501**
ComplEx+ ConvE + TransE	0.162	0.228	0.285	0.415

Table 3. Percent increase between ensemble models and individual models

Ensemble	Individual Model 2	Hits@1	Hits@10	Hits@100	Hits@1000
RESCAL + TransE	RESCAL	21%	3%	25%	12%
RESCAL + TransE	TransE	5%	−30%	−18%	−22%
ComplEx + TransE	ComplEx	27%	42%	38%	37%
ComplEx + TransE	TransE	57%	24%	10%	7%
ConvE + TransE	**ConvE**	**138%**	**44%**	**51%**	**31%**
ConvE + TransE	TransE	116%	39%	2%	25%
ComplEx+ ConvE + TransE	ComplEx	49%	29%	19%	33%
ComplEx+ ConvE + TransE	ConvE	103%	17%	39%	9%
ComplEx+ ConvE + TransE	TransE	84%	13%	−6%	4%

model (e.g. ComplEx) into an ensemble might not lead to the best ensemble model (e.g. ComplEx + TransE), even each corresponding KGE model is calibrated beforehand. Table 3 shows percent increase between ensemble models and corresponding individual models, which spells out that the percent increase can be up to 138% at the maximum.

5   Related Work

Rivas-Barragan et al. [20] proposed to apply ensemble learning on multiple KGE models for drug discovery with BioKG and OpenBioLink datasets. They also had a similar argument to us in the sense that different KGE models produce different score distributions which typically lie in different intervals. Therefore, they proposed to normalize outputs of each KGE model before doing an ensemble. Regarding the ensemble, they performed the majority voting and thus reached to the conclusion that ensemble helps link prediction. However, there is no detail how the normalization was done.

In [8], the authors proposed to use a logistic sigmoid on each individual KGE model and aggregate each output with the arithmetic mean. The logistic sigmoid was used for re-scaling outputs of KGEs to [0, 1]. This method looks similar to our proposal, though we believe that the motivations are different; ours is to match a score to a probabilistic distribution – not just for the re-scaling purpose. Thus, it can be seen as a concrete instance of our proposal. Nevertheless, there is no implementation detail on how a logistic model is fit. There is no explanation on how to handle the negative sampling. Besides, there is no use of cross-validation with the sigmoid to ensure unbiased data while fitting the model.

Tabacof and Costabello [25] revealed that popular KGE models are indeed uncalibrated, meaning that the score estimates associated to the prediction are unreliable. They proposed to calibrate with the logistic and isotonic regression. Their experiments demonstrate significantly better results than the uncalibrated models from all calibration methods. This result has inspired us to further investigate on other calibration methods for ensemble learning with KGEs.

Wan et al. [28] proposed an ensemble strategy consisting of two components: (1) knowledge graph extraction and (2) combination strategy. The former component operated re-sampling to create a new training set with high diversity and the latter one used the adaptive weighting to combine multiple models according to their prediction capability. Their experiments with three classic models (TransE, DisMult, and TransR) confirm their robustness.

Joshi and Urbani [7] argued for the necessity of defining a new task (based on the classic link prediction) that classifies potential links with a binary true/false label. Following this principle, they propose a new system called DuEL. DuEL post-processes the ranked lists produced by the embedding models with multiple classifiers, which include supervised models like LSTMs, MLPs, and CNNs and unsupervised ones that consider subgraphs and reachability in the graph. Despite their accurate results, the models were not calibrated before aggregation.

6   Conclusion

This paper introduces a calibrated KGE ensemble learning method for link prediction. We have investigated and found that popular KGE models are indeed uncalibrated. That means score estimates associated to predicted triples are thus unreliable. To address this issue, we propose a pipeline of ensemble learning with KGEs that performs the probability calibration before combining the prediction of each model. Our method is formalized as a general framework that accepts to use with any calibration methods, and we demonstrate its application with the logistic regression. To fit our calibrator, we synthesize the negative triples following the standard protocol. Experiments on ConceptNet-100K confirm that significantly improvement can be obtained through ensemble learning under the calibration. Inspired by [25], we plan to further experiment the robustness with additional calibration algorithms, e.g., isotonic regression, beta calibration, and etc. The rationale is that different calibration methods could match better with the target distributions of the triples; hence, offering better explainability and/or interpretability on the outputs of ensemble learning with KGEs.

Acknowledgement. This work was supported by JSPS KAKENHI Grant Numbers JP22K18004.

A   Appendix: Preliminary Study on ConceptNet-100K

Table 4 shows our preliminary study with ConceptNet-100K. This table together with the results shown on the top part of Table 2 have inspired us to investigate an application of probability calibration on ensembling of KGEs.

Table 4. Preliminary results obtained from Pykeen under the default training

Model	Mean rank	Mean reciprocal rank
TransE	**4112.73**	**0.0279**
TransD	9877.45	0.0125
TransF	23081.42	4.03E-03
TransH	33950.49	2.00E-03
TransR	30769.40	2.19E-02
RotatE	13502.05	1.30E-02
TorusE	5086.65	0.0064
StructuredEmbedding	38003.16	9.95E+00
SimplE	16781.44	1.80E-03
RGCN	8256.60	4.44E-05
RESCAL	**4081.01**	**0.078**
QuatE	5751.13	1.00E-04
ProjE	8419.89	0.00122
PairRE	5278.03	1.00E-03
MuRE	38645.65	6.06E-05
HolE	20254.22	1.00E-04
ERMLPE	28110.41	3.70E-03
ERMLP	41002.39	2.00E-04
DistMult	**3873.39**	**0.051**
DistMA	9037.09	3.00E-04
CrossE	8963.77	6.74E-05
ConvE	**3711.98**	**0.0938**
ComplEx	**3915.91**	**0.0748**

References

1. Ali, M., et al.: PyKEEN 1.0: a Python library for training and evaluating knowledge graph embeddings. J. Mach. Learn. Res. **22**(82), 1–6 (2021)
2. Baader, F., Calvanese, D., McGuinness, D., Patel-Schneider, P., Nardi, D.: The Description Logic Handbook: Theory, Implementation and Applications. Cambridge University Press, Cambridge (2003)
3. Bollacker, K., Evans, C., Paritosh, P., Sturge, T., Taylor, J.: Freebase: a collaboratively created graph database for structuring human knowledge. In: Proceedings of the 2008 ACM SIGMOD International Conference on Management of Data, pp. 1247–1250 (2008)
4. Bordes, A., Usunier, N., Garcia-Duran, A., Weston, J., Yakhnenko, O.: Translating embeddings for modeling multi-relational data. In: Advances in Neural Information Processing Systems, vol. 26 (2013)
5. Cambria, E., Olsher, D., Rajagopal, D.: SenticNet 3: a common and common-sense knowledge base for cognition-driven sentiment analysis. In: Twenty-Eighth AAAI Conference on Artificial Intelligence (2014)
6. Dettmers, T., Minervini, P., Stenetorp, P., Riedel, S.: Convolutional 2D knowledge graph embeddings. In: Proceedings of the AAAI Conference on Artificial Intelligence, vol. 32 (2018)
7. Joshi, U., Urbani, J.: Ensemble-based fact classification with knowledge graph embeddings. In: Groth, P., et al. (eds.) The Semantic Web: 19th International Conference, ESWC 2022, Hersonissos, Crete, Greece, 29 May–2 June 2022, Proceedings, pp. 147–164. Springer, Cham (2022). https://doi.org/10.1007/978-3-031-06981-9_9
8. Krompaß, D., Tresp, V.: Ensemble solutions for link-prediction in knowledge graphs. In: PKDD ECML 2nd Workshop on Linked Data for Knowledge Discovery (2015)
9. Kull, M., Silva Filho, T.M., Flach, P.: Beyond sigmoids: how to obtain well-calibrated probabilities from binary classifiers with beta calibration (2017)
10. Lehmann, J., et al.: DBpedia-a large-scale, multilingual knowledge base extracted from Wikipedia. Semant. Web **6**(2), 167–195 (2015)
11. Lenat, D.B.: CYC: a large-scale investment in knowledge infrastructure. Commun. ACM **38**(11), 33–38 (1995)
12. Li, X., Taheri, A., Tu, L., Gimpel, K.: Commonsense knowledge base completion. In: Proceedings of the 54th Annual Meeting of the Association for Computational Linguistics (Volume 1: Long Papers), pp. 1445–1455 (2016)
13. Lin, X., Quan, Z., Wang, Z.J., Ma, T., Zeng, X.: KGNN: knowledge graph neural network for drug-drug interaction prediction. In: IJCAI, vol. 380, pp. 2739–2745 (2020)
14. Miller, G.A.: WordNet: a lexical database for English. Commun. ACM **38**(11), 39–41 (1995)
15. Nickel, M., Murphy, K., Tresp, V., Gabrilovich, E.: A review of relational machine learning for knowledge graphs. Proc. IEEE **104**(1), 11–33 (2015)
16. Nickel, M., Tresp, V., Kriegel, H.P.: A three-way model for collective learning on multi-relational data. In: ICML (2011)
17. Niculescu-Mizil, A., Caruana, R.: Predicting good probabilities with supervised learning. In: Proceedings of the 22nd International Conference on Machine Learning, pp. 625–632 (2005)

18. Noy, N., Gao, Y., Jain, A., Narayanan, A., Patterson, A., Taylor, J.: Industry-scale knowledge graphs: lessons and challenges: five diverse technology companies show how it's done. Queue **17**(2), 48–75 (2019)

19. Reiter, R.: On closed world data bases. In: Readings in Artificial Intelligence, pp. 119–140. Elsevier (1981)

20. Rivas-Barragan, D., Domingo-Fernández, D., Gadiya, Y., Healey, D.: Ensembles of knowledge graph embedding models improve predictions for drug discovery. Briefings Bioinf. **23**(6), bbac481 (2022)

21. Shi, B., Weninger, T.: Open-world knowledge graph completion. In: Proceedings of the AAAI Conference on Artificial Intelligence, vol. 32 (2018)

22. Singh, P., Lin, T., Mueller, E.T., Lim, G., Perkins, T., Li Zhu, W.: Open mind common sense: knowledge acquisition from the general public. In: Meersman, R., Tari, Z. (eds.) OTM 2002. LNCS, vol. 2519, pp. 1223–1237. Springer, Heidelberg (2002). https://doi.org/10.1007/3-540-36124-3_77

23. Speer, R., Chin, J., Havasi, C.: ConceptNet 5.5: an open multilingual graph of general knowledge. In: Thirty-First AAAI Conference on Artificial Intelligence (2017)

24. Suchanek, F.M., Kasneci, G., Weikum, G.: Yago: a core of semantic knowledge. In: Proceedings of the 16th International Conference on World Wide Web, pp. 697–706 (2007)

25. Tabacof, P., Costabello, L.: Probability calibration for knowledge graph embedding models. arXiv preprint arXiv:1912.10000 (2019)

26. Tandon, N., De Melo, G., Suchanek, F., Weikum, G.: WebChild: harvesting and organizing commonsense knowledge from the web. In: Proceedings of the 7th ACM International Conference on Web Search and Data Mining, pp. 523–532 (2014)

27. Trouillon, T., Welbl, J., Riedel, S., Gaussier, É., Bouchard, G.: Complex embeddings for simple link prediction. In: International Conference on Machine Learning, pp. 2071–2080. PMLR (2016)

28. Wan, G., Du, B., Pan, S., Wu, J.: Adaptive knowledge subgraph ensemble for robust and trustworthy knowledge graph completion. World Wide Web **23**, 471–490 (2020)

29. Wang, Q., Mao, Z., Wang, B., Guo, L.: Knowledge graph embedding: a survey of approaches and applications. IEEE Trans. Knowl. Data Eng. **29**(12), 2724–2743 (2017)

30. Wang, Z., Zhang, J., Feng, J., Chen, Z.: Knowledge graph embedding by translating on hyperplanes. In: Proceedings of the AAAI Conference on Artificial Intelligence, vol. 28 (2014)

31. Wang, Z., Chen, T., Ren, J., Yu, W., Cheng, H., Lin, L.: Deep reasoning with knowledge graph for social relationship understanding. arXiv preprint arXiv:1807.00504 (2018)

32. Xiong, C., Power, R., Callan, J.: Explicit semantic ranking for academic search via knowledge graph embedding. In: Proceedings of the 26th International Conference on World Wide Web, pp. 1271–1279 (2017)

33. Yang, B., Yih, W.T., He, X., Gao, J., Deng, L.: Embedding entities and relations for learning and inference in knowledge bases. arXiv preprint arXiv:1412.6575 (2014)

34. Zang, L.J., Cao, C., Cao, Y.N., Wu, Y.M., Cun-Gen, C.: A survey of commonsense knowledge acquisition. J. Comput. Sci. Technol. **28**(4), 689–719 (2013)

35. Zhou, Z.H.: Ensemble learning. In: Li, S.Z., Jain, A. (eds.) Encyclopedia of Biometrics. Springer, Boston (2021). https://doi.org/10.1007/978-0-387-73003-5_293

Multi-criteria Approaches to Explaining Black Box Machine Learning Models

Jerzy Stefanowski[✉]

Poznan University of Technology, Institute of Computer Science, Poznań, Poland
jerzy.stefanowski@cs.put.poznan.pl

Abstract. The adoption of machine learning algorithms, especially in critical domains often encounters obstacles related to the lack of their interpretability. In this paper we discuss the methods producing local explanations being either counterfactuals or rules. However, choosing the most appropriate explanation method and one of the generated explanations is not an easy task. Instead of producing only a single explanation, the creation of a set of diverse solutions by a specialized ensemble of explanation methods is proposed. Large sets of these explanations are filtered out by using the dominance relation. Then, the most compromise explanations are searched with a multi-criteria selection method. The usefulness of these approaches is shown in two experimental studies carried out with counterfactuals or rule explanations, respectively.

Keywords: explaining ML models · counterfactual explanations · rules · ensemble of explainers

1 Introduction

Although Artificial intelligence (AI) and especially machine learning (ML), have made tremendous progress in recent years, many problems, including the development of explainable AI methods (XAI), remain open and challenging [12]. In particular the current successful machine learning systems are usually based on complex, *black-box* models and they do not provide direct information about the system's internal logic or how the predictions for considered cases are produced [18]. The ability to provide *explanations* in a human-understandable way would support the verification of the ML systems, help to identify potential biases in data or algorithms, enable the assessment of causes of incorrect decisions and the possibility of correcting them. Finally, it could improve human trust in AI systems and their acceptance [26,28].

Up to now numerous methods producing different kinds of explanations have been introduced, for some reviews see, e.g. [2,3,12,18]. However, applying various methods, although restricted to the same form of representation often provide quite different explanations of the single predicted instance. Which puts even experienced recipients in an ambiguous decision-making situation, especially that the assessment of explanations from the point of view of various quality measures usually leads to contradictory conclusions.

© The Author(s), under exclusive license to Springer Nature Singapore Pte Ltd. 2023
N. T. Nguyen et al. (Eds.): ACIIDS 2023, LNAI 13996, pp. 195–208, 2023.
https://doi.org/10.1007/978-981-99-5837-5_17

In this paper we limit our interest to methods designed for either counterfactual explanations or for rules. For instance, a counterfactual is expected to be a similar example for which the model prediction will be changed to more desired one [32]. Usually, a single prediction can be explained by many different counterfactuals generated by various methods. Several properties and quality measures are considered to evaluate these provided counterfactuals [10]. As the popular measures usually present contradictory views on a possible explanation, it naturally leads us to exploiting a *multiple criteria decision analysis* for comparing alternative solutions and aiding human while selecting the most compromise one.

The contributions of this paper are as follows: Firstly, the discussion of various evaluation measures for a given form of explanations is provided. Secondly, instead of proposing yet another method for generating counterfactuals or rules, we claim that the already existing methods should be sufficient to provide a diversified set of explanations. As a result we propose to use an ensemble of multiple explanation methods (explainers) to provide a richer set of explanations, each of which establishes a certain trade-off between values of different quality measures (criteria). Then, the dominance relation between pairs of explanations is used in order to construct their Pareto front and filter out dominated explanations. The final explanation is selected from this front by applying one of the multiple criteria choice methods. These considerations are illustrated by experiments performed independently for counterfactuals and rules.

2   Expected Properties of Local Explanations

Due to page limits we will briefly present some basic concepts of XAI only, for more details see e.g. [2,3,18]. For the purposes of this work, we refer to the following definition of *explanation of ML models* or XAI: *Explainable-AI explores and investigates methods to produce or complement AI models to make accessible and interpretable the internal logic and the outcome of the algorithms, making such process understandable by humans* [12]. This usually opens a discussion on practical implementations of the following research questions:

- What should be explained? the instance prediction, the model, the relation of outputs to changes in the training data, or details of the training procedure?
- What form of input and output information should be considered.
- How should we represent the explanation?
- To whom the explanation should be directed - the different forms are expected depending on the type of the users (a ML engineer, top expert, simple user or auditor, ...) and their background knowledge?
- What kind of an interaction with users is expected: static or active scenarios?

Many of these questions are still open challenges. Popular XAI methods are categorized in different ways. Here, we follow [12,18] and generally distinguish between *global* and *local* ones. The global explanations refer to studying more comprehensive interaction between input attributes and output, target variables for the complete data set, while the local explanation attempt to explain the reasons for a single predicted instance. Then, a distinction is made between solutions designed for a specific black box model and *model agnostic approaches* that can be used for different classifiers.

The authors of the paper [4] proposed three levels of the evaluation and strongly promote the role of human in it. If no humans can be included in it, then for some interpretable representations proxy tasks and measures can be used. Miller in his survey on how humans see "good" explanations [17] pointed out, among other issues, the following characteristics:

- *Constructivism* – humans usually ask why this prediction was made instead of another predictions, i.e. people may think in counterfactual cases;
- *Selectivity* – people prefer to receive few main causes of the prediction instead of the complete list of all ones;
- *Focus on the abnormal* – people focus more on abnormal cause to explain the event as their elimination more likely will change the outcome (it also refers to counterfactual faithfulness);
- *Trustful* – good explanations should predict the event as trustfully as possible and the explanation must make sense (*plausible*) and be suitable to predictions of other instances;
- *Generality* – good explanations can explain many instances.

More specific properties of local explanations are the following [25]: *Fidelity* – it is associated with how well the explanation approximates the prediction of the black box model. *Consistency* – how much does an explanation differ between two different prediction models that have been trained on the same task: *Stability* – it represent how similar are the explanations for similar instances while using the same method: *Representativeness* – it describes how many additional instances are covered by the explanation; *Importance* – whether the explanation reflects the most important features or other parts of instance description.

3   Counterfactual Explanations

3.1   Basics

Counterfactual explanations (briefly *counterfactual*) x' provide information about how some attribute values in the description of example x should be changed to achieve the other desired prediction of the black box model to y'. For instance, consider a case where someone applies for a loan and gets rejected by the system. A counterfactual explanation of this decision provides the information on minimal changes of attributes' values that will alter the decision of the system to accept this loan application instead of rejecting it.

This kind of a local explanation is appreciated, as counterfactuals are quite intuitive and may indicate to people what to do in order to achieve the desired outcome. The user could better understand how ML system works by considering scenarios "what would have happened if ...". Psychological justifications of the importance of counterfactuals in human cognition are also discussed in [15].

The works on human's understanding of the counterfactuals showed that their properties such as *valid* (really changing predictions to the desired one), *sparsity* (they change as few attributes as possible), *proximity* (the closet change in x' to the classified instance x) are expected. It is also postulated that counterfactuals need to be *distributionally faithful* i.e. they should be located in positions in the attribute space ensured their plausibility (as some methods may produce a counterfactual being out-of-distribution, and unrealistic in this space).

It is also strongly claimed that changes in the description of example x' should be limited to, so called, *actionable* attributes only, while some other attributes should be frozen (see, e.g., gender of the person, race, etc.) However, not all methods for generating counterfactual explanations are handling these action-ability constraints. In [6] a more extended set of constraints has been studied, i.e. *monocity* - indicates the preferred direction of attribute value changes (e.g. only an increase of the age is allowed), *one-hot enconding* – in case of nominal attributes changes of encoding to arbitrary real numbers are prevented and a change should preserve an appropriate zero-one encoding, *range of attribute value change* – it expresses the allowed range of changes to obtain realistic examples.

The reader is referred to [31] for a comprehensive review of algorithms used for generating counterfactual explanations. They could be categorized in main classes depending on the kind of exploiting methodological principles [12], i.e. instance-based explainers, tree approaches, optimization-based approaches or specialized heuristic search. The third category contains many popular algorithms that optimize a certain loss function, which often balances the *validity* of prediction y' with the *proximity* of counterfactual x' to instance x.

3.2   Quality Measures

The discussed properties of local explanations (Sect. 2) and characteristics of counterfactual ones (Subsect. 3.1) lead us to defining different measures evaluating of counterfactuals (see below list). The most of them exploit the *distance* between two examples $x, y \in X$ (or an example x and its counterfactual x'), where X denotes the training data.

- *Proximity(x,x')* is a *distance(x,x')*,
- *Sparsity(x,x')* is a number of attributes changed in x to get x',
- *Feasibility(x)*, also called *Implausibility*, is a distance between generated x' its k nearest real neighbors from X.
- *Discriminative Power(x')* is the ratio of the k nearest neighbors of x' which are labeled by y' to all neighbours (a reclassification rate of its k nearest neighbors),
- *Stability(x')* - similar instances x and z with the same predictions should have the similar explanations and the same predictions $b(x') = b(z') = y'$.

Independently each counterfactual should be verified whether its description satisfies the considered actionability constraints. Furthermore several other measures can be defined for a set of counterfactuals (if a method can generate several explanations), see [6,13]. The *diversity* of selected counterfactuals is the most often exploited.

3.3   Ensemble of Explanation Methods

Let us recall the experimental studies, such as e.g. [6] showing that the currently available algorithms may produce considerably different counterfactuals for the single instance that is classified by the black box model. While evaluating them with respect to the aforementioned quality measures, it should be noted that their values may be contradictory. For example, the optimization of the high proximity may result in finding the counterfactual being too close to the decision boundary. On the other hand, working with the discrimination power measure should favour explanations located more in safer groups of examples labelled by the desired class. Thus, looking for a single counterfactual which could optimize all such considered criteria is difficult. Moreover one could rather expect that several alternative counterfactual explanations will present different trade-offs between values of these criteria.

Therefore instead of looking for this single explanation method optimizing the set of criteria, we advocate for providing a richer set of diverse counterfactuals for the given instance, which could be obtained by means of an *ensemble* of different explanation methods (briefly *explainers*). This is inspired by good experiences of using multiple classifiers to boost predictive accuracy.

As counterfactuals should be diversified, they should be generated by relatively diversified explainers. Here we promote to consider methods based on completely different background, having positive literature recommendations and corresponding to stable open source implementation.

Furthermore their final selection could be done with respect to multi-criteria analysis with special filtering - what is described below:

1. Use an ensemble of different explainers to generate a diverse set of counterfactuals for a considered instance.
2. Verify whether each explanation changes the prediction of the black box model to the desired values; and reject explanations that fail this test.
3. Verify whether an explanation satisfies actionability constraints and reject ones not satisfying them.
4. Given selected multiple criteria, use the dominance relation to skip explanations dominated by others.
5. If the number of remaining counterfactual is still too high, use a multi-criteria choice method to support the user in finding the most compromise solution.

Recall that the *dominance relation* for two alternatives x and y means that the solution x is rated no worse on each criterion than the alternative y and it is better on at least one of them. The set of all non-dominated solutions constitutes the *Pareto front* of diverse explanations with respect to these considered criteria. Exploitation of these solutions restricts the decision maker's attention to the most efficient ones and allows to look for tradeoffs between criteria [5].

3.4   Multi-criteria Analysis of Counterfactuals

If the number of explanations generated by the methods within the ensemble is still high enough, the user should be supported in the analysis of this set with methods from the *multiple criteria decision analysis* (MCDA). To this aim we may consider adaptations of many MCDA methods developed in the last decades [14]. For the problem under consideration, we suggest using one of the three possible approaches:

1. An aggregation of all criteria to the single objective function.
2. A method of the reference point.
3. Dialogue methods.

The construction of the single objective function is one of the paradigms of MCDA, where the decision maker's preferences are used to model importance of criteria. Quite often an additive weight sum is exploited, however it requires additional techniques for eliciting the decision maker preferences to the trade-offs between criteria and/or estimating weight values. In case of some related explanation methods it is already realized by the composite loss function, such as Wachter's two criteria loss (the weighted sum of the validity and proximity) in the optimization approaches to look for counterfactuals [32]. It could be used to select a required subset of k explanations coming from the ensemble of explainers as an extension of two-criteria selection function (diversity of this set and proximity) [13]. However these proposals require special optimization methods for tuning weights, which may be challenging when the number of counterfactuals is too limited as a result of the filtering step in the ensemble.

The second approaches built a reference point in the criteria space, e.g. the *ideal point* having the best possible values for every criterion, and then to look for the nearest solution from the Pareto front with respect to a chosen distance measure. The Ideal Point method is recommended for the case of equally important criteria [5], which is more realistic in case of the Pareto front defined on the criteria values, being not so intuitive for humans as the original attributes.

The third group of approaches covers interactive scenarios of the dialogue with the decision maker, realized in the following steps: (1) A presentation to the decision maker the acceptable number of explanations from the Pareto Front – they could be sampled from all non-dominated explanations and are evenly distributed on the front; (2) The decision maker selects the most interesting ones, which potentially covers a smaller range of the Pareto Front. (3) In the next iterations, this zoomed range is a basis of generating the next m alternatives to be presented to the decision maker. The procedure is repeated until he/she will be satisfied with the sufficiently smaller number of alternatives or finally selects the one compromise explanation.

The above approaches definitely differ with respect to exploitation of information on relative preferences for criteria, which coul be difficult to obtain depending on the possibility of cooperation with decision makers.

3.5   Experiments with Selecting Counterfactuals

To show the usefulness of the proposed multi-criteria approach we carried our an experimental study with 9 methods for generating counterfactuals (Dice, Cadex,

Fimap, Wachter optimization, CEM, CFProto, Growing Spheres - GS, Action-ableRecourse - AR, FACE)[1] available in open source libraries and integrated in the project [29]. Where possible, generations of up to 20 counterfactuals were parametrized for each of the implemented methods. The obtained counterfactuals were evaluated with three criteria: *Proximity*, *Feasibility* (with the number of neighbors k equal to 3) and *Discrimination Power* (with $k = 9$). HEOM was used as a distance measure in each of these criteria. The neural network consisting of 3 linear layers, each of which consists of 128 neurons and ReLU activation function, was used to be a black box classifier. The approach was tested with two datasets Adult and German Credit, due to their frequent use in the evaluation of counterfactuals in the related literature. For both actionability immutable constraints were defined for sex, race or native country of the person.

Table 1. The average number of retrieved counterfactual explanations per instance in the test datasets, along with their standard deviation.

Dataset	All	Valid	Actionable	Pareto-Front
Adult	81.4 ± 2.8	67.1 ± 9.2	60.2 ± 8	7.6 ± 3.7
German	85 ± 3.8	66.5 ± 7	65.4 ± 7.3	11.9 ± 3.9

The summary of all steps of constructing the ensemble is presented in Table 1; where *Valid* and *Actionable* denote the number of counterfactual remaining after the filtering phases while *Pareto-Front* results of using the dominance relation.

It is clearly seen in Table 1 that the proposed ensemble of explainers produces, on average, over 80 counterfactual explanations per instance. Then a relatively large number of these explanations fail to meet the validity and actionability requirements, and are subsequently removed from further consideration. By using the dominance relation, we are able to identify a concise Pareto front, resulting in a sufficient number of alternatives (8–12). An additional analysis of the considered methods showed that all methods used in our ensemble may generate non-dominated counterfactuals, but they can make a varied contribution to the Pareto front (some methods are more effective [29]).

The non-dominated could be further exploited, e.g. by the ideal point method to select the single most compromise solution. The analysis presented in [29] was based on a comparative analysis of the final solution selected by the Ideal Point method. It showed that this MCDA approach has led to good values of additional evaluation measures where it was competitive with the solutions offered by the best of the single counterfactual generation method.

---

[1] The description of these methods can be found, e.g., in [10,31].

4   Rule Based Explanations

4.1   Rule Evaluation Measures

A fairly similar multi-criteria approach to the selection of explanations can also be applied to rules. Let us recall that rules are one of the popular form both for local and global explanations of ML models, see a survey [2]. They are also one of the oldest and intensively studied knowledge representation in artificial intelligence, machine learning and data mining [28,30].

Rules are typically represented as *if* <premise> *then* <conclusion> statements. The rule premise is composed of elementary conditions being some tests on attribute values, while the conclusion indicates the class assignment. We say that an example x satisfies a rule r, or r covers x, if it logically satisfies all the conditions of this rule.

The rules are already used to represent local explanations of predictions of ML black box models. ANCHOR seems to be the most known proposal [24], however there are many newer algorithms such as LORE [11], see e.g. their review in Chapter 4.2 of [2]. Furthermore rules can be also induced in a specialized way to provide a global explanation of MLP neural networks (e.g. TREPAN) or ensembles [2] or combined with Bayesian approaches to be a valid interpretable model for high stake decisions concerning humans [16].

Moreover, more numerous sets of rules could be used in XAI forming, the so-called, characteristic description of certain concepts being more global explanation [1]. Especially in the case of larger sets of rules, some of which may play an interchangeable role in relation to the coverage of similar examples in the training data [27]. So, the effective selection of a subset of rules or even a single rule is needed. This leads us to the multi-criteria selection methods discussed previously for contracts.

However, such individual rules are described by different quality measures, which we briefly discuss below. Generally speaking the evaluation of individual rules could be done with two categories of measures:

- The quantitative measures calculated on the basis of the relationships of rule parts to the training examples.
- The measures used to compare syntax of the condition parts of rules.

For the brevity of presentation, any rule r is denoted as $E \rightarrow H$. The most of the measures quantify the relationship between E and H, and the corresponding examples. These measures defined as functions of four non-negative values that can be gathered in a 2×2 contingency table (see Table 2).

For a particular data set, a is the number of examples that satisfy both the rule's premise and its conclusion, b is the number of learning examples for which only H is satisfied, etc. Based on the number of examples in this table several measures could be defined, see reviews in [7,8]. In Table 3 we list the ones which are valid for our discussion. The particular attention should be paid to a group of so called (*Bayesian*) *confirmation measures*. Their common feature is that they satisfy *the property of Bayesian confirmation* [30] (or simply: confirmation),

Table 2. An exemplary contingency table of the rule's premise and conclusion

	H	$\neg H$	Σ
E	a	c	$a + c$
$\neg E$	b	d	$b + d$
Σ	$a + b$	$c + d$	n

which can be regarded as an expectation that a measure obtains positive values
when the rule's premise increases the knowledge about the conclusion, zero when
the premise does not influence the conclusion at all, and finally, negative values
when the premise has a negative impact on the conclusion.

Table 3. Basic rule evaluation measures

Support	$sup(H, E) = a$ or $sup(H, E) = a/n$
Confidence	$conf(H, E) = a/(a + c)$
Change of confidence $D(H, E)$	$P(H\|E) - P(H) = \frac{a}{a+c} - \frac{a+b}{n} = \frac{ad-bc}{n(a+c)}$
Confirmation $M(H, E)$	$P(E\|H) - P(E) = \frac{a}{a+b} - \frac{a+c}{n} = \frac{ad-bc}{n(a+b)}$
Confirmation $S(H, E)$	$P(H\|E) - P(H\|\neg E) = \frac{a}{a+c} - \frac{b}{b+d} = \frac{ad-bc}{(a+c)(b+d)}$
Confirmation $N(H, E)$	$P(E\|H) - P(E\|\neg H) = \frac{a}{a+b} - \frac{c}{c+d} = \frac{ad-bc}{(a+b)(c+d)}$
Confirmation $C(H, E)$	$P(E \wedge H) - P(E)P(H) = \frac{a}{n} - \frac{(a+c)(a+b)}{n^2} = \frac{ad-bc}{n^2}$
Confirmation $F(H, E)$	$\frac{P(E\|H)-P(E\|\neg H)}{P(E\|H)+P(E\|\neg H)} = \frac{\frac{a}{a+b} - \frac{c}{c+d}}{\frac{a}{a+b} + \frac{c}{c+d}} = \frac{ad-bc}{ad+bc+2ac}$

Yet other measures for rule explanations concern such more general properties which may also correspond to general properties of good explanations:

- A *simplicity* of the explanation, which could be expressed as the number of
  conditions in the rule premise $Length(r)$.
- *Importance* – being the number of the most important attributes occurring in
  the condition part of the rule; It should be based on background knowledge
  on the role of attributes and it is defined as the relative ratio of the number
  of such attributes to all attributes in the condition part of rule r.

To consider other properties of rule such as their diversity or stability we
need to calculate the distance (or similarity) of condition parts of many rules. A
popular example of similarity or distance measures for a pair of two rules r_i and
r_j could be defined with Jaccard index of two sets of elementary conditions:

$$J(r_i, r_j) = \frac{|P(r_i) \cap P(r_j)|}{|P(r_i) \cup P(r_j)|}$$

where $P(r_i)$ and $P(r_j)$ are sets of the elementary conditions in rules r_i and r_j,
respectively; and IYI is a cardinality of the set Y. Such measures are also gen-

eralized for rules, where conditions are defined as subset of values, in particular for numerical attributes (see e.g. [8]).

Table 4. Characteristics of selected BRACID rules for the minority class; sup*N or sup*S denote the considered function aggregating rule support and one of the rule confirmation measure

Data set	Pruning	#Rules	Avg.sup	Avg.conf	Avg.S	Avg.N
balance-scale	none	52	2.08	0.61	0.54	0.03
	sup*N	5	4.60	0.32	0.24	0.07
	sup*S	5	2.00	0.88	0.80	0.04
breast-cancer	none	77	3.36	0.71	0.42	0.03
	sup*N	8	10.13	0.74	0.46	0.10
	sup*S	8	9.13	0.82	0.54	0.09
cmc	none	354	6.59	0.72	0.50	0.02
	sup*N	35	18.57	0.65	0.43	0.05
	sup*S	35	12.69	0.78	0.56	0.03
haberman	none	122	6.05	0.72	0.46	0.06
	sup*N	12	12.25	0.78	0.55	0.14
	sup*S	12	9.42	0.90	0.66	0.11
hepatitis	none	66	7.42	0.99	0.82	0.23
	sup*N	7	12.57	1.00	0.86	0.39
	sup*S	7	12.57	1.00	0.86	0.39
transfusion	none	161	6.36	0.67	0.44	0.03
	sup*N	16	18.50	0.68	0.46	0.08
	sup*S	16	15.56	0.77	0.54	0.07
yeast	none	155	7.432	0.905	0.875	0.145
	sup*N	16	10.689	0.915	0.877	0.209
	sup*S	16	8.875	0.944	0.915	0.174

If the considered set of rules is too large, the multi-criteria approach presented in the previous section can be used. However, it is not necessary to construct an analogous ensemble, because many algorithms naturally create more numerous sets of rules. In addition, the filtering phases should be based on different principles, because the validity and actionability test are not adequate for the evaluation of rules. Rather, the possible similarity of single rules should be tested to avoid further processing of their duplicates.

Still, given the selected rule evaluation criteria, it is worth to apply the dominance relation to eliminate dominated rules and, consequently, to build a Pareto Front. Moreover, if confirmation measures are used as rule criteria, then rules with the negative values of these measures should be discarded. Further exploitation of the rules on this front can be carried out using one of the three previously presented groups of methods.

4.2   Experiments with Selecting a Subset of Rules

To illustrate the usefulness of the aggregated multicriteria function let us consider the experiences with interpreting rule sets induced from imbalanced data sets. We will consider BRACID, being the specialized rule induction algorithm, which addresses class imbalances in data [19,20]. More precisely, it induces a hybrid representation of more general rules and specialized ones corresponding to single examples with respect to their difficulty factors characteristic for the complex distribution of minority classes [22]. Although BRACID proved to be an accurate classifier for imbalanced datasets, its potentially high number of rules may lower the possibility of analyzing the rules by humans. To address it the authors of BRACID proposed a special post-pruning strategy [21].

This strategy could be extended by the evaluation of candidate rules, which exploits the aggregation single function being a product of few criteria, mainly of a *rule support* and $S(H, E)$ or $N(H, E)$ *confirmation measure*. Additionally it is integrated with a *weighted rule covering* algorithm [23] leading to obtaining diversified rules with respect to learning examples satisfying them.

The usefulness of this approach was experimentally studied over a dozen of benchmark imbalanced datasets [23]. Some of these results of averaged rule qualities are presented in Table 4. The general observations from these experiments showed that each variant of selected rules with this approach improved the considered rule measures. Besides increasing average values of $S(H, E)$ and $N(H, E)$ measures, the average rule supports for both minority and majority classes were higher. For instance, for *cmc* data the average rule supports increased from 6.59 to 18.57 examples in the minority class, and from 7.3 to 21.0 examples in the majority class. Classification performance of selected rules did not decrease much compared to the non-pruned set of all BRACID rules.

It is possible to further select individual rules using the Ideal Point method or dialogue approach discussed in the previous chapter. Independently, it is possible to apply the author's approach to assessing relative importance of the elementary conditions in the conditional parts of these selected rules, which is based on an adaptation of the Shapley indices, where the evaluation functions relate again to the support of the rule and the confirmation; for more details see [9,28].

5   Conclusions

An evaluation of explanations produced by various XAI methods and selecting the most appropriate one is still a challenging problem. In this paper the possibilities of using multi-criteria approaches to select either a subset or single explanations by means of multi-criteria approach are discussed.

As shown in the case of counterfactuals, the use of dominance relations alone (built with multiple criteria) can significantly reduce the sets of explanations leading to a small yet quite diverse set of alternatives. Their further analysis and selection of the best compromise explanation could be done by adopting one of three types of multi-criteria methods. The choice of one of them is strongly dependent on possible cooperation with experts to obtain preferential information from them about the resolving of the tradeoff between the values of the criteria or a reliable estimate of the weights in the aggregated functions. As for criteria defined with counterfactual quality measures obtaining dialogue-based preference information from people may not be so easy, we suggest starting with methods that do not require more advanced preferential information, such as the Ideal Point Method. Some of the rule evaluation measures are more naturally interpretable, which can facilitate the creation of slightly more complex multi-criteria analysis and selection scenarios.

Finally, the discussed methodology could be generalized for other types of explanations of predictions of black box ML models such as e.g. prototypes (see their presentation in [18]).

Acknowledgements. This research was partially supported by TAILOR, a project funded by EU Horizon 2020 research and innovation programme under GA No 952215. and by PUT Statutory Funds no 0311/SBAD/0740.

References

1. Alkhatib, A., Boström, H., Vazirgiannis, M.: Explaining predictions by characteristic rules. In: Amini, MR., Canu, S., Fischer, A., Guns, T., Kralj Novak, P., Tsoumakas, G. (eds.) Machine Learning and Knowledge Discovery in Databases. ECML PKDD 2022. Lecture Notes in Computer Science(), vol. 13713, pp. 389–403. Springer, Cham (2023). https://doi.org/10.1007/978-3-031-26387-3_24
2. Bodria, F., Giannotti, F., Guidotti, R., Naretto, F., Pedreschi, D., Rinzivillo, S.: Benchmarking and survey of explanation methods for black box models. Data Min Knowl Disc (2023). https://doi.org/10.1007/s10618-023-00933-9
3. Carvalho, D.V., Pereira, E.M., Cardoso, J.S.: Machine learning interpretability: a survey on methods and metrics. Electronics **8**(8), 832 (2019)
4. Doshi-Velez, F., Kim, B.: Towards a rigorous science of interpretable machine learning (2017). https://arxiv.org/abs/1702.08608
5. Ehrgott, M.: Multicriteria Optimization. Springer-Verlag, Cham (2005)
6. Falbogowski, M., Stefanowski, J., Trafas, Z., Wojciechowski, A.: The impact of using constraints on counterfactual explanations. In: Proceedings of the 3rd Polish Conference on Artificial Intelligence, PP-RAI 2022, pp. 81–84 (2022)
7. Fürnkranz, J., Gamberger, D., Lavrac, N.: Foundations of Rule Learning. Cognitive Technologies, Springer, Cham (2012)
8. Geng, L., Hamilton, H.: Interestingness measures for data mining: a survey. ACM Comput. Surv. **38**(3), 9 (2006)
9. Greco, S., Słowiński, R., Stefanowski, J.: Evaluating importance of conditions in the set of discovered rules. In: An, A., Stefanowski, J., Ramanna, S., Butz, C.J., Pedrycz, W., Wang, G. (eds.) RSFDGrC 2007. LNCS (LNAI), vol. 4482, pp. 314–321. Springer, Heidelberg (2007). https://doi.org/10.1007/978-3-540-72530-5_37
10. Guidotti, R.: Counterfactual explanations and how to find them: literature review and benchmarking. Data Min. Knowl. Disc., 1–55 (2022)

11. Guidotti, R., Monreale, A., Ruggieri, S., Pedreschi, D., Turini, F., Giannotti, F.: Local rule-based explanations of black box decision systems. arXiv preprint arXiv:1805.10820 (2018)
12. Guidotti, R., Monreale, A., Ruggieri, S., Turini, F., Giannotti, F., Pedreschi, D.: A survey of methods for explaining black box models. ACM Comput. Surv. (CSUR) **51**(5), 1–42 (2018)
13. Guidotti, R., Ruggieri, S.: Ensemble of counterfactual explainers. In: Soares, C., Torgo, L. (eds.) DS 2021. LNCS (LNAI), vol. 12986, pp. 358–368. Springer, Cham (2021). https://doi.org/10.1007/978-3-030-88942-5_28
14. Hwang, C.L., Yoon, K.: Methods for multiple attribute decision making. In: Multiple Attribute Decision Making. Lecture Notes in Economics and Mathematical Systems, vol. 186, pp. 58–191. Springer, Berlin (1981). https://doi.org/10.1007/978-3-642-48318-9_3
15. Keane, M.T., Kenny, E.M., Delaney, E., Smyth, B.: If only we had better counterfactual explanations: Five key deficits to rectify in the evaluation of counterfactual XAI techniques. In: Proceedings of the 30th International Joint Conference on Artificial Intelligence (IJACI-21), pp. 4466–4474 (2021)
16. Letham, B., Rudin, C., McCormick, T.H., Madigan, D., et al.: Interpretable classifiers using rules and Bayesian analysis: building a better stroke prediction model. Ann. Appl. Statist. **9**(3), 1350–1371 (2015)
17. Miller, T.: Explanation in artificial intelligence: insights from the social sciences. Artif. Intell. **267**, 1–38 (2019)
18. Molnar, C.: Interpretable machine learning (2019). https://christophm.github.io/interpretable-ml-book/
19. Naklicka, M., Stefanowski, J.: Two ways of extending Bracid rule-based classifiers for multi-class imbalanced data. In: Third International Workshop on Learning with Imbalanced Domains: Theory and Applications. Proceedings of Machine Learning Research, vol. 154, pp. 90–103. PMLR (2021)
20. Napierala, K., Stefanowski, J.: BRACID: a comprehensive approach to learning rules from imbalanced data. J. Intell. Inf. Syst. **39**(2), 335–373 (2012)
21. Napierala, K., Stefanowski, J.: Post-processing of BRACID rules induced from imbalanced data. Fund. Inform. **148**, 51–64 (2016)
22. Napierala, K., Stefanowski, J.: Types of minority class examples and their influence on learning classifiers from imbalanced data. J. Intell. Inf. Syst. **46**(3), 563–597 (2016)
23. Napierała, K., Stefanowski, J., Szczęch, I.: Increasing the interpretability of rules induced from imbalanced data by using Bayesian confirmation measures. In: Appice, A., Ceci, M., Loglisci, C., Masciari, E., Raś, Z.W. (eds.) NFMCP 2016. LNCS (LNAI), vol. 10312, pp. 84–98. Springer, Cham (2017). https://doi.org/10.1007/978-3-319-61461-8_6
24. Ribeiro, M.T., Singh, S., Guestrin, C.: Anchors: high-precision model-agnostic explanations. In: Proceedings of the Thirty-Second AAAI Conference on Artificial Intelligence, pp. 1527–1535 (2018)
25. Robnik-Šikonja, M., Bohanec, M.: Perturbation-based explanations of prediction models. In: Zhou, J., Chen, F. (eds.) Human and Machine Learning. HIS, pp. 159–175. Springer, Cham (2018). https://doi.org/10.1007/978-3-319-90403-0_9
26. Samek, W., Müller, K.-R.: Towards explainable artificial intelligence. In: Samek, W., Montavon, G., Vedaldi, A., Hansen, L.K., Müller, K.-R. (eds.) Explainable AI: Interpreting, Explaining and Visualizing Deep Learning. LNCS (LNAI), vol. 11700, pp. 5–22. Springer, Cham (2019). https://doi.org/10.1007/978-3-030-28954-6_1

27. Stefanowski, J., Krawiec, K., Wrembel, R.: Exploring complex and big data. Int. J. Appl. Math. Comput. Sci. **27**(4), 669–679 (2017)
28. Stefanowski, J., Wozniak, M.: Interpretation of models learned from complex medical data. In: Medical Informatics. Seria Inzynieria Biomedyczna, Podstawy i Zastosowania [in Polish], pp. 295–314. Academic Press Exit (2019)
29. Stepka, I., Lango, M., Stefanowski, J.: On usefulness of dominance relation for selecting counterfactuals from the ensemble of explainers. In: Proceedings of the 4rd Polish Conference on Artificial Intelligence, PP-RAI (2023)
30. Szczech, I., Susmaga, R., Brzezinski, D., Stefanowski, J.: Rule confirmation measures: Properties, visual analysis and applications. In: Greco, S., Mousseau, V., Stefanowski, J., Zopounidis, C. (eds.) Intelligent Decision Support Systems: Combining Operations Research and Artificial Intelligence-Essays in Honor of Roman Słowiński, pp. 401–423. Springer (2022). https://doi.org/10.1007/978-3-030-96318-7_20
31. Verma, S., Dickerson, J.P., Hines, K.E.: Counterfactual explanations for machine learning: a review. arXiv:abs/2010.10596 (2020)
32. Wachter, S., Mittelstadt, B., Russell, C.: Counterfactual explanations without opening the black box: automated decisions and the GDPR. Harv. JL Tech. **31**, 841 (2017)

Improving Loss Function for Polyp Detection Problem

Anh Tuan Tran[✉], Doan Sang Thai, Bao Anh Trinh, Bao Ngoc Vi,
and Ly Vu[iD]

Le Quy Don Technical University, Hanoi, Vietnam
tuantva86@gmail.com

Abstract. The utilization of automatic polyp detection during endoscopy procedures has been shown to be highly advantageous by decreasing the rate of missed detection by endoscopists. In this paper, we propose a new loss function for training an object detector based on the EfficientDet architecture to detect polyp areas in endoscopic images. The proposed loss combines the features of the Focal loss and DIoU (Distance Intersection over Union) loss named as Focal-DIoU. In addition, we have also carried out some experiments to evaluate the proposed loss function. The experimental results show that our proposed model achieves higher accuracy than previous works on two public datasets.

Keywords: deep learning · EfficientDet · Focal · DIoU · polyp detection

1 Introduction

Polyps are abnormal growths that can develop in various parts of the body, including the colon, stomach, and uterus. In the context of colon health, polyps can potentially develop into cancerous tumors over time [1]. The detection and removal of colorectal polyps are widely regarded as being most effectively accomplished through an endoscopy procedure. However, polyps can be missed in an endoscopy procedure for several reasons, such as, small size of polyps, low experience of doctors and procedure speed [2]. Therefore, computer-aided systems possess significant potential by reducing missed detection rate of polyps to prevent the development of colon cancer [1,2]. Particularly deep learning approaches have shown promising results in improving the accuracy and efficiency of detecting polyps in medical images.

Deep learning models, such as convolutional neural networks (CNNs), have been widely used in medical image analysis due to their ability to automatically learn meaningful features from raw data [12]. By training a CNN-based object detection model on a large dataset of medical images, the model can learn to detect and localize polyps accurately and efficiently. Object detection problems using deep learning approaches such as Regions with CNN features (R-CNN) [15], Fast R-CNN [16], Feature Pyramid Networks (FPN) [20], Single Shot Detector (SSD) [17], You Only Look Once (YOLO) [19], and EfficientDet [27] have gained popularity in recent years. Building upon this achievement,

© The Author(s), under exclusive license to Springer Nature Singapore Pte Ltd. 2023
N. T. Nguyen et al. (Eds.): ACIIDS 2023, LNAI 13996, pp. 209–220, 2023.
https://doi.org/10.1007/978-981-99-5837-5_18

deep learning models have found extensive application in the domain of medical image analysis [12]. Between them, EfficientDet is one of the most effective models for object detection [27].

The loss function for bounding box regression (BBR) is crucial for object detection, with the ln-norm loss being the most commonly used [29]. In BBR, there exists the imbalance problem in training samples. Specifically, the number of high-quality samples (anchor boxes) with small regression errors is much fewer than low-quality samples (outliers) due to the sparsity of target objects in images, e.g., polyps. This paper introduces a novel loss function, named as Focal-DIoU, which integrates the Focal loss and the DIoU loss. The proposed loss function aims to enhance the accuracy of small objects detection while concurrently addressing imbalances in class distribution. Therefore, the detector with the proposed loss function can work well with polyp detection. Moreover, the proposed loss can be easily incorporated into multiple detection models. The experimental results show that the Focal-DIoU loss can enhance the accuracy of the EfficientDet model for the polyp detection problem.

The contribution of our work is summarized as follows:

1. Introduce the Focal-DIoU loss function that handles the imbalance between foreground and background classes together with enhancing the localization of small objects.
2. Integrate the Focal-DIoU loss to train the EfficientDet model for polyp detection.
3. Conduct the various experiments on two well-known polyp datasets. The results show that our proposed loss can enhance the accuracy for the polyp detection problem.

The rest of this paper is organized as follows. Section 2 we first gather essential reviews about our topic. Section 3 presents the backbone EfficientDet and common BBR loss functions. In Sect. 4, we present the proposed method with the new loss function called Focal-DIoU and the object detection model for polyp detection. The experimental settings are provided in Sect. 5. After that, Sect. 6 presents the experimental results and discussions. Conclusions are discussed in Sect. 7.

2   Related Work

In recent years, there have been a number of studies proposed about deep learning-based approaches to object detection such as R-CNN [3], Fast R-CNN [16], FPN [20], SSD [17], YOLO [19], and EfficientDet [27]. There are two types of object detection models, i.e., a two-stage detector and a one-stage detector [10]. The two-stage detector includes a preprocessing step for generating object detection proposals and a detection step for identifying objects. The one-stage detector has an integrated process containing both above two steps.

Two-stage detectors, e.g., Mask R-CNN [22], consist of two separated modules, i.e., a region proposal network (RPN) [3] and a detection module. The RPN generates object proposals, which are then refined by the detection module. This two-stage approach has shown to achieve better accuracy than one-stage detectors, especially in detecting small objects and handling occlusion. However, the disadvantages of the two-stage framework are the requirement of large resources for computation.

To overcome the above shortcomings, one-stage detectors have been developed recently, e.g., YOLO [19], SSD [17], CenterNet [23] and EfficientDet [27] have a simple and efficient architecture that can detect objects in a single phase. These detectors use a feature pyramid to detect objects at different scales and employ anchor boxes to handle object variability. However, they often suffer from lower accuracy compared to two-stage networks, especially in detecting small objects and handling class imbalance. Recently, EfficientDet is one of the most effective object detection model due to using a compound scaling method to balance the model's depth, width, and resolution [24].

Different types of networks can be applied in medical object detection. An object detection algorithm could detect lesions automatically and assist diagnosis during the process of endoscopic examination. Hirasawa et al. [5] used SSD to diagnose the gastric cancer in chromoendoscopic images. The training dataset consisted of 13,584 images and the test dataset included 2,296 images from 77 gastric lesions in 69 patients. The SSD performed well to extract suspicious lesions and evaluate early gastric cancer. Wu et al. [7] proposed an object detection model named ENDOANGEL for real-time gastrointestinal endoscopic examination. ENDOANGEL has been utilized in many hospitals in China for assisting clinical diagnosis. Gao et al. [6] analyzed perigastric metastatic lymph nodes of computerized tomography (CT) images using Faster R-CNN. The results showed that the Faster R-CNN model has high judgment effectiveness and recognition accuracy for CT diagnosis of perigastric metastatic lymph nodes.

The loss function for bounding box regression (BBR) is crucial for object detection, with the ln-norm loss being the most commonly used [29]. However, it is not customized to adapt to the intersection over union (IoU) evaluation metric. The IoU loss is also used in the object detection models. However, the IoU loss will always be zero when two bounding boxes have no intersection [13]. Thus, the generalized IoU (GIoU) loss [25] was proposed to address the weaknesses of the IoU loss, i.e., Recently, the Complete IoU Los (CIoU) and the distance IoU (DIoU) [26] were proposed with faster convergence speed and better performance. However, above losses seem less effective with the imbalance training data. To handle this, we propose the new loss function that combines the IoU based loss, i.e., DIoU and the Focal loss [21] for the polyp detection problem.

3   Background

In this section, we first review the detector used in this paper, i.e., EfficientDet backbone. Second, we describe the mathematical computation of some related loss functions.

3.1   EfficientDet Architecture

EfficientDet [27] is a recent object detection architecture proposed by Tan et al. in 2020. It is based on the EfficientNet backbone [24], which is a family of efficient convolutional neural networks that achieves a state-of-the-art performance on image classification tasks.

One of the key features of EfficientNet is its use of a compound scaling method to balance the model's depth, width, and resolution. This method allows the model to achieve high accuracy with fewer parameters compared to other object detection architectures [24]. Based on EffiecientNet, EfficientDet can extract features from input images effectively. Moreover, EfficientDet employs the Bidirectional Feature Pyramid Network (BiFPN) module [27] to integrate features from multiple scales to improve the accuracy of the detection results. Efficient-Det has different versions labeled from D0 to D7 with increasing depth, width, and resolution. In this paper, we use the EfficientDet-D0 backbone which is the smallest and fastest version.

3.2   Loss Function

The loss function is an important component of an object detection model. It helps to guide the training process to enhance the accuracy for the object detection problem. Here, we introduce some common loss functions for the object detection problem.

IoU Loss: The Intersection over Union (IoU) is commonly used as a metric for evaluating the performance of object detection models. It can also be used as a loss function to optimize the model during training [18]. The IoU loss measures the similarity between the predicted bounding box and the ground truth bounding box. It is defined as follows:

$$L_{\text{IoU}} = 1 - IoU, \tag{1}$$

where IoU is the intersection over union between the predicted bounding box and the ground truth bounding box. However, the IoU loss has some weakness when measuring the similarity between two bounding boxes. It does not reflect the closeness between the bounding boxes correctly [13].

Smooth L1: The Smooth L1 loss was first proposed for training Fast R-CNN [16]. The Smooth L1 loss function is defined as follows:

$$L_{\text{Smooth L1}}(x) = \begin{cases} 0.5x^2, & \text{if } |x| < 1 \\ |x| - 0.5, & \text{otherwise} \end{cases} \tag{2}$$

where x is the difference between the predicted and the ground truth bounding boxes.

This loss function is widely used in popular object detection frameworks such as Faster R-CNN and Mask R-CNN because of the smoothness and robustness to outliers. However, similar to the IoU loss, it also do not consider the distance between the bounding boxes. Moreover, the Smooth L1 loss bias to larger bounding boxes.

DIoU Loss: The Distance-IoU (DIoU) loss [26] is proposed to directly minimize the normalized distance between predicted and ground truth bounding for achieving faster convergence. The DIoU loss takes into account the aspect ratio and diagonal distance of the predicted and ground truth bounding boxes. The loss penalizes the distance between the center points of the predicted and ground truth boxes as well as the difference between their diagonal lengths. The formula for DIoU loss can be represented as follows:

$$L_{\mathrm{DIoU}}(b, b^{gt}) = 1 - \mathrm{IoU} + \frac{d(b, b^{gt})^2}{c^2}, \tag{3}$$

where b and b^{gt} denote the central points the predicted and ground truth bounding boxes, respectively; $d(.)$ is the Euclidean distance and c is the diagonal length of the smallest enclosing box covering the bounding boxes. The DIoU loss has shown to be effective to improve the accuracy of object detection models, especially with small objects or many objects in one image [26].

Focal Loss: The Focal Loss is designed to address the imbalance between foreground and background classes during training of object detection models [21]. This loss tries to down-weight easy samples and thus focus training on negative samples. Let's define p is the probability estimated by the model for positive class. Then, we define $p_t = p$ for the positive class and $p_t = 1 - p$ for the negative class. The computation of this loss is as follows:

$$L_{\mathrm{Focal}}(p_t) = -(1 - p_t)^\gamma log(p_t), \tag{4}$$

where γ is the focusing parameter that smoothly adjusts the rate for down-weighting easy samples.

4 Methods

In this section, we present the new loss function named as Focal-DIoU to improve the performance of the DIoU loss. After that, we present the polyp detection model with the EfficientNet-B0 backbone trained by the proposed loss.

4.1 The Proposed Loss

The existence of BBR losses has some drawbacks when applying to the polyp detection problem. Firstly, inspired by the improvement of convergence speed as well as the performance, the IoU-based losses, such as IoU and DIoU, still do not solve the imbalance between high-quality and low-quality anchor boxes. In other hand, the other losses based on the Focal loss are successful in tackling the imbalance problem by increasing the contribution of high-quality boxes [13]. However, these losses only work well with medium or large objects which are not suitable for polyp detection. Therefore, to tackle these above problems, we propose the Focal-DIoU loss which combines the advantages of the Focal and DIoU loss to provide a more effective and robust loss function for training polyp detection models.

Firstly, the Focal-DIoU loss integrates the Focal loss, which is originally designed for addressing the issue of class imbalance in object detection. The idea behind the Focal loss is that it assigns different weights to different samples. Specifically, higher weights are assigned to samples that are miss-classified or hard to classify. This allows the Focal-DIoU to focus more on challenging samples, such as rare objects with larger errors, leading to better optimization and improved performance on imbalanced data.

Secondly, the Focal-DIoU loss also retains the advantages of DIoU, which considers both the aspect ratio and the distance between bounding boxes. By incorporating the distance penalty term, Focal-DIoU can effectively handle variations in object size and position, making it more robust to difference of object scales and object misalignment.

We integrate the DIoU and Focal loss by re-weighting DIoU by the value of IoU, then the Focal-DIoU is computed as below:

$$L_{\text{Focal-DIoU}} = -(1 - \text{IoU})^{\gamma} log(\text{IoU}) L_{\text{DIoU}}. \tag{5}$$

In this paper, we use the Focal-DIoU loss in Eq. 5 to train the detector for the polyp detection problem. The modulating factor $(1 - \text{IoU})^{\gamma}$ intuitively decreases the loss contribution from easy samples and expands the range in which a sample obtains a low loss.

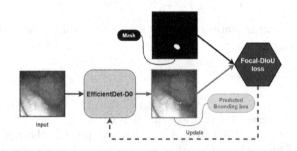

Fig. 1. Architecture of Proposed Model for Polyp Detection.

4.2 Proposed Model

As mentioned in Sect. 3.1, our proposed model is based on the EfficientDet-D0 model as introduced in Sect. 3. The input image is first resized to a fixed size (e.g., $512 \times 512 \times 3$). Then, it is passed through a backbone network EfficientNet-D0, which consists of multiple stages with different spatial resolutions. The backbone network is responsible for extracting multi-scale features from the input image.

As shown in Fig. 1, after passing through the backbone network, the output of the network is a prediction mask. The Focal-DIoU loss function is calculated based on the prediction mask and the ground truth mask with all input samples of a training batch size of the dataset. The value of loss function is used to optimize the weights of the EfficientDet-D0 network.

5 Experimental Settings

This section presents the datasets and the experimental settings used in this paper.

5.1 Datasets

This section presents two polyp datasets used in our experiments, i.e., Kvasir-SEG and CVC-ClinicDB dataset [30].

The Kvasir-SEG dataset [28] is the collection of 1,000 endoscopic images of the gastrointestinal tract, including the esophagus, stomach, duodenum, and colon, obtained from two Norwegian medical centers. The dataset contains images of different types of abnormalities, such as polyps, compression, bleeding, swelling, inflammation, white stool, tumors, and ulcers, with the resolution of $512 \times 512 \times 3$. The image samples are annotated by experts to indicate the location and type of abnormality present in each image.

The CVC-ClinicDB dataset [4] comprises 612 endoscopic images of the colon obtained from the Clinic Hospital of Barcelona in Spain. The dataset includes images of polyps, respectively, acquired using a linear endoscope and a convex endoscope. The images are annotated by experts to indicate the presence or absence of polyps.

In order to prepare the datasets for evaluating the proposed object detection model, we split the datasets into three subsets, i.e., training, validating, and testing, by the ratio as 8:1:1, respectively. The numbers of samples for training, testing, and validating are shown in Table 1.

We transform the mask images of the datasets into bounding boxes to fit with an object detection problem. Here, we apply a contour detection algorithm to the binary mask images to identify the boundaries of the objects in the images. The contours are then converted into rectangular bounding boxes that enclose the objects. This process is executed for each mask image in the datasets. The resulting bounding box annotations are used to train and evaluate our proposed object detection model.

Table 1. Dataset splitting.

Dataset	Total	Train	Test	Val
Kvasir-SEG	1000	800	100	100
CVC-ClinicDB	612	489	61	62

5.2   Parameter Settings

For each dataset, we employ two steps in the training phase. In the first step, the pre-trained EfficientDet-D0 model on the COCO dataset [14] is trained only on the last layer of the EfficientDet-D0 while freezing the rest layers. In the second step, we train all layers of the EfficientDet-D0 model and the early stopping is used to terminate the training process. In the first step, the learning rate, batch size, and the number of epochs are 0.005, 32, and 10, respectively. These values for the second step are 0.001, 16, and 200, respectively.

5.3   Experiment Setup

We conduct the experiments on a computing system with the following specifications: CPU Intel(R) Xeon(R) CPU@ 2.00 GHz, 16 GB of RAM, and a Tesla T4 GPU with 16 GB of VRAM. We use the Python programming language with the PyTorch library [8] to implement our proposed polyp detection model.

For comparison, we train the detectors with the same backbone, i.e., EfficientDet-D0 but using five different loss functions, i.e., Smooth L1 [9], IoU [18], CIoU [11], DIoU [26], and the proposed loss Focal-DIoU. These experiments are conducted on two different datasets, as mentioned in Sect. 5.1. We use COCO metrics [14] for evaluation in our experiments that are based on Average Precision (AP). Notice that, AP is averaged over all classes and we make no distinction between AP and mAP. AP is a performance evaluation metric used in object detection and recognition tasks. It calculates the model's accuracy in determining the location and classifying objects in an image. AR (Average Recall) is a similar performance evaluation metric that focuses on the model's coverage, i.e., its ability to detect all objects present in an image.

6   Results

6.1   Accuracy Comparison

As can be seen from Table 2, the detector with the Focal-DIoU loss achieves the highest AP as 0.637, with competitive performance in other metrics as well. The detector with Focal-DIoU generally outperforms those with the Smooth L1, IoU, CIoU, and DIoU loss in most of the evaluated metrics. On this dataset, the detector with IoU shows the worst performance. Notably, the detector with DIoU achieves the highest values in some specific metrics such as AP_{75} and

AP_L. However, the detector with Focal-DIoU remains the top-performing loss function in terms of overall performance with the highest AP score. The reason is that the proposed loss function helps the training detector by considering the small, medium, and the large size polyps. Thus, using our proposed loss function enhances the overall accuracy compared with the previous loss function, such as Smooth L1, IoU, CIoU, and DIoU on the Kvasir-SEG dataset for the polyp detection problem.

Table 2. The results of EfficientDet-D0 with different loss functions on Kvasir-SEG dataset.

$LossFunction$	AP	AP_{50}	AP_{75}	AP_L	AR	AR_{50}	AR_{75}	AR_L
Smooth L1	0.612	0.832	0.715	0.695	0.638	0.657	0.657	0.743
IoU	0.542	0.769	0.583	0.617	0.557	0.629	0.633	0.709
CIoU	0.619	**0.853**	0.709	0.702	0.637	0.672	0.672	0.755
DIoU	0.623	0.848	**0.727**	0.704	0.645	**0.691**	**0.691**	**0.772**
Focal-DIoU	**0.637**	0.850	0.680	**0.721**	**0.663**	0.679	0.679	0.764

Similarly, as is presented in Table 3, the detector with the Focal-DIoU loss achieves the highest AP as 0.790. The detector with the CIoU loss also delivers competitive results with the highest Recall as 50% IoU threshold (i.e., AP_{50}). Generally, the detectors with the Focal-DIoU and CIoU loss outperform others in almost all of the evaluated metrics. We can observe that Focal-DIoU achieves the highest values in most metrics except AR_{50} and AR_{50} where CIoU gets the highest score. Overall, the proposed loss function helps the detector for improving the accuracy for the polyp detection problem.

Table 3. The results of EfficientDet-D0 with different loss functions on CVC-ClinicDB dataset.

$LossFunction$	AP	AP_{50}	AP_{75}	AP_L	AR	AR_{50}	AR_{75}	AR_L
Smooth L1	0.773	0.993	0.879	0.753	0.787	0.829	0.829	0.817
IoU	0.773	0.957	0.909	0.773	0.783	0.808	0.808	0.813
CIoU	0.789	0.986	0.893	0.778	0.806	**0.835**	**0.835**	0.830
DIoU	0.776	0.949	0.886	0.767	0.790	0.808	0.808	0.803
Focal-DIoU	**0.790**	**0.996**	**0.926**	**0.805**	**0.814**	0.827	0.827	**0.840**

6.2   Visualization

To observe the results of the polyp detection visually, we show the detection results for several images of the experimental datasets in Table 4. This table shows that the detector with the Focal-DIoU loss helps to detect the polyp area more correctly in both experimental datasets. Especially, on the Kvasir-SEG dataset, the detector with the Focal-DIoU loss achieves more correctly polyp detection results compared with the detectors with other loss function. For the CVC-ClinicDB dataset, the detectors with loss functions achieves similar accuracy. Overall, the detector with the Focal-DIoU loss presents the best detection result even with very small and unevenly distributed polyps in the image.

Table 4. Visualization of polyp detection resulting from EfficientDet-D0 with different loss functions.

Dataset \ Loss func.	Original Image	Smooth L1	IoU	CIoU	DIoU	Focal-DIoU
Kvasir-SEG						
CVC-ClinicDB						

7   Summary

In this paper, we focus on studying loss functions to improve the accuracy and efficiency of the polyp detection model. We propose Focal-DIoU loss to train the effective detector, i.e., EfficientDet-D0 backbone for the polyp detection problem. The proposed loss function can help the detector consider small, medium, and large size of polyps in the training process. Thus, the detector enhances the accuracy to detect various sizes of polyps. The experimental results show that the detector with the proposed loss function achieves higher accuracy than the detectors with other loss functions for the polyp detection problem. This proves that the proposed loss function can help to improve the polyp detection problem.

References

1. Lee, S.H., et al.: An adequate level of training for technical competence in screening and diagnostic colonoscopy: a prospective multicenter evaluation of the learning curve. Gastrointest. Endosc. **67**(4), 683–689 (2008)
2. Leufkens, A., Van Oijen, M., Vleggaar, F., Siersema, P.: Factors influencing the miss rate of polyps in a back-to-back colonoscopy study. Endoscopy **44**(05), 470–475 (2012)
3. Girshick, R., Donahue, J., Darrell, T., Malik, J.: Region-based convolutional networks for accurate object detection and segmentation. IEEE Trans. Pattern Anal. Mach. Intell. **38**(1), 142–158 (2015)
4. Bernal, J., Sánchez, F.J., Fernández-Esparrach, G., Gil, D., Rodríguez, C., Vilariño, F.: WM-DOVA maps for accurate polyp highlighting in colonoscopy: validation vs. saliency maps from physicians. Comput. Med. Imaging Graph. **43**, 99–111 (2015)
5. Hirasawa, T., et al.: Application of artificial intelligence using a convolutional neural network for detecting gastric cancer in endoscopic images. Gastric Cancer **21**(4), 653–660 (2018). https://doi.org/10.1007/s10120-018-0793-2
6. Gao, Y., et al.: Deep neural network-assisted computed tomography diagnosis of metastatic lymph nodes from gastric cancer. Chin. Med. J. **132**(23), 2804–2811 (2019)
7. Wu, L., et al.: A deep neural network improves endoscopic detection of early gastric cancer without blind spots. Endoscopy **51**(06), 522–531 (2019)
8. Paszke, A., et al.: Pytorch: an imperative style, high-performance deep learning library. In: Advances in Neural Information Processing Systems, vol. 32 (2019)
9. Fu, C.Y., Shvets, M., Berg, A.C.: Retinamask: learning to predict masks improves state-of-the-art single-shot detection for free. arXiv preprint arXiv:1901.03353 (2019)
10. Du, L., Zhang, R., Wang, X.: Overview of two-stage object detection algorithms. J. Phys. Conf. Ser. **1544**(1), 012033 (2020)
11. Wang, X., Song, J.: ICIoU: improved loss based on complete intersection over union for bounding box regression. IEEE Access **9**, 105686–105695 (2021)
12. Puttagunta, M., Ravi, S.: Medical image analysis based on deep learning approach. Multimedia Tools Appl. **80**(16), 24365–24398 (2021). https://doi.org/10.1007/s11042-021-10707-4
13. Zhang, Y.F., Ren, W., Zhang, Z., Jia, Z., Wang, L., Tan, T.: Focal and efficient IOU loss for accurate bounding box regression. Neurocomputing **506**, 146–157 (2022)
14. Lin, T.-Y., et al.: Microsoft COCO: common objects in context. In: Fleet, D., Pajdla, T., Schiele, B., Tuytelaars, T. (eds.) ECCV 2014. LNCS, vol. 8693, pp. 740–755. Springer, Cham (2014). https://doi.org/10.1007/978-3-319-10602-1_48
15. Girshick, R., Donahue, J., Darrell, T., Malik, J.: Rich feature hierarchies for accurate object detection and semantic segmentation. In: Proceedings of the IEEE Conference on Computer Vision and Pattern Recognition, pp. 580–587 (2014)

16. Girshick, R.: Fast R-CNN. In: Proceedings of the IEEE International Conference on Computer Vision, pp. 1440–1448 (2015)

17. Liu, W., et al.: SSD: single shot MultiBox detector. In: Leibe, B., Matas, J., Sebe, N., Welling, M. (eds.) ECCV 2016. LNCS, vol. 9905, pp. 21–37. Springer, Cham (2016). https://doi.org/10.1007/978-3-319-46448-0_2

18. Yu, J., Jiang, Y., Wang, Z., Cao, Z., Huang, T.: Unitbox: an advanced object detection network. In: Proceedings of the 24th ACM International Conference on Multimedia, pp. 516–520 (2016)

19. Redmon, J., Divvala, S., Girshick, R., Farhadi, A.: You only look once: unified, real-time object detection. In: Proceedings of the IEEE Conference on Computer Vision and Pattern Recognition, pp. 779–788 (2016)

20. Lin, T.Y., Dollár, P., Girshick, R., He, K., Hariharan, B., Belongie, S.: Feature pyramid networks for object detection. In: Proceedings of the IEEE Conference on Computer Vision and Pattern Recognition, pp. 2117–2125 (2017)

21. Lin, T.Y., Goyal, P., Girshick, R., He, K., Dollár, P.: Focal loss for dense object detection. In: Proceedings of the IEEE International Conference on Computer Vision, pp. 2980–2988 (2017)

22. He, K., Gkioxari, G., Dollár, P., Girshick, R.: Mask R-CNN. In: Proceedings of the IEEE International Conference on Computer Vision, pp. 2961–2969 (2017)

23. Duan, K., Bai, S., Xie, L., Qi, H., Huang, Q., Tian, Q.: Centernet: keypoint triplets for object detection. In: Proceedings of the IEEE/CVF International Conference on Computer Vision, pp. 6569–6578 (2019)

24. Tan, M., Le, Q.: Efficientnet: rethinking model scaling for convolutional neural networks. In: International Conference on Machine Learning, pp. 6105–6114. PMLR (2019)

25. Rezatofighi, H., Tsoi, N., Gwak, J., Sadeghian, A., Reid, I., Savarese, S.: Generalized intersection over union: a metric and a loss for bounding box regression. In: Proceedings of the IEEE/CVF Conference on Computer Vision and Pattern Recognition, pp. 658–666 (2019)

26. Zheng, Z., Wang, P., Liu, W., Li, J., Ye, R., Ren, D.: Distance-IoU loss: faster and better learning for bounding box regression. In: Proceedings of the AAAI Conference on Artificial Intelligence, pp. 12993–13000 (2020)

27. Tan, M., Pang, R., Le, Q.V.: Efficientdet: scalable and efficient object detection. In: Proceedings of the IEEE/CVF Conference on Computer Vision and Pattern Recognition, pp. 10781–10790 (2020)

28. Jha, D., et al.: Kvasir-SEG: a segmented polyp dataset. In: Ro, Y.M., et al. (eds.) MMM 2020. LNCS, vol. 11962, pp. 451–462. Springer, Cham (2020). https://doi.org/10.1007/978-3-030-37734-2_37

29. Wang, Q., Cheng, J.: LCornerIoU: an improved IoU-based loss function for accurate bounding box regression. In: 2021 International Conference on Intelligent Computing, Automation and Systems (ICICAS), pp. 377–383. IEEE (2021)

30. Wang, F., Hong, W.: Polyp dataset (2022). https://doi.org/10.6084/m9.figshare.21221579.v2

Flow Plugin Network for Conditional Generation

Patryk Wielopolski[1]([⊠]) [iD], Michał Koperski[2], and Maciej Zięba[1,2] [iD]

[1] Faculty of Information and Communication Technology, Wroclaw University of
Science and Technology, Wybrzeze Wyspianskiego 27, 50-370 Wroclaw, Poland
{patryk.wielopolski,maciej.zieba}@pwr.edu.pl
[2] Tooploox Ltd., ul. Teczowa 7, 53-601 Wrocław, Poland
michal.koperski@tooploox.com

Abstract. Generative models have gained many researchers' attention
in the last years resulting in models such as StyleGAN for human face
generation or PointFlow for 3D point cloud generation. However, by
default, we cannot control its sampling process, i.e., we cannot generate a sample with a specific set of attributes. The current approach is
model retraining with additional inputs and different architecture, which
requires time and computational resources. We propose a novel approach
that enables generating objects with a given set of attributes without
retraining the base model. For this purpose, we utilize the normalizing
flow models - Conditional Masked Autoregressive Flow, and Conditional
Real NVP, as a Flow Plugin Network (FPN).

Keywords: Plugin networks · Normalizing Flows · Conditional Image
Generation · Deep generative models

1 Introduction

In the last years, generative models have achieved superior performance in object
generation with leading examples such as StyleGAN [8] for human face synthesis
and PointFlow [22] for 3D point cloud generation and reconstruction area. These
models perform exceptionally well in the case of the unconditional generation
process. However, by default, we cannot control its generating process, i.e., we
cannot out-of-the-box generate a sample with a specific set of attributes. To
perform such a conditional generation, we must put additional effort into creating
a new model with such functionality. In the case of images, specific solutions
were proposed, e.g., conditional modification of the well-known unconditional
generative models - Conditional Variational Autoencoders [20] and Conditional
Generative Adversarial Networks [14]. These approaches provide a good result in
conditional image generation; however, they require additional effort for training
and model design. The ideal solution would consist of a robust transformation
of the unconditional base model to a conditional one.

This work proposes a novel approach that enables a generation of objects
with a given set of attributes from an already trained unconditional generative

model (a base model) without its retraining. We assume that the base model has an autoencoder architecture, i.e., an encoder and a decoder network that enables the transformation of an object to the latent space representation and inverts that transformation, respectively. In our method, we utilize the concept of the plugin network [10], whose task is to incorporate information known during the inference time to the pretrained model and extend that approach to generative models. As the Flow Plugin Network backbone model, we use the conditional normalizing flow models - Conditional Masked Autoregressive Flow and Conditional Real NVP. Finally, we perform several experiments on three datasets: MNIST, ShapeNet, and CelebA, including conditional object generation, attribute manipulation, and classification tasks. The code for the method and experiments is publicly available.[1]

Concluding, our contributions are as follows:

- we propose a method for conditional object generation from already trained models with an autoencoder architecture, both generative and non-generative;
- we show that the proposed method could be used for non-generative variants of the autoencoder models, thus making them generative;
- we show that it also could be successfully used for classification tasks or as a tool for attribute manipulation.

The rest of the paper is organized as follows. In the next section, we describe the theoretical background. Afterward, we describe in detail our proposed method. In the following section, we shortly describe related works. Then we present the results of the experiments, and in the last section, we summarize our work and propose directions for future research.

2 Background

2.1 Autoencoders

Autoencoder [4] is a neural network primarily designed to learn an informative representation of the data by learning to reconstruct its input. Formally, we want to learn the functions $\mathcal{E}_\phi : \mathbb{R}^n \to \mathbb{R}^q$ and $\mathcal{D}_\theta : \mathbb{R}^q \to \mathbb{R}^n$ where ϕ and θ are parameters of the encoder and decoder. The model is usually trained by minimizing the reconstruction loss: $L_{rec} = \sum_{n=1}^{N} ||\mathbf{x}_n - \mathcal{D}_\theta \circ \mathcal{E}_\phi(\mathbf{x}_n)||_2^2$, assuming training $\mathcal{X}_N = \{\mathbf{x}_n\}$ is given. Some regularization terms can enrich the simple architecture of the autoencoder to obtain the generative model. For Variational Autoencoder (VAE) [9] models, it is achieved by enforcing the latent representation to be normally distributed using Kullback-Leibler divergence. For the adversarial autoencoders, [13], the assumed prior on the embeddings is obtained via adversarial training. With such extensions, the model is not only capable of representing the data on a low-dimensional manifold, but it is also capable of generating samples from an assumed prior in latent space.

---

[1] https://github.com/pfilo8/Flow-Plugin-Network-for-conditional-generation.

2.2   Normalizing Flows

Normalizing flows [18] represent the group of generative models that can be efficiently trained via direct likelihood estimation. They provide a general way of constructing flexible probability distributions over continuous random variables. Suppose we have a D-dimensional real vector $\mathbf{x}$, and we would like to define a joint distribution over $\mathbf{x}$. The main idea of flow-based modeling is to express $\mathbf{x}$ as a transformation T of a real vector $\mathbf{u}$ sampled from a base distribution $p_u(\mathbf{u})$ with a known density function:

$$\mathbf{x} = T(\mathbf{u}) \quad \text{where} \quad \mathbf{u} \sim p_u(\mathbf{u}) \tag{1}$$

It is common to chain multiple transformations $T_1, \ldots, T_K$ to obtain transformation $T = T_K \circ \cdots \circ T_1$, where each T_k transforms $\mathbf{z}_{k-1}$ into $\mathbf{z}_k$, assuming $\mathbf{z}_0 = \mathbf{u}$ and $\mathbf{z}_K = \mathbf{x}$. In practice we implement either T_k or T_k^{-1} using a neural network f_{ϕ_k} with parameters ϕ_k. Assuming given training data $\mathcal{X}_N = \{\mathbf{x}_n\}$ the parameters ϕ_k are estimated in the training procedure by minimizing the negative log-likelihood (NLL):

$$-\log p_{\mathbf{x}}(\mathbf{x}_n) = -\sum_{n=1}^{N} [\log p_{\mathbf{u}}(\mathbf{u}_n) - \sum_{k=1}^{K} \log |\det J_{T_k}(\mathbf{z}_{n,k-1})|], \tag{2}$$

where $p_u(\mathbf{u})$ is usually a Gaussian distribution with independent components. The most challenging aspect of training flow-based models is an optimization of $\log |\det J_{T_k}(\mathbf{z}_{k-1})|$, where the Jacobian $J_T(\mathbf{u})$ is the $D \times D$ matrix of all partial derivatives of T, that may be challenging to compute for high-dimensional data. That issue is usually solved by enforcing the Jacobian matrix to be triangular, for which the determinant is easy to calculate.

3   Flow Plugin Network

3.1   Overview

We consider the base pretrained model $\mathcal{M}$ with an autoencoder architecture, i.e., it has an encoder $\mathcal{E}$, a bottleneck with latent space $\mathcal{Z}$, and decoder $\mathcal{D}$, which was previously trained on dataset $\mathcal{X}$. We also assume we have a set of attributes $\mathcal{Y}$ for samples in the dataset $\mathcal{X}$. The main idea in the proposed method is to use a conditional normalizing flow model $\mathcal{F}$ to learn a bidirectional conditional mapping between a simple base distribution $\mathcal{N}$ (usually standard normal distribution) and a latent space representation which will be conditioned by some attributes (classes) from the set $\mathcal{Y}$. This approach will enable us to generate a random sample from the specific region of the latent space, which should correspond to object representation with a particular set of attributes. In the last step, using a decoder $\mathcal{D}$, we can generate an object with a requested set of attributes. The schema of the proposed method can be found in Fig. 1.

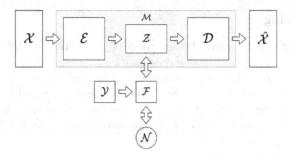

Fig. 1. High-level schema of the Flow Plugin Network method.

3.2 Training

We assume that we have a pretrained model $\mathcal{M}$, samples $\mathbf{x}$ from the dataset $\mathcal{X}$, and attributes $\mathbf{y}$ from the attribute set $\mathcal{Y}$. Our training objective is to learn a bidirectional mapping between a simple base distribution and a latent space representation. For that reason, we need to have a dataset consisting of pairs $(\mathbf{z}_i, \mathbf{y}_i)$ where $\mathbf{z}_i$ is a latent space representation of the vector $\mathbf{x}_i$. In the first step, we obtain such a dataset by encoding vectors $\mathbf{x}_i$ to vectors $\mathbf{z}_i$ using encoder $\mathcal{E}$. In the second step, we train conditional normalizing flow $\mathcal{F}$ using obtained pairs $(\mathbf{z}_i, \mathbf{y}_i)$, by minimizing the conditional negative log-likelihood function:

$$-\sum_{i=1}^{N} \log p(\mathbf{z}_i|\mathbf{y}_i) = -\sum_{i=1}^{N} \left(\log p_u(T^{-1}(\mathbf{z}_i|\mathbf{y}_i)) + \log|\det J_{T^{-1}}(\mathbf{z}_i|\mathbf{y}_i)|\right). \quad (3)$$

It is essential to highlight that weights of the base model $\mathcal{M}$ are frozen during training, and only parameters of transformation T are optimized.

As plugin networks, we use a conditional version of the normalizing flows. For MAF [17], the conditioning component is concatenated to the inputs to mean and log standard deviation functions:

$$\mu_i = f_{\mu_i}(\mathbf{x}_{1:i-1}, \mathbf{y}), \quad \alpha_i = f_{\alpha_i}(\mathbf{x}_{1:i-1}, \mathbf{y}) \quad (4)$$

For RealNVP [6] model, the same operation is applied to scaling and shifting functions for each of the coupling layers:

$$\mathbf{x}_{1:d} = \mathbf{u}_{1:d} \quad (5)$$
$$\mathbf{x}_{d+1:D} = \mathbf{u}_{d+1:D} \odot \exp\left(\mathbf{s}(\mathbf{u}_{1:d}, \mathbf{y})\right) + \mathbf{t}(\mathbf{u}_{1:d}, \mathbf{y}), \quad (6)$$

3.3 Conditional Object Generation

In this section, we will present how the proposed approach can be applied to conditional image generation. First, we generate a sample from the base distribution $\mathcal{N}$ and construct a vector that encodes the attributes $\mathbf{y}$ of the desired object that will be generated. Then, the generated sample is passed via the flow conditioned on the vector of attributes embeddings to obtain a vector in the latent space $\mathcal{Z}$ of the base model $\mathcal{M}$. After that, we process that vector to the decoder and obtain the requested object.

3.4   Attribute Manipulation

We could also use the proposed approach for object attribute manipulation, e.g., we may want to change the color of the person's hair on the image. First, to perform such an operation, we need to encode an input object to a latent space representation. In the next step, we need to pass the obtained sample through the normalizing flow $\mathcal{F}$ with attributes corresponding to the original image and obtain a sample in the base distribution domain. Subsequently, we need to change the selected attributes. Then, we go back to the latent space representation using the previously obtained sample from the base distribution and the new attributes. Finally, we decode the obtained representation and obtain the final result.

3.5   Classification

The proposed approach can be easily applied as a generative classification model that utilizes the autoencoder's feature representation $\mathcal{Z}$. Assuming that a set $\mathbf{Y}$ is a set of all possible classes, we can use the Bayes rule to calculate predictive distribution $P(\mathbf{y}|\mathbf{z})$, which is a probability of specific class $\mathbf{y}$ given the vector of the latent space representation $\mathbf{z}$: $P(\mathbf{y}|\mathbf{z}) = \frac{P(\mathbf{z}|\mathbf{y})P(\mathbf{y})}{P(\mathbf{z})}$, where $P(\mathbf{y})$ is an assumed class prior and $P(\mathbf{z}) = \sum_{\mathbf{y} \in \mathbf{Y}} P(\mathbf{z}|\mathbf{y})P(\mathbf{y})$. As a consequence, we can omit the denominator as it is a normalizing factor and perform classification according to the rule given by the formula $\hat{\mathbf{y}} = \mathrm{argmax}_{\mathbf{y} \in \mathbf{Y}} P(\mathbf{z}|\mathbf{y})P(\mathbf{y})$.

4   Related Works

The problem of conditional generation is solved using a variety of generative approaches. Some of them extend traditional VAE architectures to construct the conditional variants of this kind of model for tasks like visual segmentation [20], conditional image [21], or text generation [23]. Conditional variants of GANs (cGANs) [14] are also popular for class-specific image [16] or point cloud [15] generation. Conditioning mechanisms are also implemented in some variants of flow models, including Masked Autoregressive Flows (MAF) [17], RealNVP [6], and Continuous Normalizing Flows [7]. The conditional flows are successively applied to semi-supervised classification [3], super-resolution generation [12], noise modeling [2], structured sequence prediction [5], or 3D points generation [19]. The presented methods achieve outstanding results in solving a variety of tasks. However, they are trained in an end-to-end fashion, and any changes in the model require very expensive training procedures from scratch.

That issue is tackled by plugin models introduced in [10]. The idea behind plugins is to incorporate additional information, so-called partial evidence, into the trained base model without modifying the parameters. It can be achieved by adding small networks named plugins to the intermediate layers of the network. The goal of these modules is to incorporate additional signals, i.e., information about known labels, into the inference procedure and adjust the predicted output accordingly. Plugin models were successively applied to the classification and segmentation problems. The idea of incorporating an additional, conditional flow model into a pretrained generative model was introduced in StyleFlow model [1]. This approach was designed for StyleGAN [8], the state-of-the-art model for generating face images. MSP [11] utilizes manipulation of VAE latent space via matrix subspace projection to enforce desired features on images.

Compared to the reference approaches, we present a general framework for utilizing conditional flow for various tasks, including conditional image genera-tion, attribute manipulation, and classification. Our model can be applied to any base autoencoder with bottleneck latent space without additional requirements.

5   Experiments

This section evaluates the proposed method for three tasks: conditional object generation, attribute manipulation, and classification. We perform experiments on two domains: images (MNIST, CelebA) and 3D point clouds (ShapeNet). Our goal in the experiments is to qualitatively and quantitatively validate our method.

5.1   Conditional Object Generation

Our goal in this section is qualitative validation of conditional image and con-ditional 3D point cloud generation capabilities of Flow Plugin Network (FPN). In this experiment, we utilize the already trained non-conditional autoencoder-like models and plug in our FPN to make the whole architecture conditionally generative.

Methodology. The first part of the experiment we performed on the MNIST dataset. Due to the lack of the standard baseline VAE and AE models, we trained these models by ourselves. The model's bottleneck with latent space has a dimensionality equal to 40.

The second part of the experiments we performed on the ShapeNet dataset. It has more classes, which additionally are imbalanced, and it's an application to a different domain. We used PointFlow [22], a de-facto standard model for Point Clouds. The model has an autoencoder-like architecture and operates on a 128-dimensional latent space. We used an already trained non-generative version of the model for all 55 classes. For both datasets, we trained two models with different normalizing flow architectures for each dataset: Conditional Masked Autoregressive Flow (C-MAF) and Conditional Real NVP (C-RNVP).

Results. We started experiments by evaluating conditional generative capabilities on latent spaces. The goal of this experiment was to evaluate if the Flow Plugin Network is capable of correctly learning the conditional mapping between base distributions and latent spaces of the models. A successful experiment will inform us that our model correctly generates embeddings, and possible errors on the final object could be caused by the decoder part of the autoencoder.

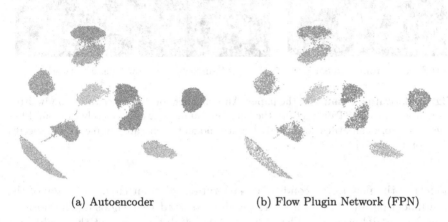

(a) Autoencoder   (b) Flow Plugin Network (FPN)

Fig. 2. Comparison of the latent spaces obtained using Autoencoder on MNIST dataset. We can observe that FPN could correctly fit and then sample the latent space distributions for all models. The results were visualized in the 2D space after UMAP transformation, and colors represent different classes.

We generated latent space of the Autoencoder model for the MNIST dataset. It was generated by encoding images using the encoder part of the models. On the other hand, the latent space for Flow Plugin Networks with MAF backbone was generated by sampling 1000 data points conditioned on the specific class. The results are presented in Fig. 2. We can observe that our method generates samples in the proper locations of the latent space, which confirms its ability to reconstruct the latent space properly.

In the next experiment, we go one step further, and we sample latent space vectors using the Flow Plugin Network model and decode them by the decoder from the base model. The experiment aims to evaluate the model in an end-to-end conditional generation process.

In this experiment, we utilize the already trained non-conditional autoencoder-like model and plug in our FPN to make the whole architecture conditionally generative. The results for the MNIST dataset are presented in Fig. 3. We can observe samples generated from the VAE model on the left - the samples' classes are entirely random. On the right, our method generated samples with specific digit classes. What is essential here is that the weights of the already trained model are not changed during the FPN training phase. Thanks to that, we showed that we could extend the standard autoencoder-like

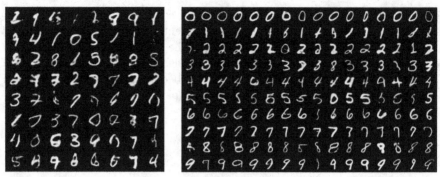

(a) Samples from Autoencoder.   (b) Samples from Flow Plugin Network.

Fig. 3. Motivating example of the paper. Autoencoder correctly generates handwritten digits from the MNIST dataset (on the left), but numbers are completely random. Flow Plugin Network can utilize an already trained model to generate samples of the specific class (on the right).

models to the new task - conditional generation, without changes to any of the base model parameters. Moreover, we also observed that choosing the normalizing flow model used as a Flow Plugin Network did not impact the qualitative results.

We also performed the same experiment for the ShapeNet dataset using only the C-MAF backbone. The resulting objects are in Fig. 4, and the model correctly generated the requested examples. The original PointFlow model was a pure autoencoder without generative capabilities. The authors created separate models for different classes. Still, we used the general model trained on all classes to generate samples from all classes, effectively turning a non-generative autoencoder model into a generative one.

5.2 Attribute Manipulation

Methodology. In this experiment, we use the CelebA dataset, Variational Autoencoder with Latent Space Factorization via Matrix Subspace Projection [11] (MSP) as a base model and Conditional Masked Autoregressive Flow as a Flow Plugin Network model.

Results. The results are presented in Fig. 5. We have the original image on the left, the middle reconstruction from the base model, and the image with changed attributes on the right. We can notice that the base model introduces some artifacts in the reconstruction. The same artifacts are visible in the images generated with our method, as our method uses the base model decoder to generate the image. However, more importantly, the qualitative results show that our proposed method can control the attributes of the reconstructed image. Our model can do it without retraining or modifying the base model.

Fig. 4. Examples of 3D Point Clouds generated from PointFlow latent space using Flow Plugin Network model.

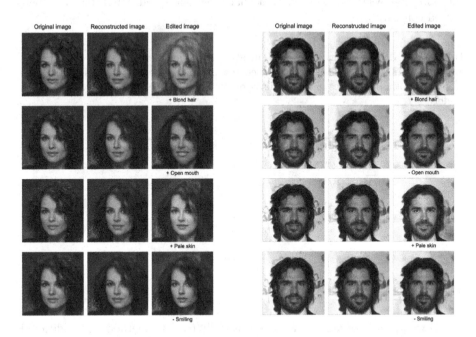

Fig. 5. Results of the feature manipulation experiment on the CelebA dataset. Legend: +/- corresponds to adding or removing a particular feature from the conditioning attributes.

5.3   Classification

In this section, we show the results of the classification experiment in which we use the proposed Flow Plugin Network model as a classifier as described in Sect. 3.5. It is a fascinating property of our model, as we can extend either generative or non-generative models (trained in an unsupervised way) to classification models.

Methodology. Similar to the Conditional Object Generation part of the experiments, we use MNIST and ShapeNet datasets. Additionally, we use the same base and Flow Plugin Network models described for Conditional Object Generation. This experiment compares our method with other classification baselines trained in a supervised way. Specifically, we compare our method to Logistic Regression, SVM with linear kernel, and SVM with RBF kernel. We performed a hyperparameter grid search for each baseline method using 5-fold cross-validation.

Results. The accuracy assessment obtained by all models is presented in Table 1. We can observe that the Flow Plugin Network models were the best-performing ones on the MNIST dataset, and the ShapeNet dataset obtained comparable results to the baseline methods. Moreover, we can observe that both normalizing flows backbones - Conditional MAF and Conditional RealNVP performed comparably.

Table 1. MNIST and ShapeNet Classification experiment results. Our Flow Plugin Network architecture extends unsupervised autoencoders to the classification task. The proposed method achieves a similar level of accuracy as the supervised classification models.

Model	MNIST	ShapeNet
Logistic Regression	0.8833	0.8443
Linear SVM	0.9164	0.8503
RBF SVM	0.9603	**0.8643**
FPN (Conditional-MAF)	**0.9677**	0.8404
FPN (Conditional-RealNVP)	0.9616	0.8370

6   Summary

In this work, we have proposed and successfully tested a method for conditional object generation based on a model with an autoencoder architecture. In the case of non-generative autoencoders, this method makes them generative. Moreover, we used a trained Flow Plugin Network for classification and attribute manipulation tasks.

During experiments, we have shown that using the proposed method, we can conditionally generate images and 3D point clouds from generative models such as Variational Autoencoder and non-generative ones such as Autoencoder or PointFlow. Moreover, we have performed a classification task on MNIST and ShapeNet datasets and compared the results with shallow machine learning models. We obtained the best results using our model on the former dataset and the latter comparable ones. Lastly, we have successfully manipulated images of human faces.

Acknowledgements. The work of M. Zieba was supported by the National Centre of Science (Poland) Grant No. 2020/37/B/ST6/03463.

References

1. Abdal, R., Zhu, P., Mitra, N.J., Wonka, P.: Styleflow: attribute-conditioned exploration of stylegan-generated images using conditional continuous normalizing flows. ACM Trans. Graph. **40**(3), 21:1–21:21 (2021)
2. Abdelhamed, A., Brubaker, M., Brown, M.S.: Noise flow: noise modeling with conditional normalizing flows. In: 2019 IEEE/CVF International Conference on Computer Vision, ICCV 2019, Seoul, Korea (South), 27 October–2 November 2019, pp. 3165–3173. IEEE (2019)
3. Atanov, A., Volokhova, A., Ashukha, A., Sosnovik, I., Vetrov, D.P.: Semi-conditional normalizing flows for semi-supervised learning. CoRR abs/1905.00505 (2019)
4. Bank, D., Koenigstein, N., Giryes, R.: Autoencoders. CoRR abs/2003.05991 (2020)
5. Bhattacharyya, A., Hanselmann, M., Fritz, M., Schiele, B., Straehle, C.: Conditional flow variational autoencoders for structured sequence prediction. CoRR abs/1908.09008 (2019)
6. Dinh, L., Sohl-Dickstein, J., Bengio, S.: Density estimation using real NVP. In: 5th International Conference on Learning Representations, ICLR 2017, Toulon, France, 24–26 April 2017, Conference Track Proceedings. OpenReview.net (2017)
7. Grathwohl, W., Chen, R.T.Q., Bettencourt, J., Sutskever, I., Duvenaud, D.: FFJORD: free-form continuous dynamics for scalable reversible generative models. In: 7th International Conference on Learning Representations, ICLR 2019, New Orleans, LA, USA, 6–9 May 2019. OpenReview.net (2019)
8. Karras, T., Laine, S., Aila, T.: A style-based generator architecture for generative adversarial networks. In: IEEE Conference on Computer Vision and Pattern Recognition, CVPR 2019, Long Beach, CA, USA, 16–20 June 2019, pp. 4401–4410. Computer Vision Foundation/IEEE (2019)
9. Kingma, D.P., Welling, M.: Auto-encoding variational bayes. In: 2nd International Conference on Learning Representations, ICLR 2014, Banff, AB, Canada, 14–16 April 2014, Conference Track Proceedings (2014)
10. Koperski, M., Konopczynski, T.K., Nowak, R., Semberecki, P., Trzcinski, T.: Plugin networks for inference under partial evidence. In: IEEE Winter Conference on Applications of Computer Vision, WACV 2020, Snowmass Village, CO, USA, 1–5 March 2020, pp. 2872–2880. IEEE (2020)
11. Li, X., Lin, C., Li, R., Wang, C., Guerin, F.: Latent space factorisation and manipulation via matrix subspace projection. In: Proceedings of the 37th International

Conference on Machine Learning, ICML 2020, 13–18 July 2020, Virtual Event. Proceedings of Machine Learning Research, vol. 119, pp. 5916–5926. PMLR (2020)

12. Lugmayr, A., Danelljan, M., Van Gool, L., Timofte, R.: SRFlow: learning the super-resolution space with normalizing flow. In: Vedaldi, A., Bischof, H., Brox, T., Frahm, J.-M. (eds.) ECCV 2020. LNCS, vol. 12350, pp. 715–732. Springer, Cham (2020). https://doi.org/10.1007/978-3-030-58558-7_42

13. Makhzani, A., Shlens, J., Jaitly, N., Goodfellow, I.J.: Adversarial autoencoders. CoRR abs/1511.05644 (2015)

14. Mateos, M., González, A., Sevillano, X.: Guiding GANs: how to control non-conditional pre-trained GANs for conditional image generation. CoRR abs/2101.00990 (2021)

15. Milz, S., Simon, M., Fischer, K., Pöpperl, M., Gross, H.-M.: Points2Pix: 3D point-cloud to image translation using conditional GANs. In: Fink, G.A., Frintrop, S., Jiang, X. (eds.) DAGM GCPR 2019. LNCS, vol. 11824, pp. 387–400. Springer, Cham (2019). https://doi.org/10.1007/978-3-030-33676-9_27

16. Odena, A., Olah, C., Shlens, J.: Conditional image synthesis with auxiliary classifier GANs. In: Proceedings of the 34th International Conference on Machine Learning, ICML 2017, Sydney, NSW, Australia, 6–11 August 2017. Proceedings of Machine Learning Research, vol. 70, pp. 2642–2651. PMLR (2017)

17. Papamakarios, G., Murray, I., Pavlakou, T.: Masked autoregressive flow for density estimation. In: Advances in Neural Information Processing Systems 30: Annual Conference on Neural Information Processing Systems 2017, 4–9 December 2017, Long Beach, CA, USA, pp. 2338–2347 (2017)

18. Papamakarios, G., Nalisnick, E.T., Rezende, D.J., Mohamed, S., Lakshminarayanan, B.: Normalizing flows for probabilistic modeling and inference. J. Mach. Learn. Res. 22, 57:1–57:64 (2021)

19. Pumarola, A., Popov, S., Moreno-Noguer, F., Ferrari, V.: C-flow: conditional generative flow models for images and 3D point clouds. In: 2020 IEEE/CVF Conference on Computer Vision and Pattern Recognition, CVPR 2020, Seattle, WA, USA, 13–19 June 2020, pp. 7946–7955. Computer Vision Foundation/IEEE (2020)

20. Sohn, K., Lee, H., Yan, X.: Learning structured output representation using deep conditional generative models. In: Advances in Neural Information Processing Systems 28: Annual Conference on Neural Information Processing Systems 2015, 7–12 December 2015, Montreal, Quebec, Canada, pp. 3483–3491 (2015)

21. Yan, X., Yang, J., Sohn, K., Lee, H.: Attribute2Image: conditional image generation from visual attributes. In: Leibe, B., Matas, J., Sebe, N., Welling, M. (eds.) ECCV 2016. LNCS, vol. 9908, pp. 776–791. Springer, Cham (2016). https://doi.org/10.1007/978-3-319-46493-0_47

22. Yang, G., Huang, X., Hao, Z., Liu, M., Belongie, S.J., Hariharan, B.: Pointflow: 3D point cloud generation with continuous normalizing flows. In: 2019 IEEE/CVF International Conference on Computer Vision, ICCV 2019, Seoul, Korea (South), 27 October–2 November 2019, pp. 4540–4549. IEEE (2019)

23. Zhao, T., Zhao, R., Eskénazi, M.: Learning discourse-level diversity for neural dialog models using conditional variational autoencoders. In: Barzilay, R., Kan, M. (eds.) Proceedings of the 55th Annual Meeting of the Association for Computational Linguistics, ACL 2017, Vancouver, Canada, 30 July–4 August Volume 1: Long Papers, pp. 654–664. Association for Computational Linguistics (2017)

Speech and Text Processing

Investigating the Impact of Parkinson's Disease on Brain Computations: An Online Study of Healthy Controls and PD Patients

Artur Chudzik[1]([envelope]) [iD], Aldona Drabik[1] [iD], and Andrzej W. Przybyszewski[1,2] [iD]

[1] Faculty of Computer Science, Polish-Japanese Academy of Information Technology, Warsaw, Poland
{artur.chudzik,adrabik,przy}@pjwstk.edu.pl
[2] Department of Neurology, University of Massachusetts Medical School, 65 Lake Avenue, Worcester, MA 01655, USA
andrzej.przybyszewski@umassmed.edu
https://pja.edu.pl , https://nd.pja.edu.pl

Abstract. Early detection of Parkinson's disease (PD) is critical for effective management and treatment. In our recent study, we collected data on brain computations in individuals with PD and healthy controls using an online platform and multiple neuropsychological tests. Using logistic regression, we achieved an accuracy rate of 91.1% in differentiating PD patients and healthy controls. However, two PD patients were classified as healthy subjects, and two healthy individuals were misclassified as PD patients. We also utilized multinomial logistic regression to predict the UPDRS3 group of patients and healthy individuals, achieving the same high accuracy. Our findings suggest that cognitive and behavioral tests can detect early changes in brain computations, potentially indicating the onset of PD before clinical symptoms appear. This has significant implications for early detection and intervention of neurological disorders, improving outcomes and quality of life for affected individuals. Overall, our study provides new insights into the utility of neuropsychological tests and statistical methods for detecting and monitoring PD.

Keywords: Parkinson's disease · Brain computations · Online testing

1 Introduction

This study focuses on the test that detects early neurological symptoms of major public health problems related to neurodegenerative diseases (ND). NDs are incurable and debilitating conditions that result in progressive degeneration and death of nerve cells. The process starts with an asymptomatic stage when the person feels fine and shows no signs of neurodegenerative disease, and clinical examination also will show no abnormalities. During the asymptomatic stage of neurodegenerative disease, individuals may not exhibit any symptoms and may not seek medical attention, leading to a lack of abnormalities detected during

N. T. Nguyen et al. (Eds.): ACIIDS 2023, LNAI 13996, pp. 235–246, 2023.
https://doi.org/10.1007/978-981-99-5837-5_20

clinical examination. All NDs are progressing relentlessly over the years and have proved to be stubbornly incurable. Thus, we believe that it is crucial to find a way to detect the early onset of NDs.

1.1 Alzheimer's and Parkinson's Disease

The most common neurodegenerative disorders are Alzheimer's disease (AD) and Parkinson's disease (PD) [4], and they are in our area of interest because of their partial similarities in symptoms. First, it is necessary to provide a comprehensive explanation of Parkinson's disease as our patients are afflicted with this condition. Parkinson's disease is a progressive neurodegenerative disorder that begins to develop approximately 20 years prior to the appearance of symptoms, during which a significant portion of the brain is already affected. It is characterized by the progressive loss of dopaminergic neurons in the brain, leading to a range of motor and non-motor symptoms, including tremors, rigidity, and cognitive impairment. Early detection of Parkinson's disease is crucial for the effective management and treatment of the condition. PD affects three fundamental systems: motor, cognitive, and emotional. The disease typically starts with motor impairments such as bradykinesia. In contrast, Alzheimer's disease initially affects cognitive abilities like mild cognitive impairment (MCI), which not all PD patients have. Although MCI is not a typical symptom of Parkinson's disease, it can occur during the disease progression as an early stage where symptoms are not severe but are detectable. As the disease advances, individuals may experience late-stage complications, including advanced cognitive and motor symptoms. It is important to note that preclinical symptoms **may vary** between Alzheimer's disease and Parkinson's disease, the presentation and progression of symptoms **can vary widely** between individuals, and **no two cases of PD are exactly alike**. In summary, both AD and PD characteristics **may include** the following common manifestation [3, 14, 18].

- Mild cognitive impairment (MCI), such as language or visuospatial perception and memory impairment (mainly in AD, not always in PD);
- affected rote memory and frontotemporal executive functions;
- depression;
- sleep problems;
- automatic response inhibition decay;
- difficulty with emotion recognition;
- motor slowness symptoms (predominantly with PD, but also associated with preclinical AD).

1.2 Digital Biomarkers

In this study, we place emphasis on the use of digital biomarkers, which are measurable and quantifiable medical signs collected through digital devices or platforms, to gain insights into brain computations. Reaction time (RT) is a widely studied digital biomarker that plays a crucial role in measuring neurological function, particularly in individuals with PD. Previous research, including our own studies [14], has demonstrated that measures such as saccadic delay and movement-related potentials can serve as reliable indicators of the state of PD. Others [10] have employed movement-related potentials in a choice reaction time task to explore the underlying causes of reaction time delay in Parkinson's disease. Movement-related potentials showed that motor processes required more time for Parkinson's disease patients making complex responses. The study also found that one or more premotor processes were slowed in Parkinson's disease patients based on delayed onset of movement-related potentials. These findings suggest that reaction time may be a valuable measure for tracking the progression of Parkinson's disease and the effectiveness of treatment. It is worth noting that reaction time is just one measure of neurological function that can be used with other measures. It may also help evaluate the impact of Parkinson's disease on the nervous system.

1.3   Digital Screening

We utilized an online platform to administer neuropsychological tests, which are widely used as the gold standard for assessing cognitive function. Our aim was to detect early changes in brain computations in individuals with PD, which could indicate the onset of the disease before the appearance of clinical symptoms. In addition to the participants' responses, we also collected additional temporal measures, including Instrumental Reaction Time (IRT) and Time-to-Submit - TTS. IRT measures the time between the screen appearing and the participant's first option selection, while TTS measures the time it takes for the participant to click the submit button. In the following text, TTS is also referred to as "response time". It is well recognized that neuropsychological testing has great diagnostic and screening power, but it requires proper training, tools, and time. Our goal is to evaluate if a single online tool can support these operations, and thus provide a cross-sectional set of neuropsychological examinations that will contribute to the overall understanding of the patient's psychophysical state. To ensure the validity of our results, we recruited both a group of individuals diagnosed with PD and a group of healthy controls for evaluation. While it is well-established that reaction time generally decreases with age, with previous studies estimating an average decrease of 4-10 ms per year [2,17], our observations of instrumental reaction times in our study were higher than expected based on age-related decline alone. Despite an average age difference of approximately 47 years between the groups, as detailed in the "Results" section, our findings suggest that factors beyond age-related decline may have contributed to the observed differences in reaction time between the PD and healthy control groups.

2   Methods

We intended to create an online method of neuropsychological assessment. The implementation in the form of a computer test started with the requirements gathering and prototype. First, we asked trained psychologists and neurologists to create a general overview of the battery of tests used in their practice, being a gold standard. We decided to use a multi-tiered approach to assessment, including the tests described below.

2.1 GDS-15

Geriatric Depression Scale - is a short version (15 questions) of the test developed in '86 by Sheikh and Yesavage [16]. The short version contains 15 of the 30 questions from the extended version that showed the most significant correlation with signs of depression. Out of 15 items, 10 indicate the presence of depression when given a positive answer, while the remaining items (questions 1, 5, 7, 11, 13) indicate depression when given a negative answer. Scores 0–4 are considered "normal" depending on age and education; 5–8 indicate mild depression; 9–11 indicate moderate depression and 12–15 indicate severe depression. The GDS has 92% sensitivity and 89% specificity as assessed by diagnostic criteria [8]. A validation study comparing long and short GDS forms to self-assessment of depressive symptoms successfully distinguished adults with depression from non-depressed people with a high correlation (r = 0.84, p < 0.001) [16]. The online implementation of the study in our version consists of 15 questions, displayed individually, with a single choice option between "yes" or "no". The sample question from this set is: "Have you dropped many of your activities and interests?" Every question is consistent with the official translation of the test in a selected language version. The test in its standard form consists of questions and answers printed on a single A4 sheet, and therefore it is possible to resolve it non-linearly. Our version shows each question separately, which allows us to measure both reaction (instrumental reaction time - IRT) and response (time-to-submit - TTS) time.

2.2 TMT A&B

Trail Making Test - is a neuropsychological test for visual attention and task switching, developed in '55 by Reitan [15]. It consists of two parts. In both, instruction to the subject is to connect a set of dots as quickly as possible while maintaining accuracy. The test can provide information on visual search speed, scanning, processing speed, mental flexibility, and executive functioning. The TMT A and B results are as high as the number of seconds to complete the task; higher scores follow the level of impairment. In part A with 25 dots - a healthy person can finish it on average in 29 s, and a patient with deficiencies in more than 78 s. In part B with 25 dots - a healthy person can finish it on average in 75 s, and a patient with deficiencies in more than 273 s. The standard form test asks to combine tracks 1-2-3- (version A) or 1-A-2-B- (version B) on the paper with a pen on the paper tray. We ask patients to select circles in a given order three times in the online version. First test: version A relies on 15 circles, and this part focuses mainly on examining cognitive processing speed. Second: version B (short) consists of 10 circles (5 with letters and 5 with numbers), and version B (long) is 20 circles (10 for both letters and numbers). These versions assess executive functioning. Each time we allocate circles randomly with uniform distribution on the screen. It is worth mentioning that there is no record of the error rate in the pen and paper version of the test. Because the online version is self-assessed, we had to implement this feature and notify the user of making a mistake by marking the circle in red and the correct connection displayed as a green circle. This mechanism gives the users feedback to get on the right path themselves. The completion, however, might be longer than in the standard version since there is no supervisor. Here, we record the error rate for each part of that task, IRT and TTS.

2.3   CDT and CCT

The Clock Drawing Test - is used to screen cognitive disorders in clinical practice. The origins of this test are not clear, but the probable precursor was Sir Henry Head [5,6]. There are many ways to conduct this test, but a common task is to draw a clock with a face, all numbers, and hands showing a given time. One way is to draw two lines perpendicular to each other, obtaining four quadrants of the clock's face. Then, we can count the number of digits in each quadrant, and if the quadrant is correct while it contains three numbers (error score is between 0 and 3 for each quadrant). The standard score is below 4 points. In the original study, a score over 4 revealed a sensitivity of 87% and a specificity of 82% for identifying dementia. Another test related to drawing tasks might be CCT - the Cube Copying test, valid (yet limited) for routine clinical dementia screening. As presented in Maeshima et al. [11], quantitatively scored cube copying can estimate cognitive dysfunction in dementia patients. The execution of both tasks in the digital form relied on the area of the screen divided by opaque lines, mimicking a standard notebook. Participants drew a figure with a cursor or a finger on mobile devices. We were concerned about the performance of older patients who were not fluent with computer technology because drawing on the computer screen introduces a novelty factor. Also, this interface lacks the naturalness of the pen-and-paper method. However, most users completed both assignments. Those tasks were not time-restricted. Nevertheless, we recorded IRT and TTS as well as the paintings.

2.4   MoCA

Montreal Cognitive Assessment - is a screening test developed in '05 by Nasreddine et al. [12] is a cognitive test to detect MCI. The test checks language, memory, visual and spatial thinking, reasoning, and orientation. MoCA scores range from 0 to 30, and 26 or more is considered normal. In the original study, people without cognitive impairment scored 27.4 (average); subjects with mild cognitive impairment (MCI) scored 22.1 (average); people with Alzheimer's disease scored 16.2 (average). The test has a 90.0–93.0% sensitivity and a specificity of 87.0% in the MCI assessment. MoCA implements three earlier tasks: TMT B, CCT, and CDT. The following tasks are related to language fluency. First: "Name this animal". We depict a cow, horse, and a lion, and we ask the participant to type the name of the presented animal into the text field. Additionally, our task depicts the lion with the incorrect number of legs because this disturbance seems to be a response time delay factor in a patient who suffers from AD. The second task from this series is a repetition of two syntactically complex sentences: "I only know that John is the one to help today". and "The cat always hid under the couch when dogs were in the room". We asked patients to replicate both sentences in a written form, disabling the copy-paste option. The third language fluency task was to write as many English words as possible that start with the letter F. The patient had 60 s for execution. Considering that older participants are less fluent in typing, we introduced two mechanisms that could align their chances. First, we delayed the countdown by the number of seconds calculated as (number of words in the task * 60/average reading speed per minute), assuming that the lower boundary of the average reading speed is 200 words per minute. Next, each keystroke stopped the countdown, allowing writing the whole word even at a slow pace. Lack of the keystroke during the next consequent 3 s was starting the countdown again. The next group of tasks focuses on attention and concentration. First, we display one letter per second, and a person has to click the button each time the letter "A" shows on the screen; next, we ask about the serial subtraction starting at 100. Likewise, we present two sets of numbers; each time, the subject must repeat them by writing in the forward or backward order to evaluate the working memory. We measure the error rate and average response time for all tasks. Also, we assess the abstract reasoning by a describe-the-similarity task. We ask about what two pairs of words have in common (in a single word): watch + ruler and train + bike, and we evaluate answers alongside limited dictionaries of means of transportation, traveling, measuring, and instruments. Here also we measure the error rate, IRT, and TTS. The next part focuses on short-term memory. We involved two learning trials of five nouns and delayed recall after approximately five minutes. For the first trial, the patient must write words and receive visual cues if they are correct. If not, it is possible to rewind and see them again. If this operation fails more than twice, we save this fact into the database, skipping into the next question. We display this task again at the end of the MoCA part. Each time, we count the error rate, IRT, and TTS. Finally, we evaluate the spatio-temporal orientation by asking the subject for the date and where the test occurs. We validate the provided year, month, exact date, and day of the week with the system clock, counting the number of errors. The place is scored manually after the test. For each part, we measure the instrumental reaction time and Time-to-Submit. We are aware of an essential, fundamental difference in switching from a verbal task (hears -> speaks) to a written form (sees -> writes), especially when taking into consideration motor problems (writing), leading to a field of uncertainty that we must treat with utmost meticulousness.

2.5   Epworth

Epworth Sleepiness Scale - is an eight questions test that focuses on daytime sleepiness, created in '91 by Johns [7]. On a 4-point scale (0-3), subjects are asked to rate the likelihood of falling asleep during eight different activities throughout the day. The Epworth score (sum of 8 responses, each scored on a 0-3 scale) can range from 0 to 24. The higher the Epworth score, the higher the average tendency to "daytime sleepiness". The test showed 93.5% sensitivity and 100% specificity in the narcolepsy diagnosis study [9]. In our online version of this test, we ask participants to determine the likelihood of falling asleep in multiple situations, such as "Sitting and reading" or "Watching TV". Possible answers are: "zero probability of falling asleep", "unlikely to fall asleep", "average probability of falling asleep", and "high probability of falling asleep". Each answer has 0-3 points accordingly. Here, we display each question with possible answers separately, each time measuring IRT and TTS.

2.6 FER

Facial Emotion Recognition - is a set of tests dedicated to recognizing emotions conveyed through different channels, where one of them is to match a label with a given emotional expression. Multiple studies suggest that the results of patients with PD are performing significantly worse than that of healthy controls [1]. The link between facial expression and FER impairment reveals since the earliest studies on FER in recall embodied simulation theory, suggesting that disturbed motor processing can lead to deficiency in emotion recognition. We decided to implement this task with six faces expressing particular emotions, alongside six radio buttons with emotions' names. We presented each face separately and obtained all of them from the "Warsaw set of emotional facial expression pictures" (WSEFEP) [13]. Each face presented anger, disgust, fear, happiness, sadness, or surprise. We selected those pictures with the highest recognition marks (e.g., accuracy with intended display) from independent judges. The test evaluates the correctness of the answer, IRT, and TTS for each displayed expression.

2.7 Online Study

To conclude, we distinguished 66 questions requiring various forms of responses, and we implemented them as web application components. Computer assessment allowed us to extend classical metrics: each question could hold a precise IRT and TTS along with the answer. Measuring time on the client-side is crucial for assessing the performance of participants. One widely used method for measuring these metrics is the JavaScript method `performance.now()`, which provides a high-resolution timestamp in milliseconds. Unlike other methods that rely on the system clock, `performance.now()` is not affected by changes to the system clock and provides a more accurate representation of the time it takes for code to execute. `performance.now()` returns a DOMHighResTimeStamp value that represents the number of milliseconds elapsed since the performance timing origin, which is typically the time the page was loaded or refreshed. This method is often used in conjunction with other JavaScript functions, such as `setTimeout()` and `requestAnimationFrame()`, to measure the time it takes for code to execute and to optimize performance. In our study, we utilized this method to measure the time to first selection (IRT), and Time-to-Submit (TTS). The user interface of the application was implemented using the React JavaScript library, which is widely used for building modern, scalable, and interactive web applications.

3 Results

The present study aimed to investigate the effects of Parkinson's Disease (PD) on brain computations using an online platform. Temporal values (IRT and TTS) were recorded in milliseconds, but for improved legibility and comprehension, the results are presented in seconds. Both IRT and TTS are averages (calculated without outliers) based on partial measurements of single questions. To determine the statistical significance of group differences, p-values were calculated and comparisons were considered statistically significant if the p-value was less than 0.05. Statistical analyses were conducted using SPSS 29.

3.1 Comparison of Cognitive and Sleep-Related Measures

A total of 45 participants were recruited for this study, with 15 PD patients (8 females, 7 males) with a mean age of 70.8 years (standard deviation [SD] = 5.93) and 30 healthy controls (3 females, 27 males) with a mean age of 24 years. The selection of participants was based on the availability of individuals who met the criteria for each group. While the age difference between the PD and healthy control groups was noticeable, it is important to note that age was not utilized as a variable in the machine learning analysis, making it less relevant to the study objectives. The severity of motor symptoms in patients with Parkinson's disease was assessed using the Movement Disorder Society-Unified Parkinson's Disease Rating Scale (UPDRS) Part III. Patients were grouped into five categories based on their UPDRS3 scores: Group 0 (score 0–9) n = 0, Group 1 (10–19) n = 5, Group 2 (20–29) n = 3, Group 3 (30–39) n = 3, Group 4 (40+) n = 4. All healthy controls were classified into Group 0 (n = 30). First, we found that the PD patients had a slightly lower mean MOCA score (24.67, SD = 3.519) than healthy controls (26.27, SD = 1.202), but the difference was not significant (p = 0.107).

Similarly, the PD patients had a slightly higher mean Epworth score (5.13, SD = 1.685) than healthy controls (4.13, SD = 1.776), but the difference was still not significant (p = 0.077). Also, we found that the mean GDS15 score for PD patients (5.60, SD = 1.056) was slightly lower than that of healthy controls (6.43, SD = 1.305), but again the difference was not significant (p = 0.38). Finally, the mean FER score for both groups was also not significantly different (PD patients: 6.87, SD = 0.352; healthy controls: 6.77, SD = 0.626) (p = 0.57). These findings suggest that there was no significant impairment in facial expression recognition in the PD group compared to the healthy control group. We found, however, that there are significant differences between the scores of TMT B (PD patients: 4.54, SD = 7.70; healthy controls: 0.67, SD = 1.20) (p = 0.043).

3.2 Temporal Results in Cognitive Tests

We also measured the participants' IRT and TTS for each cognitive tests' question. In the MoCA test, the mean instrumental reaction time for the healthy group was significantly faster (3.62 s) than the clinical group (5.90 s) (p < 0.001). Similarly, the healthy group also had a significantly faster Time-to-Submit (8.00 s) compared to the clinical group (13.67 s) (p < 0.001). The same pattern was observed in the Epworth Sleepiness Scale test, where the healthy group had a significantly faster mean instrumental reaction time (4.70 s) and TTS (6.45 s) compared to the clinical group (8.57 s and 10.45 s, respectively) (p < 0.001).

In contrast, there was no significant difference between the healthy and clinical groups in instrumental reaction time and response time in the Geriatric Depression Scale (GDS-15) test. The mean instrumental reaction time for the healthy group was 4.57 s, and for the clinical group, it was 5.82 s. The Time-to-Submit for the healthy group was 6.58 s, and for the clinical group, it was 7.18 s. We also measured the participants' IRT and TTS in the Facial Expression Recognition (FER) task. The mean instrumental reaction time for the healthy group was significantly faster (3.49 s) than the clinical group (5.23 s) (p < 0.001). However, there was no significant difference between the healthy and clinical groups in the TTS (6.06 s for the healthy group and 6.74 s for the clinical group).

3.3   Predicting Health Status with Cognitive and Emotional Measures

In our study, logistic regression was employed as the statistical method to predict the binary outcome of a patient's health status based on cognitive and emotional measures. Logistic regression models the probability of the binary outcome by applying a logistic function, which transforms a linear combination of the predictor variables. It is a widely used method in machine learning and particularly suitable when the dependent variable is categorical. The logistic regression model in our study utilized default parameter values, including the probabilities of inclusion (PIN = 0.05) and exclusion (POUT = 0.10), as well as a tolerance value (TOLERANCE = 0.0001) to assess multicollinearity. The PIN represents the probability that a variable will be included in the model, while the POUT represents the probability of excluding a variable. The tolerance value indicates the degree of multicollinearity, with a lower value indicating a higher degree of correlation among predictor variables, which can affect the interpretation of regression coefficients.

The results of our experiment showed promising findings in terms of differentiating PD patients and healthy controls based on cognitive and behavioral tests. Our initial attempt to detect healthy controls using only the MOCA score resulted in a 77.8% accuracy rate. However, when we included additional tests such as the Epworth Sleepiness Scale, and Geriatric Depression Scale, the accuracy dropped to 73.3%. Moreover, adding FER score parameter had no impact on this value. It is noteworthy that the inclusion of instrumental reaction time measurements in the MOCA test resulted in a significant increase in accuracy rate to 84.4%, indicating their potential in PD detection. Additionally, combining the results of all tests with IRT for MoCA resulted in a high accuracy rate of 91.1% with a sensitivity of 86.67% and a specificity of 93.33%, underscoring the significance of employing a combination of cognitive and behavioral tests in conjunction with IRT to enhance accuracy and establish a possible digital biomarker for early detection of the disease.

Table 1. Classification Results of Multinomial Logistic Regression using TMT B, IRT, and TTS Measures.

Observed	Predicted					
	G0	G1	G2	G3	G4	% Correct
G0	30	0	0	0	0	100.0
G1	1	4	0	0	0	80.0
G2	1	1	1	0	0	33.3
G3	0	0	0	3	0	100.0
G4	1	0	0	0	3	75.0
Overall Percentage	**73.3**	**11.1**	**2.2**	**6.7**	**6.7**	**91.1**

3.4   Predicting Parkinson's Disease Severity with TMT B and Temporal Measures

We utilized multinomial logistic regression to predict the UPDRS3 group of both PD patients and healthy controls based on their TMT B scores, IRT, and TTS measures. Multinomial logistic regression is a statistical method used to predict categorical outcomes with more than two categories. In our case, patients were grouped into five categories based on their UPDRS3 scores, with healthy controls classified as Group 0. The model was implemented with maximum iterations set to 100, maximum step halving set to 5, and log-likelihood and parameter convergence set to 0. Our analysis showed that using only TMT B score and IRT, we achieved an accuracy of 82.2% in predicting the UPDRS3 group. However, when TTS was added to the model, the overall accuracy increased to 91.1% (Table 1). These results suggest that TMT B error rate, IRT, and TTS might be reliable measures for predicting the UPDRS3 group of patients with PD.

4   Discussion

The primary goal of our research group is to investigate new and innovative ways to detect and diagnose neurodegenerative diseases, such as Parkinson's Disease and Alzheimer's Disease, as early as possible. Early detection is essential because it allows for timely interventions, potentially leading to improved outcomes and quality of life for affected individuals.

In our latest study, we investigated the effects of PD on brain computations using an online platform. We collected cognitive and behavioral data from PD patients and healthy controls, measuring IRT and TTSs, as well as performance on a battery of cognitive tests. Our findings suggest that cognitive and behavioral tests can be used to detect early changes in brain computations, potentially indicating the onset of PD before clinical symptoms appear. There was no significant difference in the mean Montreal Cognitive Assessment score between the PD patients and the healthy controls. The mean Epworth Sleepiness Scale score was slightly higher in the PD group than in the healthy group, although the difference was not significant. Our study also revealed that the mean Geriatric Depression Scale (GDS-15) score in the PD group was only marginally lower than in the healthy group, and the difference was not significant. Moreover, we measured the participants' IRT and TTS for each cognitive tests' question. It is worth noting that the PD patients in our study were undergoing treatment with medications which have positive impact on brain computations. Our findings suggest that IRT and TTSs were significantly slower in the PD group compared to the healthy group, particularly in the MoCA and Epworth tests. Interestingly, we found no significant difference between the groups in IRT and TTSs in the GDS-15 test. In the next step we performed a logistic regression analysis to evaluate the effectiveness of our cognitive and behavioral tests in differentiating PD patients and healthy controls, being a first step for early disease detection based on online testing approach. The initial attempt to detect healthy controls using only the MoCA score resulted in a 77.8% accuracy rate. However, when additional tests such as the Epworth Sleepiness Scale and Facial Expression Recognition task were included, together only with MoCA IRT, the accuracy rate increased to 91.1%. This result suggests that a combination of cognitive and behavioral tests may be more effective in identifying early changes in brain computations associated with PD. As a next part of analysis, we performed a multinomial logistic regression analysis to evaluate the effectiveness of our cognitive and behavioral tests in differentiating PD patients and healthy controls. Our decision to focus on the TMT B test was based on its widespread use as a neuropsychological test that has demonstrated high sensitivity in detecting cognitive impairments in PD patients, particularly in attention and executive function domains. The first experiment included only the TMT B score and IRT, resulting in an accuracy rate of 82.2%. We then added TTS to the model, resulting in an increased accuracy rate of 91.1%. These results suggest that adding temporal measures such as IRT and TTS to cognitive tests such as TMT B can improve the accuracy of predicting UPDRS3 group classification. Of course, as with any study, there are limitations to our research. One limitation is the small sample size, which could impact the generalizability of our findings. Furthermore, we only included a limited set of cognitive and behavioral tests in our study. Future research should explore the use of additional tests to improve the accuracy of early detection of PD. Despite these limitations, our study provides evidence that cognitive and behavioral tests can be used to detect early changes in brain computations associated with PD. In extrapolating the results of our study, it is plausible to apply the findings to other neurodegenerative diseases, such as Alzheimer's disease. Similar to PD, early detection of AD is crucial for timely interventions and improved outcomes. Cognitive and behavioral tests, along with measures such as IRT and TTS, can potentially serve as digital biomarkers to detect early changes in brain computations associated with AD. However, further research is necessary to validate the effectiveness of these tests specifically for AD and explore the potential integration of cognitive and behavioral tests with innovative technologies like chatbots to enhance the assessment process. By leveraging digital biomarkers and innovative approaches, we can advance early detection and diagnostic strategies for various neurodegenerative diseases, ultimately improving patient outcomes and quality of life.

References

1. Argaud, S., Vérin, M., Sauleau, P., Grandjean, D.: Facial emotion recognition in Parkinson's disease: a review and new hypotheses. Mov. Disord. **33**(4), 554–567 (2018)
2. Deary, I.J., Der, G.: Reaction time, age, and cognitive ability: longitudinal findings from age 16 to 63 years in representative population samples. Aging Neuropsychol. Cogn. **12**(2), 187–215 (2005)
3. Goldman, J.G., Aggarwal, N.T., Schroeder, C.D.: Mild cognitive impairment: an update in Parkinson's disease and lessons learned from Alzheimer's disease. Neurodegener. Dis. Manag. **5**(5), 425–443 (2015)
4. Hansson, O.: Biomarkers for neurodegenerative diseases. Nat. Med. **27**(6), 954–963 (2021)
5. Hazan, E., Frankenburg, F., Brenkel, M., Shulman, K.: The test of time: a history of clock drawing. Int. J. Geriatr. Psychiatry **33**(1), e22–e30 (2018)
6. Head, H.: Aphasia and Kindred Disorders of Speech. Cambridge University Press, Cambridge (2014)
7. Johns, M.W.: A new method for measuring daytime sleepiness: the epworth sleepiness scale. Sleep **14**(6), 540–545 (1991)
8. Koenig, H.G., Meador, K.G., Cohen, H.J., Blazer, D.G.: Self-rated depression scales and screening for major depression in the older hospitalized patient with medical illness. J. Am. Geriatr. Soc. **36**(8), 699–706 (1988)
9. Kumar, S., Bhatia, M., Behari, M.: Excessive daytime sleepiness in Parkinson's disease as assessed by Epworth sleepiness scale (ESS). Sleep Med. **4**(4), 339–342 (2003)
10. Low, K.A., Miller, J., Vierck, E.: Response slowing in Parkinson's disease: a psychophysiological analysis of premotor and motor processes. Brain **125**(9), 1980–1994 (2002). https://doi.org/10.1093/brain/awf206
11. Maeshima, S., et al.: Usefulness of a cube-copying test in outpatients with dementia. Brain Inj. **18**(9), 889–898 (2004)
12. Nasreddine, Z.S., et al.: The Montreal cognitive assessment, MoCA: a brief screening tool for mild cognitive impairment. J. Am. Geriatr. Soc. **53**(4), 695–699 (2005)
13. Olszanowski, M., Pochwatko, G., Kuklinski, K., Scibor-Rylski, M., Lewinski, P., Ohme, R.K.: Warsaw set of emotional facial expression pictures: a validation study of facial display photographs. Front. Psychol. **5**, 1516 (2015)
14. Przybyszewski, A.W., Sledzianowski, A., Chudzik, A., Szlufik, S., Koziorowski, D.: Machine learning and eye movements give insights into neurodegenerative disease mechanisms. Sensors **23**(4), 2145 (2023)
15. Reitan, R.M.: The relation of the trail making test to organic brain damage. J. Consult. Psychol. **19**(5), 393 (1955)
16. Sheikh, J.I., Yesavage, J.A.: Geriatric depression scale (GDS): recent evidence and development of a shorter version. Clin. Gerontol. J. Aging Mental Health (1986)
17. Thompson, J.J., Blair, M.R., Henrey, A.J.: Over the hill at 24: persistent age-related cognitive-motor decline in reaction times in an ecologically valid video game task begins in early adulthood. PLoS ONE **9**(4), e94215 (2014)
18. Zokaei, N., Husain, M.: Working memory in Alzheimer's disease and Parkinson's disease. In: Processes of Visuospatial Attention and Working Memory, pp. 325–344 (2019)

Speech Enhancement Using Dynamical Variational AutoEncoder

Hao D. Do[(⊠)] [iD]

FPT University, Ho Chi Minh City, Vietnam
haodd3@fpt.edu.vn

Abstract. This research focuses on dealing with speech enhancement via a generative model. Many other solutions, which are trained with some fixed kinds of interference or noises, need help when extracting speech from the mixture with a strange noise. We use a class of generative models called Dynamical Variational AutoEncoder (DVAE), which combines generative and temporal models to analyze the speech signal. This class of models makes attention to speech signal behavior, then extracts and enhances the speech. Moreover, we design a new architecture in the DVAE class named Bi-RVAE, which is more straightforward than the other models but gains good results. Experimental results show that DVAE class, including our proposed design, achieves a high-quality recovered speech. This class could enhance the speech signal before passing it into the central processing models.

Keywords: speech enhancement · dynamical variational autoEncoder · generative model

1 Introduction

In speech processing, there is a big gap between research and real-world applications. This results from the difference between research data, and real speech signal [27]. In reality, the speech is mixed with a lot of interference signals, including background noise, traffic, music, fan sounds, and many other kinds of sounds. These elements make the spectral of the recorded speech different from the clean speech used in the research.

This research aims to improve speech signal quality via the generative learning approach. To do that, we used a class named Dynamical Variational AutoEncoder (DVAE) [3,15] as the primary model. This is the combination of a generative model and a temporal model. It can model the temporal behavior of speech signals in the time domain and the spectral distribution in the frequency domain. After that, DVAE extracts the speech signal from the mixture. Finally, the extracted speech is enhanced to emphasize the spectral distribution. The speech recovered with DVAE is not only so clean but also clear and normalized its amplitudeí.

N. T. Nguyen et al. (Eds.): ACIIDS 2023, LNAI 13996, pp. 247–258, 2023.
https://doi.org/10.1007/978-981-99-5837-5_21

In the experiment, we re-implemented many particular architectures in DVAE class for enhancing speech. Moreover, we also proposed a new design, called Bi-RVAE, in this class with more simple architecture while not reducing the model's performance. Experimental results show that DVAE class, including our new design Bi-RVAE, is good enough for applications.

Our main contributions in this research are as follows:

- Demonstrating the potential of DVAE class in speech enhancement problems.
- Designing a new architecture in DVAE class, more simple and more effective than many previous models.

The remaining of this paper is structured with 4 main sections. Section 2 mentions and briefly summarizes some studies related to this work. Next, the paper details the proposed model, including the essential elements, via Sect. 3. We present the mathematical formula system, model design, and the main phases for training. After that, Sect. 4 gives some experiments to evaluate our proposed method. Finally, the paper ends with Sect. 5, including a summary and conclusion for our research.

2   Related Works

During the development of speech enhancement, there are many pieces of research related to how to model the speech signal. Most traditional approaches focus on processing the speech signal in a particular context. Generally, these methods are reasonable solutions for some situations, but their applicability is not high. A fixed model with various background sounds or noises is challenging to adapt in the real world. The modern approaches turn to model the human speech distribution, then extract and enhance the human speech only. The following paragraphs briefly mention and describe some main strategies for this problem.

The first group uses the digital transforms [7,8,20] or digital filters [21] to model and process the signal, including human speech. Some kinds of noise or interference distribute in a fixed frequency range, so many filters can be applied to remove them from the mixture. This general approach can be applied to many cases of speech. In real applications, there are many more complicated interference sounds, this method cannot work perfectly, but it has a low computational cost and high reliability. Advanced extensions such as wavelet transform or filter bank can be applied in some cases with a more complicated problem.

The second group uses the machine learning approach as the main method to model the speech and the interference signal. In this group, Independent component analysis (ICA) [16,17,26] and its variants [18,25,30] form the main trend. Particularly, these models try to separate the mixed signal into a collection of simple elements, then extract and post-process the speech signal. This is a powerful approach because it can represent a signal into many pieces and then choose some of them to extract. However, ICA is not a perfect solution. In the computing aspect, ICA uses a matrix with a limited capacity to present the signal. So one ICA cannot model the speech mixed with so many backgrounds. On the other hand, if the ICA matrix has a large capacity, it can cause a high computation cost and reduce the ability of real applications.

Unlike these two traditional approaches, the third group uses a deep neural network to model the speech signal distribution. Simon Leglaive et al. [24] use VAE and its extension to extract and present the speech signal. The key idea is to use the probability distribution to show the behavior of speech signal in the frequency domain [22,23]. This approach focuses on speech and extracts this kind of signal out of the mixture, and it does not depend on the mixture or the background sounds. In this work, we follow VAE and extend this class to develop a more robust architecture for speech enhancement.

3   Dynamical Variational AutoEncoder for Speech Enhancement

3.1   Variational AutoEncoder

Variational AutoEncoder (VAE) is a deep generative model using latent variables to represent the main content in the speech signal [9–11]. This model includes a generating model and an inferring model. In reality, VAE uses two neural networks: an Encoder for inferring model and a Decoder for generating model.

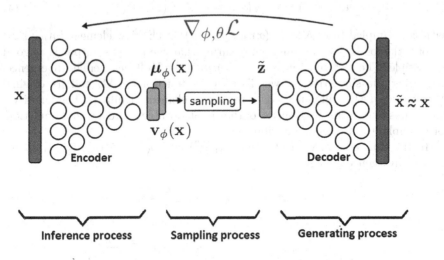

Fig. 1. Main components in a Variational AutoEncoder

Let x, z denote two random variables, a generative model, which generates x from z, is defined by:

$$p_\theta(x) = \int p_\theta(x|z)p(z)dz, \tag{1}$$

with θ is the hidden parameter of $p(x|z)$.

If $p(z) = \mathcal{N}(z, 0, 1)$, the likelihood function is:

$$p_\theta(x|z) = \mathcal{N}(x, \mu_\theta(z), diag\{v_\theta(z)\}), \tag{2}$$

with μ, v denote the mean and variance of $p(x)$ after the process of generating.

In the inferring model, the approximate posterior is defined as:

$$q_\phi(z|x) = \mathcal{N}(z, \mu_\phi(x), diag\{v_\phi(x)\}), \tag{3}$$

with ϕ is the parameter of $q(z|x)$.

Generally, the operation process in VAE (Fig. 1) includes three phases:

- Inferring process: Encoder, corresponding to the probability q with parameter ϕ, represents input mixture signal x into a distribution, described by μ, v
- Sampling process: From the distribution, take a sample to get the real value representing the latent value z
- Generating process: Decoder, corresponding to p with parameter θ, receives latent variable z and then estimates the value data x.

Training, or learning a VAE, is the process that updates the model parameters ϕ and θ to optimize the objective function or loss function. Different from many models for classification, the objective function of VAE is the Lower Bound function, defined by:

$$\mathcal{L}(x, \phi, \theta) = ln(p_\theta(x|\tilde{z})) - \mathcal{KL}(q_\phi(z|x)||p(z)), \tag{4}$$

with $\tilde{z}$ is sampled from $\mathcal{N}(z, \mu_\phi(x), diag\{v_\phi(x)\})$. The first element $ln(p_\theta(x|\tilde{z}))$ denotes the log-likelihood function, ensuring that the model's output is as good as possible. The remaining element $\mathcal{KL}(q, p)$ is the Kullback Leibler divergence, measuring the similarity of two distribution functions. VAE and its original AutoEncode (AE) difference come from this element. Kullback Leibler divergence aims to cast the distribution of the latent variable z into a wanted function, for example, the Gaussian distribution.

In the simple case, with $q(\cdot) \sim \mathcal{N}(\mu_1, \sigma_1^2)$ and $p(\cdot) \sim \mathcal{N}(\mu_2, \sigma_2^2)$, Kullback Leibler divergence becomes:

$$\mathcal{KL}(q||p) = \int_{-\infty}^{\infty} (q(z) \times \log \frac{q(z)}{p(z)}) dx, \tag{5}$$

or:

$$\mathcal{KL}(q||p) = \int_{-\infty}^{\infty} (q(z) \times \log \frac{\frac{1}{2\sqrt{2\pi\sigma_1^2}} \exp - \frac{(z-\mu_1)^2}{2\sigma_1^2}}{\frac{1}{2\sqrt{2\pi\sigma_2^2}} \exp - \frac{(z-\mu_2)^2}{2\sigma_2^2}}) dz. \tag{6}$$

Then the Eq. (6) is reduced to:

$$\mathcal{KL}(q||p) = \int_{-\infty}^{\infty} q(z) \times \left(\log \sqrt{\frac{\sigma_2^2}{\sigma_1^2}} + \frac{(z-\mu_2)^2}{2\sigma_2^2} - \frac{(z-\mu_1)^2}{2\sigma_1^2} \right) dz. \tag{7}$$

If $p(\cdot) \sim \mathcal{N}(0,1)$, Eq. (7) becomes:

$$\mathcal{KL}(q||p) = -\frac{1}{2}(\sigma_1^2 + \mu_1^2 - \log \sigma_1^2 - 1). \tag{8}$$

Although much more flexible than the original AE, VAE includes three main disadvantages as follows:

- Using a static distribution in the latent space.
- Only performing well with static data, such as an image or a short frame of a speech signal.
- Could not take advantage of temporal data or time series data. This aspect makes VAE difficult to be used in speech processing or video processing.

These limits could be improved a lot with the combination of VAE and temporal architecture. This motivates the existence of a more robust model, which is discussed in the next section.

3.2 Dynamical Variational AutoEncoder

DVAE is the combination of a VAE model and a temporal model, such as RNN, LSTM, GRU [2,4,15]. This approach exploits and represents data in an informative latent space with VAE. On the other hand, with the temporal connection, DVAE will take advantage of temporal properties in data, focusing on using the sequential order in data, such as speech and video, to capture more helpful information.

Different from VAE, the generating process in DVAE is defined as:

$$p_\theta(x_{1:T}|z_{1:T}) = \Pi_{t=1}^T p_\theta(x_t|x_{1:t-1}, z_{1:T}), \tag{9}$$

and the inferring process is defined based on the following equation:

$$q_\phi(z_{1:T+1}|x_{1:T}) = \Pi_{t=1}^T q_\phi(z_{t+1}|z_{1:t}, x_{1:T}), \tag{10}$$

with $1:T$ denotes the sequence of T nodes in the data sequence. It means that the Encoder and Decoder networks in DVAE receive a sequence of input nodes, not a single vector. In this context, a node represents a feature vector. DVAE can be seen as a sequence of VAE network, which receives a sequence of input vectors, then processes and yields a sequence of output.

Figure 2 shows some combination of VAEs, briefly described by z and x nodes, with different temporal orders. Corresponding with each combination, a DVAE model can be designed to strengthen the traditional VAE. Depending on the approaches to design, modify, extend, and simplify the sequential relationship in the combination, the probabilities in Eqs. (9) and (10) will be renovated to form a new architecture for DVAE.

In a DVAE architecture, the temporal models, including RNN, LSTM, and GRU, can be used anywhere in the design. They can exist in Encoder, latent code layer, Decoder, or their connections. Based on the data properties or the

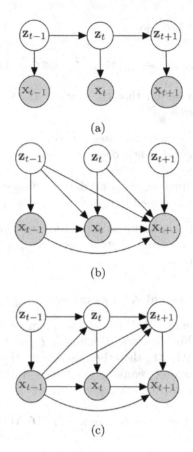

Fig. 2. combination of a temporal model with VAE. Figure 2a is a simple case for sequential relationships in the DVAE model, similar to Markov models. Figures 2b and 2c are the more complicated sequential models.

research target, the architecture should be designed reasonably for the particular problem.

With this approach, there are many different designed architectures for DVAE, such as:

- Deep Kalman Filter (DKF) [19,31]
- Stochastic Recurrent Networks (STORN) [1]
- Variational Recurrent Auto-Encoder (VRAE) [12]
- Recurrent Variational AutoEncoder (RVAE) [22]
- Stochastic Recurrent Neural (SRNN) [13]

3.3   Our DVAE Architecture for Speech Enhancement: Bi-RVAE

We combine a VAE and a temporal model for speech enhancement to form our DVAE, named Bi-RVAE. Notably, we apply Uni-directional RNN in the Encoder

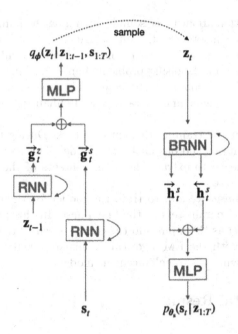

Fig. 3. Bi-RVAE - Our design for DVAE uses Uni-directional RNN for Encoder and Bi-directional RNN for Decoder.

and Bi-directional RNN in the Decoder (Fig. 3). We developed our model based on the idea from RVAE. In this model, Leglaive used BRNN for both Encoder and Decoder networks to get the bi-directional properties of speech signals. This is not necessary. We keep BRNN at Decoder and replace BRNN at Encoder with an RNN. In the end, with BRNN at Decoder, we can capture directly two sides of information from latent variable z, and input indirectly.

In our opinion, using Bi-directional RNN for both Encoder and Decoder is unnecessary. This is because the natural hearing mechanism of humans is hearing one time and thinking many times. Human ears only hear once during the time flow, particularly with human speech or other sounds. This process cannot be reversed, so it is not reasonable if the DVAE contains a Bi-directional in the Encoder. On the other side, after hearing, the human brain can think many times before returning to the result or the conclusion. This motivates us to use the Bi-directional connection for Decoder.

3.4 Two Phases for Training DVAE for Speech Enhancement

Speech enhancement is a particular problem. In this task, the input and the output are similar and different. The input signal contains human speech and other sounds, while the expected output contains human speech only. The difference comes from the other sounds in the input, which are called interference sounds.

There is no information about these sounds, including the number of sounds and the contribution of each sound to the input signal.

DVAE is a generative model, not a denoising model. While speech enhancement is more similar to a denoising problem than generating data, the training process is modified to accomplish the goal. This process includes learning to capture the speech behavior and extracting and enhancing human speech in a mixture signal.

The first phase is the training Re-synthesis task. During this phase, we use clean speech as the input and output for the model. This step aims to train the DVAE network to learn to extract the main contents of the clean speech and represent them in a latent space.

In the second phase, we aim to train the model for the enhancement task using the pre-trained parameters in the first phase. In this phase, we push the mixture to the model as the input and the clean speech as the output. Then we re-train the model with the EM algorithm. This step could tune and transfer the model from re-synthesis to enhancement model.

4   Experimental Result

4.1   Dataset

To evaluate our proposed method for speech enhancement, we designed an experiment with data including human speech and noise signals. We used the TIMIT dataset for human speech [14]. This dataset is a standard English dataset in many speech processing applications. TIMIT includes 630 speakers, and each speaker has recorded 10 utterances. TIMIT contains 6300 utterances with 5.4 hours in duration.

Noise signals used in this research came from the well-known dataset QUT-NOISE [5,6]. It contains a lot of everyday noises such as "home", "street", "car", "cafe", etc. In this work, we used these noises to create the mixture signals. We randomly mixed noise from QUT-NOISE to human speech from TIMIT with SDR among $\{-5, 0, 5\}$ dB. In this way, the mixture contained human speech as the primary content and everyday noises with a moderate amplitude, similar to the natural environment.

Then, the data for the two training phases were organized as follows:

- Re-synthesis phase: (input, output) = (speech, speech)
- Enhancement phase: (input, output) = (mixture, speech)

The networks were trained 20 epochs for each phase. After that, the results were reported in Sect. 4.3.

4.2   Evaluation Method

We used these two measures to evaluate the proposed method for the speech enhancement problem: Perceptual evaluation of speech quality (PESQ) [28] and Extended short-time objective intelligibility (ESTOI) [29].

While PESQ considers the quality of output speech, ESTOI focuses on the intelligibility of speech. Score for PESQ distribute in $[-0.5, 4.5]$), when the score for ESTOI belongs to $[0, 1]$. With these two measures, the higher numbers imply better results.

4.3   Experiments and Results

In this section, we re-implemented these models: DKM, STORN, VRNN, RVAE, and SRNN, to compare with our proposed model Bi-RVAE. Both of the five models above come from DVAE class with different designs in detail. The numerical results are displayed in Tables 1 and 2.

Table 1. comparison the quality in enhanced human speech

Model	PESQ
DKM	2.063
STORN	3.417
VRNN	3.457
RVAE	2.232
SRNN	3.531
Bi-RVAE (our proposed model)	3.233

Table 1 shows the PESQ score for our model and the others. With PESQ, the higher score, the better model. The best result is SRNN with 3.511, higher approximately 0.3 than our model. Although our model Bi-RVAE only achieves 3.233, it provides good quality for the restored sounds. Mainly, with PESQ score larger than 3.2, the model can return a good sound. Compared with the original RVAE, our Bi-RVAE gains a much better score, with more than 1.0 in improvement.

Table 2. Comparison the intelligibility in enhanced human speech

Model	ESTOI
DKM	0.802
STORN	0.937
VRNN	0.940
RVAE	0.855
SRNN	0.944
Bi-RVAE (our proposed model)	0.938

The intelligibility of speeches yielded by different models is shown in Table 2. Similar to PESQ, the ESTOI score is as higher as better. The highest score is from SRNN with 9.44 while our Bi-RNN ranks third with 9.38. Except for DKM and RVAE, the remaining models return perfect results with more than 0.9 for ESTOI.

4.4 Discussion

Via these two experiments, there are at least two crucial observations. The first is the general performance for DVAE class. The second is improving our proposed model Bi-RVAE compared to the original RVAE.

Most DVAE models, including STORN, VRNN, SRNN, and our model Bi-RVAE, achieve outstanding results in both PESQ and ESTOI. This means that the DVAE class shows applicable results in general. Because DVAE includes a generative model and a temporal model, it combines the strengths of these two models. It leads to the fact that DVAE is an appropriate methodology for speech processing.

On the other hand, although the design for Bi-RVAE is a modified form of the original RVAE, our model performs a much better result than RVAE. Bi-RVAE simulates the working process of humans, so the rightness is higher than the RVAE with a more complicated design. The better results from our model imply that a reasonable structure is better than mixing many models without any strategy.

5 Conclusion

This research focuses on speech enhancement and aims to improve speech quality before recognition. Our approach mainly uses and develops Dynamical Variational AutoEncoder (DVAE) to model the speech signal. We also propose a particular design in DVAE class named Bi-RVAE.

Generally, DVAE combines a Variational AutoEncoder (VAE) and a temporal model. Via this approach, the latent variable and latent space in VAE play an essential role in representing the behavior of human speech. Moreover, to take advantage of sequential properties in human speech, a temporal model is combined with VAE to form a more general form of a generative model for sequential data.

Experimental results show that DVAE is a good design for speech enhancement. On the other hand, our design Bi-RVAE, with Uni-directional RNN for Encoder and Bi-directional RNN for Decoder, yields a good result with high-quality restored sounds and could be a potential approach for speech enhancement problems.

Acknowledgement. Hao D. Do was funded by Vingroup JSC and supported by the Ph.D Scholarship Programme of Vingroup Innovation Foundation (VINIF), Institute of Big Data, code VINIF.2022.TS.037.

References

1. Bayer, J., Osendorfer, C.: Learning stochastic recurrent networks (2014). https://doi.org/10.48550/ARXIV.1411.7610
2. Bie, X., Girin, L., Leglaive, S., Hueber, T., Alameda-Pineda, X.: A benchmark of dynamical variational autoencoders applied to speech spectrogram modeling, pp. 46–50 (2021). https://doi.org/10.21437/Interspeech.2021-256
3. Bie, X., Leglaive, S., Alameda-Pineda, X., Girin, L.: Unsupervised speech enhancement using dynamical variational auto-encoders (2021)
4. Bie, X., Leglaive, S., Alameda-Pineda, X., Girin, L.: Unsupervised Speech Enhancement using Dynamical Variational Auto-Encoders (2021). https://hal.inria.fr/hal-03295630. Working paper or preprint
5. Dean, D., Kanagasundaram, A., Ghaemmaghami, H., Rahman, M.H., Sridharan, S.: The QUT-NOISE-SRE protocol for the evaluation of noisy speaker recognition (2015). https://doi.org/10.21437/Interspeech.2015-685
6. Dean, D., Sridharan, S., Vogt, R., Mason, M.: The QUT-NOISE-TIMIT corpus for the evaluation of voice activity detection algorithms, pp. 3110–3113 (2010). https://doi.org/10.21437/Interspeech.2010-774
7. Do, H.D., Chau, D.T., Tran, S.T.: Speech representation using linear chirplet transform and its application in speaker-related recognition. In: Nguyen, N.T., Manolopoulos, Y., Chbeir, R., Kozierkiewicz, A., Trawinski, B. (eds.) ICCCI 2022. LNCS, vol. 13501, pp. 719–729. Springer, Cham (2022). https://doi.org/10.1007/978-3-031-16014-1_56
8. Do, H.D., Chau, D.T., Tran, S.T.: Speech feature extraction using linear chirplet transform and its applications. J. Inf. Telecommun. 1–16 (2023). https://doi.org/10.1080/24751839.2023.2207267
9. Do, H.D., Tran, S.T., Chau, D.T.: Speech separation in the frequency domain with autoencoder. J. Commun. **15**, 841–848 (2020)
10. Do, H.D., Tran, S.T., Chau, D.T.: Speech source separation using variational autoencoder and bandpass filter. IEEE Access **8**, 156219–156231 (2020)
11. Do, H.D., Tran, S.T., Chau, D.T.: A variational autoencoder approach for speech signal separation. In: International Conference on Computational Collective Intelligence (2020)
12. Fabius, O., van Amersfoort, J.R.: Variational recurrent auto-encoders (2014). https://doi.org/10.48550/ARXIV.1412.6581
13. Fraccaro, M., Sønderby, S.K., Paquet, U., Winther, O.: Sequential neural models with stochastic layers. In: Proceedings of the 30th International Conference on Neural Information Processing Systems, NIPS 2016, pp. 2207–2215. Curran Associates Inc., Red Hook (2016)
14. Garofolo, J., et al.: Timit acoustic-phonetic continuous speech corpus. Linguistic Data Consortium (1992)
15. Girin, L., Leglaive, S., Bie, X., Diard, J., Hueber, T., Alameda-Pineda, X.: Dynamical variational autoencoders: a comprehensive review. Found. Trends® Mach. Learn. **15**, 1–175 (2021). https://doi.org/10.1561/2200000089
16. Hao, X., Shi, Y., Yan, X.: Speech enhancement algorithm based on independent component analysis in noisy environment, pp. 456–461 (2020). https://doi.org/10.1109/ICAIIS49377.2020.9194905
17. Jancovic, P., Zou, X., Kokuer, M.: Speech Enhancement and Representation Employing the Independent Component Analysis, pp. 103–113 (2011). https://doi.org/10.2174/978160805172411101010103

18. Kemp, F.: Independent component analysis independent component analysis: principles and practice. J. Roy. Stat. Soc. Ser. D (Stat.) **52** (2003). https://doi.org/10.1111/1467-9884.00369_14

19. Krishnan, R.G., Shalit, U., Sontag, D.: Deep kalman filters (2015). https://doi.org/10.48550/ARXIV.1511.05121

20. Kumar, C., ur Rehman, F., Kumar, S., Mehmood, A., Shabir, G.: Analysis of MFCC and BFCC in a speaker identification system. In: 2018 International Conference on Computing, Mathematics and Engineering Technologies (iCoMET), pp. 1–5 (2018). https://doi.org/10.1109/ICOMET.2018.8346330

21. Kumar, C., Ur Rehman, F., Kumar, S., Mehmood, A., Shabir, G.: Analysis of MFCC and BFCC in a speaker identification system (2018). https://doi.org/10.1109/ICOMET.2018.8346330

22. Leglaive, S., Alameda-Pineda, X., Girin, L., Horaud, R.: A recurrent variational autoencoder for speech enhancement, pp. 371–375 (2020). https://doi.org/10.1109/ICASSP40776.2020.9053164

23. Leglaive, S., Girin, L., Horaud, R.: A variance modeling framework based on variational autoencoders for speech enhancement (2018). https://doi.org/10.1109/MLSP.2018.8516711

24. Leglaive, S., Girin, L., Horaud, R.: Semi-supervised multichannel speech enhancement with variational autoencoders and non-negative matrix factorization (2019). https://doi.org/10.1109/ICASSP.2019.8683704

25. Li, Y., Ye, Z.: Boosting independent component analysis. IEEE Signal Process. Lett. **29**, 1–5 (2022). https://doi.org/10.1109/LSP.2022.3180680

26. Nakhate, S., Singh, R., Somkuwar, A.: Speech enhancement using independent component analysis based on entropy maximization, pp. 154–158 (2010). https://doi.org/10.1142/9789814289771_0030

27. Park, S., Lee, J.: A fully convolutional neural network for speech enhancement, pp. 1993–1997 (2017). https://doi.org/10.21437/Interspeech.2017-1465

28. Rix, A., Beerends, J., Hollier, M., Hekstra, A.: Perceptual evaluation of speech quality (PESQ)-a new method for speech quality assessment of telephone networks and codecs. In: 2001 IEEE International Conference on Acoustics, Speech, and Signal Processing. Proceedings (Cat. No. 01CH37221), vol. 2, pp. 749–752 (2001). https://doi.org/10.1109/ICASSP.2001.941023

29. Van Kuyk, S., Kleijn, W., Hendriks, R.: An evaluation of intrusive instrumental intelligibility metrics. IEEE/ACM Trans. Audio Speech Lang. Process. (2018). https://doi.org/10.1109/TASLP.2018.2856374

30. Wei, H., Shi, X., Yang, J., Pu, Y.: Speech independent component analysis, pp. 445–448 (2010). https://doi.org/10.1109/ICMTMA.2010.604

31. Xue, Q., Li, X., Zhao, J., Zhang, W.: Deep Kalman filter: a refinement module for the rollout trajectory prediction methods (2021)

Vietnamese Multidocument Summarization Using Subgraph Selection-Based Approach with Graph-Informed Self-attention Mechanism

Tam Doan-Thanh[1], Cam-Van Thi Nguyen[2(✉)], Huu-Thin Nguyen[2], Mai-Vu Tran[2], and Quang Thuy Ha[2]

[1] Viettel Group, Hanoi, Vietnam
tamdt9@viettel.com.vn

[2] VNU University of Engineering and Technology (VNU-UET), Hanoi, Vietnam
{vanntc,18021221,vutm,thuyhq}@vnu.edu.vn

Abstract. In Multi-Document Summarization (MDS), the exceedingly large input length is a significant difficulty. In this paper, we designed a graph-based model with Graph-informed Self-attention to capture inter-sentence and inter-document relations in graph. Our models represent the MDS task as a sub-graph selection problem, where the candidate summaries are its subgraphs and the source documents are thought of as a similarity a graph. We add a pair of weight matrix learning parameters that represent the relationship between sentences and the relationship between documents in the cluster, thereby seeing the close relationship and the role of relation types on the graph. Empirical results on the MultiNews and Vietnamese AbMuSu dataset demonstrate that our approach substantially outperforms some baseline models.

Keywords: Vietnamese extractive summarization · subgraph selection · graph-informed self-attention

1 Introduction

The exponential rise in amount of data due to technology and media development makes it challenging to comprehend and analyze text documents. Information summarization involves conveying the document's main idea without requiring a full read-through. This task heavily relies on the estimation of language similarity, with both extractive and abstractive techniques explored for automatic text summarization. Abstractive summarization interprets and analyzes the text using sophisticated methods to recreate the key content in a fresh manner. In contrast, extractive summarization identifies key sections and produces a subset of original sentences. The goal of summarization is to present data while considering the significance of each sentence and its elements. Multi-Document

N. T. Nguyen et al. (Eds.): ACIIDS 2023, LNAI 13996, pp. 259–271, 2023.
https://doi.org/10.1007/978-981-99-5837-5_22

Summarization (MDS) is the process of extracting the pertinent details from a set of documents and filtering out the irrelevant information to create a concise text representation of the documents [12]. MDS can produce more succinct, insightful, and cohesive summaries of several documents by using graphs that capture relationships between textual units.

Graph-based approaches have been successfully applied in many domains such as computer vision, recommendation systems, information extraction, etc. Recent research used graph-based technique for extractive summarization and have seen considerable results compared to other methods. Prior techniques focused on extracting important textual units from documents using graph representations such as a similarity graph based on lexical similarities [5] and discourse graph based on discourse relations [2]. LexRank [5] was introduced by Erkan and Radev (2014) to determine the relative significance of sentences based on the eigenvector centrality in the connectivity graph of cosine similarity between sentences. Recently, Yasunaga et al. (2017) enhance sentence salience prediction by expanding on the rough discourse graph model and accounting for macro-level aspects in sentences. Wang et al. (2019) [16] also provide a graph-based neural sentence ordering model that makes use of an entity connecting graph to represent the interconnectedness of documents globally. Li et al. (2020) [8] based on a unique graph-informed self-attention technique, embed explicit graph representations into the neural architecture. However, these studies have primarily focused on the graph structure of source documents, overlooking the importance of graph structures in producing coherent and insightful summaries.

A well-known deep neural network technique and a graph neural networks (GNNs) are combined in the GraphSum model [3] (2020). It employs a variational autoencoder, which learns latent themes through encoding-decoding, to model the subjects included in a given text. The topic nodes and sentence nodes of a graph are supplied to the GNN. BERT [4] creates the sentence node embeddings while the autoencoder creates the topic node embeddings. Most recently, a new MDS framework called SgSum is proposed by Moye Chen et al. (2021) [1] which formulates the MDS task as a problem of sub-graph selection. In order to extract sub-graphs from encoded documents, the SgSum model employs graph pooling. In other words, it first turns the documents into a big graph and then, using pooling and convolution, creates a number of sub-graphs. Next, a summary is chosen from among these sub-graphs based on ranking.

In this paper, in order to learn inter-document relations that are better and richer in the same cluster, as well as inter-sentence relations, in the same document, we designed a graph-based model with Graph-informed Self-attention to capture these relation in a model. Leverage by [1], we formulates the MDS task as a sub-graph selection problem. To summarize, our contribution is two-fold:

- We have proposed a subgraph selection-based approach to the Vietnamese multi-document summarization problem. This strategy takes use of both the relationships between sentences within the same document (inter-sentence) as well as those within the same cluster of documents (inter-document).

– Extensive experiments on MultiNews, Vietnamese AbMuSu dataset validate the performance of our approach over baseline models including LexRank and SgSum while also surpassing the extractive baseline established by the VLSP2022-AbMuSu competition.

2 Method

This section contains a detailed presentation of method proposed in this paper. The overall framework of our approach is shown in Fig. 1.

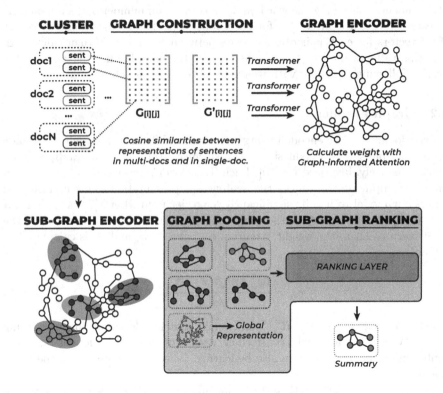

Fig. 1. The overall framework of our approach.

2.1 Graph Construction

A cluster with N documents can be presented as undirected graph $\mathcal{G} = (\mathcal{V}, \mathcal{E})$, where $\mathcal{V}$ denoted sentence nodes representation of each sentence in the input documents and $\mathcal{E}$ is a set of relationships between nodes. Sentences are the basic information units and represented as nodes in the graph. The relations between sentences are represented as edges. We assume that all sentences in a document will be connected to each other. Therefore, any two nodes in the same documents in the same cluster are connected in the graph. Let $\mathbf{G}$ denotes a graph representation matrix of the input documents, where $\mathbf{G}_{ij}$ indicates the tf-idf cosine similarity between sentence S_i and S_j. Specifically, we first represent each document d_i as a bag of words. Then, we calculate the tf-idf value v_{ik} for each token t_{ik} in a document:

$$v_{ik}(t_{ik}) = N_w(t_{ik})log(\frac{N_d}{N_{dw}(t_{ik})}) \tag{1}$$

where $N_w(t)$ is the count of word t in the document, $N_{dw}(t)$ is the total number of documents containing the word, and N_d is the total number of documents. As a result, we get a tf-idf vector for each document. The cosine similarity of the tf-idf vectors for each pair of documents between S_i and S_j is then calculated, and this value is used as the weight G_{ij} for the edge linking the pair on the graph. Following [1], edges with weights less than 0.2 are eliminated.

2.2   Graph Encoder

Source documents are encoded using a Hierarchical Transformer, which is made up of multiple shared weight single Transformers [15] that each handle a document separately, insprised by [10], Each Transformer receives a tokenized document as input and produces the sentences representation. This architecture allows our model to handle significantly lengthier input. Let x_i^{l-1} is the output of the sentence vector S_i in the graph encoding layer $(l-1)$, where x_i^0 is the input vector. Then, the weight attention e_{ij} between S_i and S_j is calculated as follow:

$$e_{ij} = \frac{(x_i^{l-1}W_Q)(x_j^{l-1}W_K)^T}{\sqrt{d_{head}}} \tag{2}$$

where W_Q, W_K and $W_V \in \mathbf{R}^{d \times d}$ are weight matrices learning during training phase. With a Hierarchical Transformer, each sentence will no longer be represented independently, but can also be learned from other sentences in the input document.

In order to capture both the relationship between any two sentences in a document cluster and the relationship between any two sentences in the same document, we propose a new weighting calculation between any two sentences S_i and S_j. Each sentence in the cluster documents can learn information from any other sentence in the cluster while also learning information from other sentences in the same document.

The Gaussian deviation of the weight matrix $\mathbf{G}$ between any two sentences on cluster documents is represented by R_{ij}, while the Gaussian deviation of the weight matrix $\mathbf{G}'$ is represented by R'_{ij}. In order to establish the best parameter pair for each data set, θ and β are modified during training.

$$R_{ij} = -\frac{(1 - \mathbf{G}_{ij}^2)}{2\sigma^2}$$
$$R'_{ij} = -\frac{(1 - \mathbf{G'}_{ij}^2)}{2\sigma^2} \tag{3}$$

In which, σ is the standard deviation, representing the influence of the graph structure. The gaussian bias $R_{ij} \in (-inf, 0]$ measuring the distance between document S_i and S_j. Gaussian bias approximates to multiplying the latent attention distribution by a weight $\in (0, 1]$ because of the exponential operation in the softmax function. R'_{ij} is also setup simmilar to R_{ij}. The new attention score is calculated by the formula:

$$\alpha_{ij} = Softmax(e_{ij} + \theta \times R_{ij} + \beta \times R'_{ij}) \tag{4}$$

This way, our model can learn better and richer inter-document relations in same, and also learn inter-sentence relations in the same document, specifically: (i) Learning from itself, through self-attention weights; (ii) Learning from other sentences through the Hierarchical Transformer, also through self-attention weights; (iii) Learning from other sentences by similarity graph of any two sentences across the entire cluster documents; (iv) Learn from other sentences by graphing similarity between any two sentences in the same document.

2.3 Sub-graph Encoder

Moye Chen et al. [1] proposed a multi-document summarization framework named SgSum which transforms summarization into the problem of sub-graph selection. They claim that the sub-graph structure can reflect the quality of candidate summaries. Using the gold summary, we first get the ROUGE score for each sentence. In order to create candidate summaries, we combine the top M scoring sentences. The sentences in each candidate's summary make up a sub-graph. Leverage by the idea of SgSum model [1], in order to model each sub-graph, we use a sub-graph encoder that has the same architecture as the graph encoder in Sect. 2.2, which means it is also encoded through a Hierarchical Transformer. The best sub-graph is then chosen as the final summary after each sub-graph has been scored in a sub-graph ranking layer. A multi-document dataset has hundreds of sentences, which means that thousands of sub-graphs must be taken into account.

2.4 Graph Pooling and Sub-graph Ranking Layer

In this study, we implement a multi-head weighted pooling procedure similar to Liu and Lapata (2019) [10], to capture the global semantic information of source texts based on the graph representation of the documents. The weight distributions across tokens are computed for each document, enabling the model to adaptively encode documents in different representation subspaces by paying attention to certain words. For each document D_i, for head $z \in 1, \ldots, n_{head}$, attention scores a_i and value vectors b_i are transformed from the input vectors x_i of sentence S_i. Finally, based on a_i, we determine a probability distribution $\hat{a}_i^z$ across tokens inside the document for each head as follow:

$$\hat{a}_i^z = exp(a_i^z)/ \sum_{i=1}^{n} exp(a_i^z) \tag{5}$$

where $a_i^z = W_a^z x_i$ and $b_i^z = W_b^z x_i$; $W_a^z \in \mathbb{R}^{1*d}$ and $W_b^z \in \mathbb{R}^{d_{head}*d}$ are weights; $d_{head} = d/n_{head}$ is the dimension of each head; n is the number of tokens in S_i. Next, we apply the weighted summation with another linear transformation and layer normaliation to produce the $head_z$ vector for the document:

$$head_z = LayerNorm(W_c^z \sum_{i=1}^{n} \hat{a}_i^z b_i^z) \tag{6}$$

where $W_c^z \in \mathbb{R}^{d_{head}*d_{head}}$. The model can adaptably accommodate numerous heads, with each document having multiple attention distributions, focused on various input perspectives. Finally, the model joins all the heads and applies a linear transformation to obtain the overall representation D of the document. $D = W_d[head_1 || \ldots head_2]$ represents the entire input document after learning all the information of the document. Therefore, we will proceed to compare candidate summaries with D in the form of a comparison of two graphs.

Using a greedy approach, we first choose a number of important sentences as candidate nodes before combining them to create candidate sub-graphs. It is easy to produce candidate sub-graphs based on the key sentences since they often serve to assemble important sub-graphs. Then we calculate cosine similarities between all sub-graphs with the global document representation D, and decide which paragraph will serve as the final summary based on its score.

2.5   Joint Optimization

The loss function for our model training includes two parts; the summary-level loss $\mathcal{L}_{summary}$, and the candidate-level loss $\mathcal{L}_{sub-graph}$:

$$\mathcal{L}_{sub-graph}^1 = max(0, cosine(K_j, K) - cosine(K_i, K) + \gamma)(i < j) \tag{7}$$

$$\mathcal{L}_{sub-graph}^2 = 1 - consine(D, K) \tag{8}$$

$$\mathcal{L}_{summary} = - \sum_{i=1}^{n} (y_i^* log(\hat{y}_i) + (1 - y_i^*)log(1 - \hat{y}_i)) \tag{9}$$

$$\mathcal{L} = \mathcal{L}_{summary} + \mathcal{L}_{sub-graph}^1 + \mathcal{L}_{sub-graph}^2 \tag{10}$$

In which, K represents gold summary, K_i represents the candiate summary ranked i và γ is hyperparameter. Specifically, $\mathcal{L}_{sub-graph}^1$ is the loss function between each pair candidate summary. The candidate pairs with more ranking deviation will have higher margin. $\mathcal{L}_{sub-graph}^2$ is the loss between candidate summaries K and input documents representation D. $\mathcal{L}_{summary}$ is the loss function that learn more accurate extracted sentence and summary representation, where a label $y_i \in \{0,1\}$ indicates whether the sentence S_i should be extracted summary sentence.

3 Experiments and Results

This section contains our experiments and results on two datasets. The results show that our model' results are better than the baseline models at ROUGE-1, ROUGE-2 and ROUGE-L.

3.1 Corpus Description

To verify effectiveness of our proposed model, we conduct experiments on two dataset, including MultiNews [6] and VLSP2022-AbMuSu [14]. Table 1 shows the data distribution of MultiNews and VLSP2-22-AbMuSu.

The MultiNews dataset is a commonly used dataset in multi-document summaries with extremely large amounts of data, introduced by Alexander Frabbri et al. in 2019 [6]. The collection includes 56,216 document clusters, each of which consists of two to ten documents on the same topic and is accompanied by a reference summary created by human specialist.

VLSP2022-AbMuSu is a Vietnamese multi-documents summaries dataset [14], including a variety of documents on topics such as economic, social, cultural, scientific and technological. The dataset is divided into three parts: training, validation, and testing. Each dataset is made up of several document clusters, each of which has three to five documents on the same topic. Each document cluster in the test and validation dataset is accompanied with a manually created reference summary. The test dataset has the same structure as the other two, but there is no reference summary associated with each cluster.

Table 1. Data distribution

	MultiNews			VLSP2022-AbMuSu		
	Train	Validation	Test	Train	Validation	Test
Number of clusters	1000	300	300	200	100	300
Number of documents	2926	879	845	621	304	914
Average of documents	3	3	3	3	3	3

3.2   Experimental Setup

Data Preprocessing: The data included in the preprocessing consists of two text files. The first file has several clusters' documents on various topics. Each cluster on a line is a collection of two or more documents, separated by conventional notation. The second file contains golden summaries, one on each line, corresponding to cluster of documents in the first file. After preprocessing, the files are merged into a single json file, with different fields. The fields after data preprocessing include:

- **SRC:** Containing the source document's tokens.
- **TGT:** Containing the gold summary document's tokens.
- **SRC_STR:** Containing segmented source document.
- **TGT_STR:** Contraining segmented gold summary document.
- **SIM_GRAPH:** Matrix representing the relationship between any two sentences across the entire cluster document, calculated based on the cosine similarity tf-idf between the two sentences.
- **SIM_GRAPH_SINGLE:** Matrix representing the relationship between any two sentences in the same document, calculated based on the tf-idf cosine similarity between the two sentences.
- **CLS_IDS:** The starting position of a sentence.
- **SEP_IDS:** The ending position of a sentence.

Evaluation Metrics: ROUGE was firstly introduced by [9]. It includes metrics for automatically evaluating a summary's value by contrasting it with other summaries produced by humans. Between the computer-generated summaries to be assessed and the ideal summaries authored by humans, the measures count the number of overlapping units, such as n-grams, word sequences, and word pairs. In our experiments, we evaluated summarization quality automatically using ROUGE-2 score. Formulas are used to determine the ROUGE-2 Recall (R), Precision (P), and F_1 between the predicted summary and reference summary are presented as follows:

$$P = \frac{|n - grams|_{Matched}}{|n - grams|_{PredictedSum}} \quad ; \quad R = \frac{|n - grams|_{Matched}}{|n - grams|_{ReferencedSum}} \tag{11}$$

$$F_1 = \frac{(2 \times P \times R)}{(P + R)}$$

Experiment Scenario. For both datasets, we conduct preprocessing then train with LexRank [5] and SgSum [1] models as base models. We perform experiments with our model with the following steps:

- *Step 1*: Preprocess the dataset to add a weight matrix $\mathbf{G'}$.
- *Step 2*: Change the parameters θ và β so that $0 \leq \theta, \beta \leq 1$ và $\theta + \beta = 1$.
- *Step 3*: Save the result of the model with the parameter set θ, β of *Step 2*.
- *Step 4*: Repeat *Step 2* until the most optimal set of parameters is selected.

Configuration. We use the base version of Roberta [11] for word segmentation and model initialization. The optimization function used is Adam [7] with first momentum of gradient equals to 0.9 and second momentum of gradient equals to 0.999; learning rate is 0.03; during training, gradient clipping with a maximum gradient norm of 2.0 is also used. Before all linear layers, dropout with a probability of 0.1 is applied. In our models, the feed-forward hidden size is 1,024; the number of heads is 8; and the number of hidden units is set at 256. Transformer encoding layers are set at 6 and graph encoding layers are set at 2, respectively. All the models are trained on Google Colab including 1 GPU (Tesla K80) with 5 epochs.

3.3 Results

MultiNews Dataset. Experimental results on the MultiNews dataset are presented in Table 2. After running the experiment, we found that, with the parameter set $\theta = 0.85$ and $\beta = 0.15$, all three scores ROUGE-1, ROUGE-2 and ROUGE-L reached the best value.

Table 2. ROUGE F_1 evaluation results on the MultiNews.

Model	θ	β	ROUGE-1	ROUGE-2	ROUGE-L
LexRank	1	0	0.4027	0.1263	0.3750
SgSum	1	0	0.4222	0.1418	0.3843
Our model	0.5	0.5	0.4296	0.1525	0.3914
	0.7	0.3	0.4299	0.1523	0.3912
	0.8	0.2	0.4268	0.1494	0.3874
	0.85	**0.15**	**0.4315**	**0.1544**	**0.3925**
	0.9	0.1	0.4277	0.1514	0.3898

Although the results do not differ too much from the SgSum model and LexRank model, the experimental results give us a new perspective on the importance of the coefficient β. Experiments show that, no matter how β changes, the results we get are always better than the baseline models at ROUGE-1, ROUGE-2 and ROUGE-L.

VLSP2022 AbMuSu Dataset. After running the proposed model on the VLSP2022 dataset, the results were evaluated by AbMusu VLSP2022 competition [14]. The experimental results are presented in Table 3. Experimental results show that the highest ROUGE-2 F1 result is 0.28 with parameters $\theta = 0.85$ and $\beta = 0.15$. With this result, the model entered the Top 10 in the AbMusu VLSP2022 competition [14], surpassing the *extractive_ baseline* of the organizers. Passing extractive_baseline means that our model passes all other VLSP2022 baselines, including rule_baseline, anchor_baseline and abstractive_baseline that are shown in Table 5. The results of the teams who passed the *extractive_baseline* are presented in Table 4. Team **minhnt2709** choose important sentences and phases from the original documents with TextRank and PageRank models [13], then they applies abstractive summarization to rewrite summaries with new sentences and phases. The disparity between their method and ours lies in the way they use pretrained language models in Vietnamese for abstractive approach.

Table 3. ROGUE F_1 score on the VLSP 2022 dataset with different parameters.

θ	β	ROUGE-2	ROUGE-1	ROUGE-L
0.5	0.5	0.2397	0.4642	0.4278
0.6	0.4	0.2552	0.4761	0.4418
0.7	0.3	0.2573	0.4748	0.4422
0.8	0.2	0.2811	0.4911	0.4585
0.9	0.1	0.2733	0.4847	0.4523
1.0	0.0	0.2715	0.4842	0.4513
0.85	**0.15**	**0.2823**	**0.4919**	**0.4614**

Table 4. Teams passed *extractive_baseline* on public test set VLSP2022

Rank	Teams	R-2 F_1	R-2 P	R-2 R
1	thecoach_team	0.3150	0.2492	0.4652
2	minhnt2709	0.3149	0.2566	0.4577
3	TheFinalYear	0.2931	0.2424	0.4137
4	TungHT	0.2875	0.2978	0.2989
5	ngtiendong	0.2841	0.2908	0.2949
6	**ThinNH (Our team)**	**0.2823**	**0.2440**	**0.3782**
7	nhanv	0.2752	0.2859	0.2827
8	vingovan	0.2680	0.1988	0.4696
9	extractive_baseline	0.2650	0.2465	0.3268

After conducting experiments with continuously changing weights, the highest result obtained R-2 F_1 is 0.2823, with parameters $\theta = 0.85$ and $\beta = 0.15$. This result is higher than extractive_baseline (1,065 times higher). Not only that, the results obtained are even higher than the unimproved SgSum model (0.2823 vs. 0.2715 respectively). This proves the experimental results show that the improved model with the new formula is initially effective on the Vietnamese dataset.

Table 5. Compare the experimental results of the proposed model with some baseline models

Model	ROUGE-2	ROUGE-1	ROUGE-L
extractive_baseline	0.2650	0.4826	0.4421
rule_baseline	0.2582	0.4640	0.4284
anchor_baseline	0.1931	0.4381	0.3928
abstractive_baseline	0.1457	0.3129	0.2797
Our model	**0.2823**	**0.4919**	**0.4614**

Table 6. Experimental results of choosing the number of summary sentences

Number of summary sentences	ROUGE-2	ROUGE-1	ROUGE-L
5	0.2436	0.4554	0.4191
6	0.2593	0.4714	0.4357
7	0.2731	0.4842	0.4505
8	0.2747	0.4846	0.4528
9	**0.2823**	**0.4919**	**0.4614**

After finding the optimal set of parameters θ and β, we conducts experiments to find the most suitable number of sentences to form a subgraph for the model. Experimental results show that, choose 9 sentences to build a sub-graph are the best. The comparison results when conducting the experiment are shown in Table 6 above.

Conclusion

We provide a subgraph selection-based approach for the Vietnamese multi-document summarization task that makes use of a graph-informed self-attention mechanism. This method makes advantage of both the connections between sentences inside a single document and those within a cluster of related documents. Experimental results on two MDS datasets including Vietnamese dataset demonstrated that our approach is significantly superior to some reliable baselines. Future research will focus on using knowledge graphs and other more informative graph representations to further enhance the quality of the summary. Using GNNs is also an idea that we are considering in encoding representational information between documents.

Acknowledment. Cam-Van Nguyen Thi was funded by the Master, PhD Scholarship Programme of Vingroup Innovation Foundation (VINIF), code VINIF.2022.TS143.

References

1. Chen, M., Li, W., Liu, J., Xiao, X., Wu, H., Wang, H.: SgSum: transforming multi-document summarization into sub-graph selection. arXiv preprint arXiv:2110.12645 (2021)
2. Christensen, J., Soderland, S., Etzioni, O., et al.: Towards coherent multi-document summarization. In: Proceedings of the 2013 Conference of the North American Chapter of the Association for Computational Linguistics: Human Language Technologies, pp. 1163–1173 (2013)
3. Cui, P., Hu, L., Liu, Y.: Enhancing extractive text summarization with topic-aware graph neural networks. In: Proceedings of the 28th International Conference on Computational Linguistics, Barcelona, Spain (Online), pp. 5360–5371. International Committee on Computational Linguistics (2020)
4. Devlin, J., Chang, M.-W., Lee, K., Toutanova, K.: BERT: pre-training of deep bidirectional transformers for language understanding. In: Proceedings of the 2019 Conference of the North American Chapter of the Association for Computational Linguistics: Human Language Technologies, Volume 1 (Long and Short Papers), Minneapolis, Minnesota, pp. 4171–4186. Association for Computational Linguistics (2019)
5. Erkan, G., Radev, D.R.: Lexrank: graph-based lexical centrality as salience in text summarization. J. Artif. Intell. Res. **22**, 457–479 (2004)
6. Fabbri, A.R., Li, I., She, T., Li, S., Radev, D.R.: Multi-news: a large-scale multi-document summarization dataset and abstractive hierarchical model. In: Proceedings of the 57th Annual Meeting of the Association for Computational Linguistics, Florence, Italy, pp. 1074–1084. Association for Computational Linguistics (2019)
7. Kingma, D.P., Ba, J.: Adam: a method for stochastic optimization. arXiv preprint arXiv:1412.6980 (2014)
8. Li, W., Xiao, X., Liu, J., Wu, H., Wang, H., Du, J.: Leveraging graph to improve abstractive multi-document summarization. In: Proceedings of the 58th Annual Meeting of the Association for Computational Linguistics, pp. 6232–6243. Association for Computational Linguistics (2020)
9. Lin, C.-Y.: Rouge: a package for automatic evaluation of summaries. In: Text Summarization Branches Out, pp. 74–81 (2004)
10. Liu, Y., Lapata, M.: Hierarchical transformers for multi-document summarization. In: Proceedings of the 57th Annual Meeting of the Association for Computational Linguistics, Florence, Italy, pp. 5070–5081. Association for Computational Linguistics (2019)
11. Liu, Y., et al.: Roberta: a robustly optimized BERT pretraining approach. arXiv preprint arXiv:1907.11692 (2019)
12. Mani, K., Verma, I., Meisheri, H., Dey, L.: Multi-document summarization using distributed bag-of-words model. In: 2018 IEEE/WIC/ACM International Conference on Web Intelligence (WI), pp. 672–675. IEEE (2018)
13. Nguyen, T.-M., et al.: LBMT team at VLSP2022-abmusu: hybrid method with text correlation and generative models for Vietnamese multi-document summarization. arXiv preprint arXiv:2304.05205 (2023)
14. Tran, M.-V., Le, H.-Q., Can, D.-C., Nguyen, Q.-A.: VLSP 2022 - abmusu challenge: Vietnamese abstractive multi-document summarization. In: Proceedings of the 9th International Workshop on Vietnamese Language and Speech Processing (VLSP 2022) (2022)

15. Vaswani, A., et al.: Attention is all you need. In: Advances in Neural Information Processing Systems, vol. 30 (2017)

16. Wang, D., Liu, P., Zheng, Y., Qiu, X., Huang, X.: Heterogeneous graph neural networks for extractive document summarization. In: Proceedings of the 58th Annual Meeting of the Association for Computational Linguistics, pp. 6209–6219. Association for Computational Linguistics (2020)

NegT5: A Cross-Task Text-to-Text Framework for Negation in Question Answering

Tao Jin, Teeradaj Racharak$^{(\boxtimes)}$ (ID), and Minh Le Nguyen (ID)

School of Information Science,
Japan Advanced Institute of Science and Technology, Nomi, Japan
{morgan,racharak,nguyenml}@jaist.ac.jp

Abstract. Negation is a fundamental grammatical construct that plays a crucial role in understanding QA tasks. It has been revealed that models trained with SQuAD1 still produce original responses when presented with negated sentences. To mitigate this issue, SQuAD2.0 incorporates a plethora of unanswerable questions to enable pre-trained models to distinguish negative inquiries. In this study, we assess the performance of the model on answerable and unanswerable questions that incorporate negative words and find out that the model's performance on unanswerable negative questions surpasses the baseline. However, the model's performance on answerable negative questions falls short of the baseline. This outcome prompts us to surmise that SQuAD2.0 includes a substantial number of unanswerable questions, but the pattern of these questions is typically limited to the addition of negative adverbs such as "never" and "not". As a result, the trained model tends to produce "unanswerable" responses when confronted with questions that contain negative expressions. To address this issue, we propose a novel framework, called *NegT5*, which adopts the text-to-text multi-task fine-tuning principle introduced in T5 for making the model able to deal with negation in QA.

Keywords: Negation · SQuAD2.0 · T5 · Dual Fine-tuning · Cross-labels

1 Introduction

The expression of sentences used by the question and answering (QA) sessions in daily conversation is generally not fixed. For example, "Where can watermelons be grown?" and "Where can watermelons not be grown?" are straightforward examples of sentence expression. Unsurprisingly, a model trained on the SQuAD1 [9] dataset yields the same answer from these two queries because the dataset does not include negative examples. To address this issue, SQuAD2.0 [10] incorporates a plethora of unanswerable questions, such as "Who did not invent the computer?", enabling the model to learn to distinguish negation in sentences.

N. T. Nguyen et al. (Eds.): ACIIDS 2023, LNAI 13996, pp. 272–285, 2023.
https://doi.org/10.1007/978-981-99-5837-5_23

Table 1. The amount of negative questions with (answerable) and without answers (unanswerable) contained in well-known QA datasets; The statistical results show that the number of question-answer pairs containing negatives is very small.

Dataset	Answerable	Unanswerable	Total
SQuAD2.0	1728	4692	6420
TriviaQA [3]	1012	0	1012
NQ [7]	1350	0	1350
QuAC [1]	346	77	423
NewsQA [12]	1626	758	2384

In SQuAD2.0, negative data is prevalent, comprising 85% of questions containing "n't" and 89% of questions containing "never", which reverse the sentence's meaning. However, in real-world scenarios, sentences frequently employ multiple negatives with diverse forms and contextual positions that they could affect.

This research is based on two primary observations: **(1)** In SQuAD2.0, since almost all unanswerable questions contain negatives, models trained on SQuAD2.0 will tend to give "unanswerable" responses when confronted with negations. Furthermore, the model's performance decreases when it encounters more complex negatives. **(2)** We also investigate five other well-known QA datasets (cf. Table 1) and find out that the data containing negative questions is quite limited and the sentence types are quite homogeneous. Regarding this point, it is relatively easy to generate questions containing negation by adding negation to the original dataset, yet the answer may not be included in the context due to the change in semantics of the question. Generating or collecting QA data specifically for many complex questions with negation is cost-prohibitive.

Therefore, we propose a cross-task text-to-text framework following the training scheme used by T5 [8] to solve this problem. Our framework, called *NegT5*, extends T5 with new negation tasks that could let the model handle negation in QA. The rationale behind NegT5 is that the model's ability to deal with negation can be improved by fine-tuning T5 in cross tasks enriched with our collected negation corpus. We conduct extensive experiments to research tasks that can be unified to handle negation in sentences. Our main contributions are twofold: (1) the NegT5 framework and (2) our negation dataset collected to train NegT5.

In our experiments with T5small, we extract the question and answer data containing negation words from SQuAD2.0 for an initial test with T5small. The results, as shown in Table 2, show the bias of the model's inference on answerable negative questions. We propose NegT5 with a fine-tuning scheme to overcome this problem. Indeed, we consider the negation cue detection task and negation scope resolution task as our primary tasks, and apply different data pre-processing methods proposed by Raffel et al. [8]. Our approach significantly improves the model's accuracy for answering questions containing negation.

2   Related Work

2.1   Handling Negation in QA

Comprehending negation is a significant challenge in the realm of QA. Recently, [4] manually incorporated negation into 305 question sentences in the SQuAD1.0 dataset and found that BERT largely overlooked negation. We also further examine five popular QA datasets and find that these datasets also overlooked negation (cf. Table 1). The notable exception is SQuAD2.0, which performed commendably in regard to negation. Nonetheless, SQuAD2.0 has a pronounced bias towards negation, in particular 85% of questions contain "n't" and 89% contain "never" [10]. Hence, the model cannot learn to distinguish negation effectively from the datasets. In marked contrast to other work, our work highlights the fact that the SQuAD2.0 dataset addresses the inadequacy of the SQuAD 1 model in recognizing negation by incorporating negation into question sentences. Despite the aforementioned improvement, the negation data remains insufficient and the sentence structure too rudimentary, with an over-reliance on a limited range of negative words. This not only leads to a substantial bias in the model but also exacerbates the paucity of QA task data that contains negation and is answerable. In response to this conundrum, we propose a solution of pre-training the model with negative task data, thereby elevating its ability to handle negative questions and mitigate the impact of the aforementioned limitations.

Table 2. Test results with T5small on the negative questions sampled from SQuAD2.0 show that the model does not perform well for questions containing negatives

Test Set	EM	F1
Unanswerable negative question	96.6	96.6
Answerable negative question	**47.4**	**51.6**
Question not contain negatives	66.7	69.8
All data	68.4	71.3

2.2   Negation Dataset

The BioScope [11] comprises a collection of medical and biological texts that have been annotated for negation, speculation, and their linguistic scope, with the aim of comparing the development and resolution of negation/hedging detection systems. The authors have compiled a wealth of negations and identified the ranges affected by these negations. The SFU reviews [6] for negation and speculation annotations have also amassed a corpus of book, movie, and consumer product reviews, which have been divided into categories such as cars, cookware, books, phones, computers, hotels, music, and movies. Unlike SQuAD2.0, which simply adds negatives such as "not", "never", BioScope and SFU review datasets have more comprehensive negatives with more complex syntaxes.

Table 3. Primary negation tasks

Negation Cue Detection and Scope Resolution
Sample: I've vowed never to live through a holiday season when I don't watch the animated How the Grinch Stole Christmas!   • Here, "never" inverts the meaning of "to live through a holiday season"   • "never" is the negation cue   • "to live through a holiday season" is the scope of the negation cue "never"
Negative Words
The negation words are not only "not" and "never" but also different kinds of negation words included in the negation cue detection task.   • An affix: (im)perfect, (a)typical, ca(n't)   • A single word: not, no, failed, lacks, unable, without, neither, instead, absent, . . .   • A set of consecutive words or discontinuous words: neithor . . . nor . . .

Negation Tasks

There are two main tasks for negation: negation cue detection and negation scope resolution. Negation cue detection is the task of identifying negation words in the text. Negation scope resolution is finding the words affected by the negation cue in text. As shown in Table 3, the sentence *"I've vowed never to live through a holiday season when I don't watch the animated How the Grinch Stole Christmas!"* Here, "never" inverts the meaning of "to live through a holiday season", and "never" is the negation cue. Also, "to live through a holiday season" is the scope of the negation cue "never".

Hence, for the negation cue detection task, an input sentence *"I've vowed never to live through a holiday season when I don't watch the animated How the Grinch Stole Christmas!"* should output *"never"*. In the negation scope resolution task, the input is the same as the negation cue detection task, but the output should be *"to live through a holiday season"*. It is also worth mentioning that the types of negatives in negation cue detection are classified according to the negatives contained in the training dataset. For the data annotation method of BioScope and SFU review datasets in the negation tasks, we follow the method of [5] to define four categories of negation cue labels: (1) label 0 stands for affix, (2) label 1 stands for normal cue, (3) label 2 stands for part of a multiword cue, and (4) label 3 stands for not a cue.

3 Methodology

To handle negation in QA, this paper introduces a novel two-stages fine-tuning method, called *NegT5* (cf. Fig. 1). In the first stage, we extend the training scheme of T5 with primary negation tasks (cf. Table 3) to reinforce the model to recognize negation. With the introduced training scheme, we *firstly* fine-tune T5 using our curated datasets (cf. Sect. 4). We also experiment this stage by fine-tuning under the negation tasks plus other auxiliary tasks in our ablation

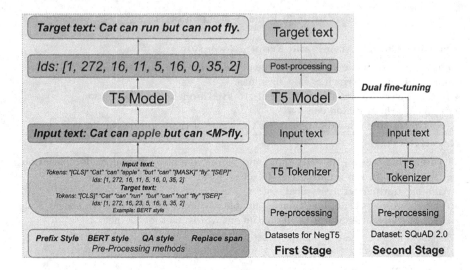

Fig. 1. Our proposed NegT5; The model is fine-tuned in two stages, the first stage is to fine-tune the original T5 model based on cross-tasks learning with primary negative tasks. The second stage is to fine-tune on the SQuAD2.0 dataset.

study (cf. Sect. 4). Since this work aims to address negation in QA, therefore we propose to perform *secondly* fine-tuning with SQuAD2.0 in the second stage to enhance the model with its capability of handling negation in QA tasks. We call our fine-tuning scheme as the *dual fine-tuning*.

More specifically, as illustrated in Fig. 1, we pre-process input text samples from the Bioscope and SFU review datasets for the first fine-tuning. We employ four different ways of pre-processing (cf. Subsects. 3.1–3.4), the Prefix style, BERT style, QA style, and Replace span, to the samples. Then, we tokenize each pre-processed sample, input the processed token into the T5 model for processing to output the token id of each word in the sentence. Finally, we convert the token back to a natural language text. For the second stage, we save our fine-tuned NegT5 model, tokenize text samples from the pre-processed SQuAD2.0 dataset, input each processed token to the NegT5 for processing for getting answer tokens, and finally convert the tokens back to natural language text. The results of this dual fine-tuning yield a novel model that can handle negation in QA.

3.1 Pre-processing Method: Prefix Style

Prefix Style is a kind of multitask learning that prefixes each task before it, and is used in T5. It makes part of its content visible in the encoder. As shown in [8] the prefix style allows the model to apply a fully visible mask to the input. Causal masking with prefixes allows self-attentive mechanisms to use fully visible masking on a part of the input sequence.

Table 4. The structures of input and output of the prefix style

Task name	Input	Output
Negation Cue Detection	Negation cue detection: sentence </s>	negative words
Negation Scope Detection	Negation scope resolution: sentence </s>	words affected by negation
Sentiment Classification	Sentiment classification: sentence </s>	positive & negative
Question Answering	question: sentence context: sentence </s>	answer

Table 5. The structures of input and output of BERT style

Task name	Input	Output
Negation Cue Detection	cat can *apple* but can <M> fly	cat can run but can not fly
Negation Scope Detection	cat can *apple* but can not <M>	cat can run but can not fly
Sentiment Classification	this movie is boring </s> <M>	this movie is boring </s> negative

Though we refer to the method of [8], the difference is that we have changed the output to the part we are concerned about, instead of outputting the original sentence. As shown in the output part of Table 4, in the negation cue detection task, our output is the negation word. In the negation scope resolution task, our output is the word affected by the negation word. Lastly, in the sentiment classification, we only output positive or negative.

3.2 Pre-processing Method: BERT Style

BERT Style refers to BERT's "Masked Language Modeling" (MLM) [2]. MLM takes an input text and corrupts 15% of its tokens. Ninety percent of the corrupted tokens are replaced with special mask tokens and 10% are replaced with random tokens. Inspired by this method, we also corrupt the tokens of a part of the sentence randomly. Instead of randomly masking as in BERT, we directly mask output labels in the main tasks: negation cue detection and negation scope resolution (cf. Table 5). For the sentiment classification task, we add output tags to the input sentences, split them with "/s", and finally mask the output tags. The output is the original sentence.

3.3 Pre-processing Method: QA Style

QA Style is employed to better handle the domain shift problem in our NegT5. Our main task (Negation) and our target task (QA) have different forms of

Table 6. The structures of input and output of QA style

Task name	Input	Output
Negation Cue Detection	question: what cue.. context: cat can..</s>	negative words
Negation Scope Detection	question: what scope.. context: cat can..</s>	words affected by negation
Sentiment Classification	question: what senti.. context: this movie..</s>	positive & negative

Table 7. The questions of QA style

Task name	Question
Negation Cue Detection	Which word is negative?
Negation Scope Detection	Which words are affected by negative words?
Sentiment Classification	What sentiment is contained in the sentence?
Sentiment Classification - invert the answer	What sentiment is not contained in the sentence?

inputs and outputs. Hence, it is important to pay our attention on solving the domain shift problem. We set all the tasks as QA tasks and then observe whether the model gives better results on SQuAD2.0 or not.

Since negation cue detection and negation scope resolution are problem- and context-free, we turn the original input into a context and then set up a unified question. As shown in Table 7, we set the questions of the negation cue detection task uniformly as: "*Which word is negative?*". The task of negation scope resolution is set uniformly as: "*Which words are affected by negative words?*". In particular, we reverse 50% of the labels of the sentiment classification task and set it up as a positive and negative question: "*What sentiment is contained in the sentence?*" and "*What sentiment is not contained in the sentence?*" (Table 6).

3.4   Pre-processing Method: Replace Span

Replace Span is a special data preprocessing format. In our experiments, we combine the negation cue detection and negation scope resolution into one task. We remove both the negation and the negation-affected words from the sentence, and let the model predict the words in the removed part. For example, we process a sentence "*Cat can run but can not fly.*". The words "*not*" and "*fly*" are negative word and the word affected by the negative word, respectively. Each corrupted token is replaced by special tokens (i.e. <X> and <Y>, resp.). In the example, the input will become "*Cat can run but can <X> <Y>.*" and we want to use these tokens to bring the model's attention to the sentence structure we urge the model to learn. The output sequence then consists of the deleted spans, bounded by sentinel tokens used to replace them in the input, plus the final ending token <Z>. That is, the output will be "<X> *not* <Y> *fly* <Z>.".

4 Experimental Setup and Results

4.1 Setting

Our experimental environment is Google Colab Pro configured to use TPU acceleration. We implemented NegT5 with the huggingface's Transfomers framework and Tokenizer, trained with 500 epochs and used the early stop technique. The batch size is set to 16 and the learning rate to 1e-4, trained by eight TPU cores. The data size for each task based on its corresponding dataset is shown on Table 8.

4.2 Prefix Style

SQuAD2.0. As shown in Table 9, NegT5 under the prefix style performs the worst under the negation tasks. In particular, the task with sentiment analysis performs better. Since the original dataset is much more than our primary negation tasks, we also adjust the sentiment analysis dataset to see if the amount of data from the auxiliary task would affect the model performance. As a result, the model perform better when we adjust the data to 4500 (i.e. consistent with the amount of data for Negation cue detection and Negation scope resolution).

Negation Test Set Extracted from SQuAD2.0. we further test on the negation test set extracted from SQuAD2.0 (cf. Table 10). The results show that our models improve significantly in answerable questions except for NegT5 fine-tuned with other pre-processing. NegT5small-prefix (sentiment-4500+negation) performs the best among other baselines in all test sets.

Table 8. The data size of each task

Dataset	Task Name	Data Size
BioScope + SFU	Negation cue detection	4497
BioScope + SFU	Negation scope resolution	4497
GLUE SST-2 (4500)	Sentiment Classification	4500
GLUE SST-2 (full)	Sentiment Classification	67349
Answerable negative QA	Question Answering	3851
Test dataset for Negation	Task name	Data size
SQuAD2.0 (Have answers)	Question Answering	152
SQuAD2.0 (No answer)	Question Answering	764

Table 9. Evaluation with the SQuAD2.0 test set; different data pre-processing methods lead to different results on each fine-tuned model. The bold parts show our results with respect to each baseline. The names in parentheses < > represent the pre-processing methods. For example, Prefix stands for the preprocessing method "Prefix style".

Model	EM	F1
T5small <baseline-*SQuAD1*>	76.7	85.6
NegT5small (Prefix)	76.7	85.6
NegT5small (BERT)	**77.0**	**85.8**
T5base <baseline-*SQuAD2.0*>	78.6	82.0
NegT5base (BERT)	**78.9**	**82.3**
T5small <baseline-*SQuAD2.0*>	68.4	71.3
NegT5small (Prefix)	68.0	71.1
NegT5small (Prefix) + sent-4500	**69.1**	**72.5**
NegT5small (Prefix) + sent-full	**68.2**	**71.7**
NegT5small (Prefix) + answerableQA	**68.4**	**72.1**
NegT5small (BERT)	**69.4**	**72.7**
NegT5small (BERT) + sent-4500	**69.0**	**72.4**
NegT5small (BERT) + sent-full	**68.8**	**72.3**
NegT5small (QA) + sent-4500	**68.5**	**72.1**
NegT5small (Replace-span)	**68.5**	**72.0**

4.3 BERT Style

SQuAD2.0. As shown in Table 11, NegT5 under the BERT-style setting has the best performance. Indeed, the models under these settings perform better than those with other auxiliary tasks. Also, NegT5small-BERTstyle (sentiment-4500+negation) with a controlled amount of data for the sentiment classification task outperform NegT5 (sentiment-full+negation). We test that 15% of all data are masked randomly without leaving 10% unmasked and find that the performance of the model dropped significantly.

Negation Test Set Extracted from SQuAD2.0. Our model performs slightly worse in the unanswerable questions (cf. Table 12), but performs much better than baselines in the answerable questions. In particular, we observe that NegT5small (sentiment-full+negation) is lower than the other experimental settings in both baseline and unanswerable negative questions, but is higher than the other models in answerable questions. This means that more positive data allows NegT5 to perform better in answerable questions. But, it performs worse in unanswerable questions.

Table 10. Evaluation with answerable and unanswerable question pairs from the test set of SQuAD2.0; we describe earlier that the simple addition of negatives to questions in SQuAD2.0 result in mostly unanswerable questions for the data. This makes the model biased. Therefore, we test the scores of answerable negative questions and unanswerable negative questions separately.

Model	Negative Questions	EM	F1
T5base <baseline>	No answer	96.2	96.2
NegT5base (BERT)	No answer	**96.6**	**96.6**
T5small <baseline>	No answer	96.6	96.6
NegT5small (Prefix)	No answer	96.1	96.1
NegT5small (Prefix) + sent-4500	No answer	**96.7**	**96.7**
NegT5small (Prefix) + sent-full	No answer	94.8	94.8
NegT5small (Prefix) + answerableQA	No answer	95.0	95.0
NegT5small (BERT)	No answer	95.8	95.8
NegT5small (BERT) + sent-4500	No answer	96.5	96.5
NegT5small (BERT) + sent-full	No answer	94.5	94.5
NegT5small (QA) + sent-4500	No answer	94.8	94.8
NegT5small (Replace-span)	No answer	95.4	95.4
T5base <baseline>	Have answers	61.8	68.2
NegT5base (BERT)	Have answers	61.8	67.3
T5small <baseline>	Have answers	47.4	51.6
NegT5small (Prefix)	Have answers	44.0	47.7
NegT5small (Prefix) + sent-4500	Have answers	**52.6**	**57.1**
NegT5small (Prefix) + sent-full	Have answers	**53.3**	**57.6**
NegT5small (Prefix) + answerableQA	Have answers	**52.6**	**58.5**
NegT5small (BERT)	Have answers	**53.3**	**57.7**
NegT5small (BERT) + sent-4500	Have answers	**52.7**	**57.3**
NegT5small (BERT) + sent-full	Have answers	**55.3**	**59.1**
NegT5small (QA) + sent-4500	Have answers	**55.3**	**60.1**
NegT5small (Replace-span)	Have answers	**51.3**	**52.2**

4.4   QA Style

SQuAD2.0. As aforementioned, the goal of employing the QA Style is to unify all data pre-processing formats into QA formats, allowing the models to improve their performance on the QA task (which is our main task). Table 13 shows that the performance is improved as expected.

Negation Test Set Extracted from SQuAD2.0. As shown in Table 14, there is a small drop in model performance in the non-answerable negative questions. But, in the answerable negative questions, the EM score is as high as 55.3 and the F1 score is as high as 60.1, a very large improvement in the performance.

Table 11. NegT5 with the BERT style pre-processing measured on the test set.

Model	2nd Fine-Tune	EM	F1
T5small \<baseline>	SQuAD1.1	76.7	85.6
NegT5small-BERTstyle (negation)	SQuAD1.1	**77.0**	**85.8**
T5small \<baseline>	SQuAD2.0	68.4	71.3
NegT5small-BERTstyle (negation)	SQuAD2.0	**69.4**	**72.7**
NegT5small-BERTstyle (negation)-all-mask	SQuAD2.0	68.9	72.2
NegT5small-BERTstyle (sentiment-full+negation)	SQuAD2.0	68.8	72.3
NegT5small-BERTstyle (sentiment-4500+negation)	SQuAD2.0	69.0	72.4
T5 \<baseline>	SQuAD2.0	78.6	82.0
NegT5-BERTstyle (negation)	SQuAD2.0	**78.9**	**82.3**

Table 12. NegT5 with the BERT style pre-processing; performance is measured on negative questions with answers and no answers on QA pairs.

Model	Negative Questions	EM	F1
T5small \<baseline>	No answer	96.6	96.6
NegT5small-BERTstyle (negation)	No answer	95.8	95.8
NegT5small-BERTstyle (negation)-all-mask	No answer	95.8	95.8
NegT5small-BERTstyle (sentiment-4500+negation)	No answer	96.5	96.5
NegT5small-BERTstyle (sentiment-full+negation)	No answer	94.5	94.5
T5small \<baseline>	Have answers	47.4	51.6
NegT5small-BERTstyle (negation)	Have answers	**53.3**	**57.7**
NegT5small-BERTstyle (negation)-all-mask	Have answers	**50.7**	**55.0**
NegT5small-BERTstyle (sentiment-4500+negation)	Have answers	**52.7**	**57.3**
NegT5small-BERTstyle (sentiment-full+negation)	Have answers	**55.3**	**59.1**

Table 13. NegT5 with the QA style pre-processing measured on the test set.

Model	2nd Fine-Tune	EM	F1
T5small \<baseline>	SQuAD2.0	68.4	71.3
NegT5small-QAstyle (sentiment-4500+negation)	SQuAD2.0	**68.5**	**72.1**

Table 14. NegT5 with the QA style pre-processing; performance is measured on negative questions with answers and no answers QA pairs.

Model	Negative Questions	EM	F1
T5small <baseline>	No answer	96.6	96.6
NegT5small-QAstyle (sentiment-4500+negation)	No answer	94.8	94.8
T5small <baseline>	Have answers	47.4	51.6
NegT5small-QAstyle (sentiment-4500+negation)	Have answers	**55.3**	**60.1**

Table 15. NegT5 with the replace span pre-processing measured on the test set.

Model	2nd fine-tune	EM	F1
T5small <baseline>	SQuAD2.0	68.4	71.3
NegT5small-replace-span (negation)	SQuAD2.0	**68.5**	**72.0**

Table 16. NegT5 based on replace span pre-processing; performance is measured on negative questions with answers and no answers QA pairs.

Model	negative questions	EM	F1
T5small <baseline>	No answer	96.6	96.6
NegT5small-replace-span (negation)	No answer	95.4	95.4
T5small <baseline>	Have answers	47.4	51.6
NegT5small-replace-span (negation)	Have answers	**51.3**	**52.2**

4.5   Replace Span

SQuAD2.0. Replace span is also a special data pre-processing format. In this experiment, we combine the negation cue detection and the negation scope resolution into one task by removing both the negation and the negation-affected words from the sentence and letting the model to predict words in the removed parts. Our experiments (Table 15) show that the replace span pre-processing also help the models performs better than baselines.

Negation Test Set Extracted from SQuAD2.0. In contrast to other pre-processing methods, Table 16 shows that the improvement of NegT5 under the replace span pre-processing is not much for answerable negative questions and is slightly lower than baseline for non-answerable negative questions.

5   Result Analysis

Key results are shown in Tables 9 and 10, where selected pre-processing is shown in the parentheses, "sent" stands for "sentiment", and "answerableQA" stands for "Answerable negative QA". The results reveal three interesting points.

(1) NegT5 prefers to give "unanswerable" in negative questions in SQuAD2.0. Almost all unanswerable questions in the dataset contain "'t" and "never", causing the bias for negative questions. (2) NegT5 improve the overall performance – not only the answerable negative questions. (3) NegT5 performs better in a context that has diverse negatives and contains long and complex sentences. Long sentences with multiple negatives are difficult to label manually and our approach that fine-tunes across tasks can solve this challenge very well; it performs well even in more extended contexts and questions containing more negatives.

6   Conclusion

Negation is a critical grammatical for the QA task to understand. This paper proposes a cross-labels dual fine-tuning method based on the T5 model, called NegT5, in order to improve the performance of the QA task that contains negation in sentences. In this study, we have made the following contributions. (1) Our NegT5 improves the overall performance. Indeed, it improve the performance in answerable negatives by 8%. Our extensive experiments show that different ways of the data pre-processing offer different performance of the fine-tuned model. (2) Long sentences with multiple negations and multiple negatives are difficult to label manually and our approach of training the model across labels with the negation tasks solves this problem very well. Our model performs better in more extended contexts and questions that contain more negatives.

References

1. Choi, E., et al.: QuAC: question answering in context. In: EMNLP (2018)
2. Devlin, J., Chang, M.W., Lee, K., Toutanova, K.: BERT: pre-training of deep bidirectional transformers for language understanding. In: NAACL-HLT (2019)
3. Joshi, M., Choi, E., Weld, D.S., Zettlemoyer, L.: TriviaQA: a large scale distantly supervised challenge dataset for reading comprehension. In: ACL (2017)
4. Kassner, N., Schütze, H.: Negated and misprimed probes for pretrained language models: birds can talk, but cannot fly. In: ACL (2020)
5. Khandelwal, A., Sawant, S.: NegBERT: a transfer learning approach for negation detection and scope resolution. In: LREC (2020)
6. Konstantinova, N., De Sousa, S.C., Díaz, N.P.C., López, M.J.M., Taboada, M., Mitkov, R.: A review corpus annotated for negation, speculation and their scope. In: LREC, pp. 3190–3195 (2012)
7. Kwiatkowski, T., et al.: Natural questions: a benchmark for question answering research. Trans. Assoc. Comput. Linguist. 7, 453–466 (2019)
8. Raffel, C., et al.: Exploring the limits of transfer learning with a unified text-to-text transformer. J. Mach. Learn. Res. 21(1), 5485–5551 (2020)
9. Rajpurkar, P., Zhang, J., Lopyrev, K., Liang, P.: SQuAD: 100,000+ questions for machine comprehension of text. In: EMNLP (2016)
10. Sen, P., Saffari, A.: What do models learn from question answering datasets? In: EMNLP (2020)

11. Szarvas, G., Vincze, V., Farkas, R., Csirik, J.: The bioscope corpus: annotation for negation, uncertainty and their scope in biomedical texts. In: Proceedings of the Workshop on Current Trends in Biomedical Natural Language Processing, pp. 38–45 (2008)

12. Trischler, A., et al.: NewsQA: a machine comprehension dataset. In: RepL4NLP (2017)

An Empirical Study on Punctuation Restoration for English, Mandarin, and Code-Switching Speech

Changsong Liu$^{(\boxtimes)}$, Thi Nga Ho, and Eng Siong Chng

Nanyang Technological University, Singapore, Singapore
{liuc0062,ngaht,aseschng}@ntu.edu.sg

Abstract. Punctuation restoration is a crucial task in enriching automated transcripts produced by Automatic Speech Recognition (ASR) systems. This paper presents an empirical study on the impact of employing different data acquisition and training strategies on the performance of punctuation restoration models for multilingual and codeswitching speech. The study focuses on two of the most popular Singaporean spoken languages, namely English and Mandarin in both monolingual and codeswitching forms. Specifically, we experimented with in-domain and out-of-domain evaluation for multilingual and codeswitching speech. Subsequently, we enlarge the training data by sampling the codeswitching corpus by reordering the conversational transcripts. We also proposed to ensemble the predicting models by averaging saved model checkpoints instead of using the last checkpoint to improve the model performance. The model employs a slot-filling approach to predict the punctuation at each word boundary. Through utilizing and enlarging the available datasets as well as ensemble different model checkpoints, the result reaches an F1 score of **76.5%** and **79.5%** respectively for monolingual and codeswitch test sets, which exceeds the state-of-art performance. This investigation contributes to the existing literature on punctuation restoration for multilingual and code-switch speech. It offers insights into the importance of averaging model checkpoints in improving the final model's performance. Source codes and trained models are published on our Github's repo for future replications and usage.(https://github.com/charlieliu331/Punctuation_Restoration)

Keywords: Punctuation Restoration · Multilingual · Codeswitching · Automatic Speech Recognition · Singaporean Speech

1 Introduction

Punctuation is an essential aspect of language since it helps to convey meaning and structure clearly. As shown in Fig. 1, an ASR system first takes in raw audio files and outputs transcripts. However, transcripts generated from most ASR systems generally remain unpunctuated, resulting in significant difficulties

N. T. Nguyen et al. (Eds.): ACIIDS 2023, LNAI 13996, pp. 286–296, 2023.
https://doi.org/10.1007/978-981-99-5837-5_24

for human and machine processing of the transcripts. This includes reducing readability, impacting comprehension, as well as disturbing and deteriorating downstream tasks' performance [1], such as machine translation [2], sentence dependency parsing [3], and sentiment analysis [4]. Hence, we need Punctuation Restoration (PR) to address this problem. Some existing approaches treat PR as a sequence labeling task [5], while others treat it as a slot-filling task [6], which is what we employ in this paper. Our objective is to investigate the impact of using in-domain and out-domain datasets, and model averaging on the multilingual and codeswitching punctuation restoration model in the speech transcript through an empirical study. We employed the same network architecture proposed by [6]. Our motivation is to produce a punctuation restoration model that can punctuate codeswitching speech, to provide insights on the impact of data preparation and model ensemble on the performance of the model, and thus to advance the state-of-the-art. This in turn will contribute to cross-cultural and cross-language communications and the performance of downstream tasks. We will present our method and analyze the impact of using in-domain and out-of-domain data, strategies for data acquisition, and model ensemble on the performance of a punctuation restoration model in subsequent sections.

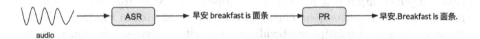

Fig. 1. ASR process diagram.

2 Related Work

Restoring punctuations for speech transcripts generally involves utilizing prosodic features and lexical features. Prosodic features, derived from audio, consist of characteristics such as pause duration, and fundamental frequency(F0). There are studies like [7,8] that use LSTM to punctuate texts with prosodic features including pause durations. Subsequent studies, such as [9], have modified the approach by replacing LSTM with bidirectional LSTM cells and incorporating an attention mechanism. However, using prosodic features can be problematic due to individual deviations in speech patterns such as tone and pitch. Lexical features, on the other hand, have been seen to be the sole input sources to various models such as HMM [10], Conditional Random Field (CRF) [11], word embedding [12], and attention-based deep learning [13]. These lexical-based approaches are commonly easier to implement than prosody-based methods. This is because they only require processing text inputs and do not require aligning audio with text. Some studies have shown that combining both prosodic and lexical features can yield the best performance. For example, in [14], the authors combined both lexical and prosodic features by integrating word-based and prosody-based networks to compensate for the deficiencies of each approach. Within the scope of this paper, we employed only lexical features acquired from texts for simplification.

For the task of multilingual punctuation restoration, recent research has focused on the utilization of multilingual language models (MLM), including Multilingual T5 [15], Multilingual-BERT [16], and XLM-RoBERTa [17], etc. These models can generate multilingual embeddings that are customizable for a wide range of subsequent tasks, including punctuation restoration. They come as pre-trained and are suitable for natural language processing (NLP) tasks. Incorporating MLMs into punctuation restoration can provide a more robust and accurate approach to handling diverse language varieties, code-switching, and multilingual texts. Previous works such as [16,17] are able to handle multilingual punctuation restoration, however, to our best knowledge there is no existing work in handling code-switching speech.

3   Methodology

3.1   Network Architecture

This work employs the off-the-shelf model architecture proposed by [6] for multilingual punctuation which consists of a pre-trained XLM-RoBERTa (XLM-R) model followed by a classifier. The architecture is illustrated in Fig. 2. The XLM-R model generates multiple embeddings in a self-supervised way and contains multiple self-attention transformer-encoder layers. The input data is tokenized using the multilingual tokenizers, i.e. SentencePiece and Jieba, and subsequently the <mask> token is added after every word. The classifier comprising two full-connected (FC) layers predicts punctuation for each masked token. The output contains tokens that represent commas, periods, or spaces. To explore the impact of data preparation and acquisition on training a Punctuation Restoration (PR) model, only this network architecture has been employed to train different PR models using datasets that have been prepared and acquired in different ways.

3.2   Datasets

Multilingual and Codeswitching Datasets. We aim to build a punctuation restoration model that can handle both multilingual speech, i.e. speakers either speak English or Mandarin, and code-switch, i.e. speakers speech English and Mandarin interchangeably either within a conversation or even within a sentence. To serve this purpose we utilized the IWSLT2012 TED Talks dataset and our in-house English-Mandarin codeswitching dataset. The IWSLT2012 TED Talk dataset, named IWSLT2012 or TED, is the same as the dataset used in [6,13]. IWSLT2012 contains the English transcripts extracted from EN-FR parallel machine translation tracks and Mandarin transcripts extracted from the ZH-EN parallel tracks. IWSLT2012 is considered a multilingual dataset in which its topic of discussion are varied and its speakers are from anywhere in the world. The in-house dataset is recorded by our team with recording devices and manually transcribed by human transcribers, we named it NTU-EnMan or EnMan in

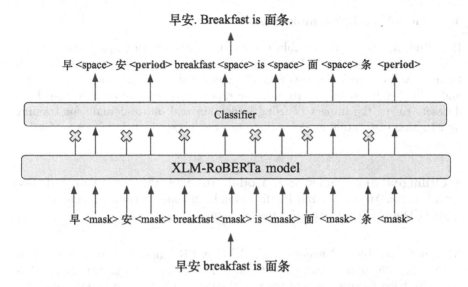

Fig. 2. Model architecture diagram.

short. NTU-EnMan contains conversational speech between two Singaporeans. The speakers may speak in English, Mandarin, or codeswitch between the two languages naturally during the entire conversation. The topics of discussion varied from daily life conversations such as hobbies, personal opinions on specific events, or road direction. The conversation between any pair of speakers was recorded in separate audio channels and transcribed accordingly into separate transcript files.

Data Acquisition. With regard to enlarging the training dataset, we acquire additional variations of NTU-EnMan by employing two preprocessing strategies for the NTU-EnMan dataset. Specifically, we first combine all the transcripts from each speaker in conversational order, however, we let the interrupting speech be kept ordered according to its start time. Meaning in a conversation between speaker A and speaker B if speaker A has not finished her sentence but speaker B interrupted, we keep the conversation order as it is. For the second strategy, we kept transcripts from speaker A's and speaker B's speeches separate. That means the conversational context between speaker A and speaker B has been lost. We named this extra set of data sampling NTU-EnMan-extra and used it together with TED and NTU-EnMan for the model training. Overall, our experiments on data preprocessing and combination aim to comprehensively evaluate the performance of our model under different data conditions, including various combinations of languages, conversation structures, and conversation orderings.

3.3 Models and Ensemble Model

Baseline Models. We established our baselines by employing the network architecture described in 3.1 to train punctuation restoration models for multilingual and codeswitching speech. The models are trained separately on TED and NTU-EnMan datasets. Results are reported on evaluation sets from both datasets to see the impact of using in-domain and out-of-domain for training punctuation restoration models. These two models stand as baselines for further experiments.

Multilingual and Codeswitch Model. To enable the capacity of predicting both English, Mandarin, and English-Mandarin codeswitching speech, we used both TED and NTU-EnMan for model training.

Model Ensemble. Conventionally, the best PR model is usually selected at the training checkpoint that gains the best results on development sets. In this work, we tried to average multiple model checkpoints to yield an overall better model. We first save the checkpoint of the model at each training epoch in every training experiment. We then compare the performance of each training epoch on development sets to choose the top-performance checkpoints. These checkpoints are then averaged to further improve the final models.

4 Experiments

In this section, details of the datasets are described, and the experimental setup and results are presented and discussed.

4.1 Data Preprocessing

Dataset preprocessing is a critical step in preparing data for machine learning models. In this section, we describe the preprocessing steps we used both IWSLT2012 and NTU-EnMan datasets.

Text Normalization. For NTU-EnMan, which is stored in TextGrid files, we first extract the text from each file and combine the resulting text files into a single complete text corpus.

We then perform text normalization to remove unnecessary symbols such as brackets, "–EMPTY–", and non-speech noises.

For the IWSLT2012 dataset, we used dual-lingual datasets prepared by [6], in which separate English and Chinese tracks are combined. For all the mentioned datasets, punctuations are converted or removed to match three classes: period, comma, and space. We convert question marks ('?') and exclamation marks ('!') to periods ('.') to indicate the end of a sentence.

Dataset Splitting. After preprocessing the datasets, we split them into training, validation, and test sets. For the IWSLT2012 dataset, we used the provided train, validation, and test splits. For NTU-EnMan, we chose an 80:10:10 split for train, validation, and test sets. It's worth noting that when splitting the data into train, validation, and test sets, we performed the splitting in a conversation-based manner to ensure that the model learns to predict punctuation within the context of a complete conversation. To provide insights into the characteristics of the preprocessed datasets, we present statistics on the number of words, periods, and commas in each dataset in Table 1.

Table 1. Number of words, commas, and periods in train, valid, and test sets.

Dataset	train			valid			test		
	word	comma	period	word	comma	period	word	comma	period
IWSLT2012	4.6M	350K	145K	36.4K	2.5K	1.5K	47.7K	3.3K	2.2K
NTU-EnMan	1.1M	67.4K	82K	144K	9.9K	10.3K	132K	8.5K	10K

Overall, our dataset preparation process aims to standardize the datasets, ensure they are suitable for machine learning models and provide a consistent evaluation methodology across all datasets. These statistics can be used to assess the distribution of punctuation in the datasets and to identify potential biases or imbalances in the data.

4.2   Experiment Setup

The experiments were set with parameters presented as follows:

- The network architecture was modified so that the classifier comprises two linear layers with dimensions of 1568 and 3, respectively. This is to reduce the output layer from 4 to 3 which corresponds to the prediction of 3 output classes, i.e. periods, commas, and spaces.
- Learning rates were set at 1e-4 for the classification layers and 3e-5 for the XLM-R model which follows the setting in the original model [6].
- For loss function we use Negative Log-Likelihood Loss, a commonly used loss function, particularly for classification tasks.
- We use the RAdam [18] optimizer, and a warm-up phase with 300 warm-up steps to gradually increase the learning rate from a small value to the optimal value.
- The maximum number of epochs was set at 10 and batch size was set at 4.
- Precision, Recall, and F1-score metrics were used to evaluate the models' performance on predicting three punctuation classes: period, comma, and space.

4.3 Results and Discussion

Multilingual and Codeswitching Models. To establish baselines for the model's performance, we initiated training it on each dataset. Subsequently, we train the model using the combination of both datasets expecting that the model is able to predict punctuations for both English, Mandarin and English-Mandarin codeswitching speech. The results are presented in Table 2 for models trained at 1 and 5 epochs. We named the models by their corresponding data sets, i.e. TED, EnMan and TED + EnMan. It can be seen that among the baselines models, which is TED and EnMan, the models performed well on their corresponding validation sets, however, degraded significantly when tested with the other validation set. For example, looking at the results when models were trained at 1 epoch, TED model achieved 74.9% F1-score when being tested with TED validation set, however, it only achieved 66.3% on EnMan validation set, resulting in 8.3% absolute reduction. The gap is even larger for the EnMan model, where the reduction was 25.2% from 77.0% to 51.8% in F1-score. This is possible because the size of training data in EnMan is significantly more than TED, and also EnMan dataset contains codeswitching sentences that further confused the model. The results were further biased when the models trained at 5 epochs, in which the gap between F1-score on TED and EnMan validation sets when tested on the TED model increased to 11.4% and 28.2% with EnMan model. These gaps, however, were reduced significantly when both the data were used to train TED+EnMan model. Performance of the model at 1 epoch was 77.3% on EnMan val., and 74.2% on TED val., and at 5 epochs was 78.6% and 74.4% respectively. It is also interesting to see that the best performance on EnMan val. was achieved at 5 epochs of training for TED+EnMan model, however, the model performance on TED val. is slightly reduced, from 74.9% with TED model to 74.4% with TED+EnMan model.

Table 2. Performance of baseline models and codeswitching model.

Models	1 epoch		5 epoch	
	PR/RC/F1	PR/RC/F1	PR/RC/F1	PR/RC/F1
	TED Val.	EnMan Val.	TED Val.	EnMan Val.
TED	**71.4/78.8/74.9**	59.8/74.5/66.3	**70.8/79.4/74.9**	55.8/73.8/63.5
EnMan	46.4/63.0/51.8	74.2/81.0/77.0	43.9/61.4/50.4	**76.3/81.1/78.6**
TED + EnMan	70.0/79.0/74.2	**75.2/79.5/77.3**	70.9/78.2/74.4	76.0/81.3/78.6

Data Acquisition. Next, we add NTU-EnMan-extra to the training set, and retrain the model resulting in a model name TED+EnMan+EnMan-extra. Results of this model in comparison to results of TED+EnMan model are presented in Table 3. Unfortunately, the results showed that the model with extra data from NTU-EnMan-extra does not help to improve the model performance on the validation sets. In fact, there were slight reductions in most of the cases, except on the EnMan val. when tested on the TED+EnMan+EnMan-extra model trained at 1 epoch. This probably suggested that acquiring additional training data by resampling or re-ordering the existing data doesn't help in improving the model performance.

Table 3. Performance on two datasets combined.

Models	1 epoch		5 epochs	
	PR/RC/F1	PR/RC/F1	PR/RC/F1	PR/RC/F1
	TED Val.	**EnMan Val.**	**TED Val.**	**EnMan Val.**
TED+EnMan	**70.0/79.0/74.2**	75.2/79.5/77.3	**70.9/78.2/74.4**	**76.0/81.3/78.6**
TED+EnMan+EnMan-extra	70.6/78.0/74.0	**75.4/81.7/78.1**	69.6/78.4/73.7	76.0/80.9/78.4

Model Ensemble

Performance of Models at One to Ten Epochs. We extended the training of models for single and combined datasets up to 10 epochs. Figure 3 shows the performance results of each model with results reporting on the TED val. EnMan val. and weighted F1-score combining TED val. + EnMan val. It can be seen TED model performs best at the 5th epoch, EnMan performs best at the 4th epoch and TED+EnMan performs best at the 3rd epoch. However, results are not much different from the 3rd epoch to the 7th epoch on all models. It seems that after 7 epochs, the performance of all models shows a tendency of degrading because of overfitting.

Performance of Averaged Models. Inspired by work reported in [18], the best model is formed by averaging a selected range of training epochs, we found that this applied to the punctuation restoration task, too. Our empirical experiments showed that averaging from the 4th epoch to the 6th epoch produced the best results on TED, EnMan, and TED+EnMan models. Table 4 shows the F1-score obtained on the best-performing model from the 1st to 10th epochs, models that were averaged from the 4th to 6th epochs, and models that were averaged from the 5th to 7th epochs. Similar results were observed when we evaluated these models on test sets from TED and EnMan datasets as showed in Table 5. This confirms our hypothesis that ensemble model works best for the task. Our final model averaged from epochs 4th to 6th achieved **75.8%** and **76.0%** on the TED validation and test sets, which is higher than results reported on [6]. Furthermore, our model has the capacity of predicting punctuations for codeswitching speech with F1 scores of **78.8%** for validation set and **78.4%** for test set, which significantly outperforms the model that only trained on English, Mandarin multilingual speech corpus, i.e. TED.

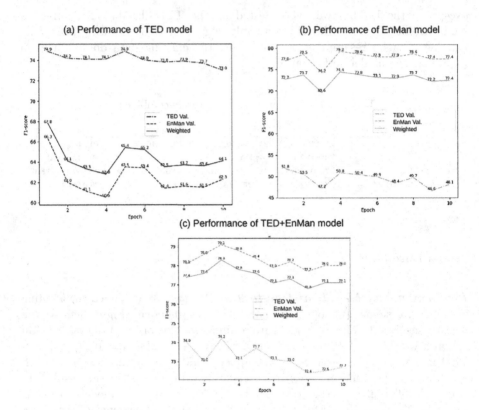

Fig. 3. F1-score on evaluation sets when trained models up to 10 epochs

Table 4. Performance best epoch and averaged epochs on validation sets.

Models	best epoch		averaged 4-5-6		averaged 5-6-7	
	TED Val.	EnMan Val.	TED Val.	EnMan Val.	TED Val.	EnMan Val.
TED	**74.9**	66.3	75.3	63.2	74.9	63.4
EnMan	50.8	**79.2**	51.6	**79.5**	51.1	**79.5**
TED + EnMan	74.4	78.6	**75.8**	78.8	**75.4**	79.1

Table 5. Performance best epoch and averaged epochs on test sets.

Models	best epoch		averaged 4-5-6		averaged 5-6-7	
	TED Test	EnMan Test	TED Test	EnMan Test	TED Test	EnMan Test
TED	74.3	66.1	**76.5**	63.2	**76.1**	63.5
EnMan	52.5	**78.3**	53.2	**78.7**	52.9	78.4
TED + EnMan	**75.2**	78.1	76.0	78.4	75.6	**78.5**

5 Conclusion and Future Work

In this study, we have investigated the importance of using the right dataset and checkpoint combinations for improving the performance of a multilingual and codeswitching punctuation restoration model. Our study offers insights into the challenges and opportunities of working with multilingual and codeswitching transcripts, highlighting the importance of developing tools and technologies that can support communication across languages and cultures. We believe that the methodology and findings presented in this paper can be applied to other areas of NLP. While our investigation has shown promising results, there are several limitations and areas for future research to consider. For example, we can explore different model architectures or incorporate additional datasets. This could yield further improvements in performance. Additionally, further research can be conducted on more languages. Overall, our study contributes to the existing literature on punctuation restoration and demonstrates the importance of dataset combination and epoch selection for improving model performance. We hope that our findings will inspire further research in this area, supporting the development of more accurate and efficient punctuation restorations.

Acknowledgements. This research is supported by the National Research Foundation, Singapore under its AI Singapore Programme (AISG Award No: AISG2-100E-2022-102). We would like to acknowledge the High Performance Computing Centre of Nanyang Technological University Singapore, for providing the computing resources, facilities, and services that have contributed significantly to this work.

References

1. Żelasko, P., Szymański, P., Mizgajski, J., Szymczak, A., Carmiel, Y., Dehak, N.: Punctuation prediction model for conversational speech. arXiv preprint arXiv:1807.00543 (2018)
2. Peitz, S., Freitag, M., Mauser, A., Ney, H.: Modeling punctuation prediction as machine translation. In: Proceedings of the 8th International Workshop on Spoken Language Translation: Papers, pp. 238–245 (2011)
3. Spitkovsky, V.I., Alshawi, H., Jurafsky, D.: Punctuation: making a point in unsupervised dependency parsing. In: Proceedings of the Fifteenth Conference on Computational Natural Language Learning, pp. 19–28 (2011)
4. Nagy, A., Bial, B., Ács, J.: Automatic punctuation restoration with bert models. arXiv preprint arXiv:2101.07343 (2021)
5. De Lima, T.B., et al.: Sequence labeling algorithms for punctuation restoration in brazilian portuguese texts. In: Intelligent Systems: 11th Brazilian Conference, BRACIS 2022, Campinas, Brazil, November 28-December 1, 2022, Proceedings, Part II, pp. 616–630. Springer, Cham (2022). https://doi.org/10.1007/978-3-031-21689-3_43
6. Rao, A., Thi-Nga, H., Eng Siong, C.: Punctuation restoration for singaporean spoken languages: English, malay, and mandarin. In: 2022 Asia-Pacific Signal and Information Processing Association Annual Summit and Conference (APSIPA ASC), pp. 546–552. IEEE (2022)

7. Baptista, F., Caseiro, D., Mamede, N., Trancoso, I.: Recovering punctuation marks for automatic speech recognition. In: Eighth Annual Conference of the International Speech Communication Association (2007)

8. Tilk, O., Alumäe, T.: Lstm for punctuation restoration in speech transcripts. In Sixteenth Annual Conference of the International Speech Communication Association (2015)

9. Tilk, O., Alumäe, T.: Bidirectional recurrent neural network with attention mechanism for punctuation restoration. In: Interspeech, pp. 3047–3051 (2016)

10. Hasan, M., Doddipatla, R., Hain, T.: Multi-pass sentence-end detection of lecture speech. In: Fifteenth Annual Conference of the International Speech Communication Association (2014)

11. Lu, W., Ng, H.T.: Better punctuation prediction with dynamic conditional random fields. In: Proceedings of the 2010 Conference on Empirical Methods in Natural Language Processing, pp. 177–186 (2010)

12. Che, X., Wang, C., Yang, H., Meinel, C.: Punctuation prediction for unsegmented transcript based on word vector. In: Proceedings of the Tenth International Conference on Language Resources and Evaluation (LREC 2016), pp. 654–658 (2016)

13. Makhija, K., Ho, T.-N., Chng, E.-S.: Transfer learning for punctuation prediction. In: 2019 Asia-Pacific Signal and Information Processing Association Annual Summit and Conference (APSIPA ASC), pp. 268–273. IEEE (2019)

14. Szaszák, G., Tündik, M.A.: Leveraging a character, word and prosody triplet for an ASR error robust and agglutination friendly punctuation approach. In: INTERSPEECH, pp. 2988–2992 (2019)

15. Xue, L., et al.: mt5: a massively multilingual pre-trained text-to-text transformer. arXiv preprint arXiv:2010.11934 (2020)

16. Pires, T., Schlinger, E., Garrette, D.: How multilingual is multilingual bert? arXiv preprint arXiv:1906.01502 (2019)

17. Conneau, A., et al.: Unsupervised cross-lingual representation learning at scale. arXiv preprint arXiv:1911.02116 (2019)

18. Liu, L., et al.: On the variance of the adaptive learning rate and beyond. arXiv preprint arXiv:1908.03265 (2019)

Novel Topic Models for Parallel Topics Extraction from Multilingual Text

Kamal Maanicshah⬤, Narges Manouchehri⬤, Manar Amayri⬤,
and Nizar Bouguila$^{(\boxtimes)}$⬤

Concordia Institute for Information Systems Engineering, Concordia University,
Montreal, QC, Canada
{kamalmaanicshah.mathinhenry,narges.manouchehri}@mail.concordia.ca,
{manar.amayri,nizar.bouguila}@concordia.ca

Abstract. In this work, we propose novel topic models to extract topics from multilingual documents. We add more flexibility to conventional LDA by relaxing some constraints in its prior. We apply other alternative priors namely generalized Dirichlet and Beta-Liouville distributions. Also, we extend finite mixture model to infinite case to provide flexibility in modelling various topics. To learn our proposed models, we deploy variational inference. To evaluate our framework, we tested it on English and French documents and compared topics and similarities by Jaccard index. The outcomes indicate that our proposed model could be considered as promising alternative in topic modeling.

Keywords: Multi-lingual topic models · Generalized Dirichlet distribution · Beta-Liouville distribution · mixture allocation

1 Introduction

Natural language processing (NLP) as a joint branch of science including linguistics, statistics and computation has grabbed lots of attention. This exciting domain of research has diverse fields such as semantic analysis and topic modeling. One of the main assumptions that could not be generalized is that the language of resources is English. Lots of models that have been conventionally used were designed to model monolingual contexts only and work with monolingual resources [2]. Considering the ongoing increase in technology specially in using online resources, more content from other languages besides English are becoming available. However, translating these valuable documents to English and using them in NLP algorithms that just work with one language is a great challenge and it is so costly and needs lots of time. Thus, there is a growing interest in finding solutions which could help scientists and industries to work with language-independent text mining tools without needing any translation resources. To tackle this issue, multilingual NLP has been introduced and helped scientists to extract information regarding topics from various data sources and documents [6,8]. In this method, various languages are tied together which helps

© The Author(s), under exclusive license to Springer Nature Singapore Pte Ltd. 2023
N. T. Nguyen et al. (Eds.): ACIIDS 2023, LNAI 13996, pp. 297–309, 2023.
https://doi.org/10.1007/978-981-99-5837-5_25

in discovering the connections in the languages of interest and building coherent topics across them. This helps us in indexing similar topics across multiple languages which helps in multilingual document retrieval [12,14]. Also, we don't have any linguistic assumption about documents or data that we intend to model. This capability empowers our model to relax the constraint of modeling just one language and identifies similar patterns across multiple corpora in various languages. Such models with their power of inference on documents could be interesting in many applications [13].

There are several methods for topic modelling. Latent Dirichlet Allocation (LDA) is one of the most well-known and popular methods for topic modelling [2,7]. In this work, we propose a novel version of LDA in which we assume that the priors are generalized Dirichlet and Beta-Liouville distributions. Conventionally, the prior for LDA is Dirichlet distribution which has a negative covariance [4]. Using generalized Dirichlet (GD) lets us to relax this constraint. However, GD requires twice the number of parameter compared to Dirichlet to be estimated. In order to avoid this we propose to use Beta-Liouville (BL) distribution [3] as prior which also allows more general covariance. Also, we assume that our model has a nonparametric structure which provides us considerable flexibility to model several topics in multiple languages. To do so, we use Dirichlet process (DP) [5]. This elegant method helps to address another task which is defining model complexity. To learn our model, we applied variational learning [1]. To evaluate the model performance, we applied it to a real world dataset with two languages, English and French. We measure the quality of topics by comparing the coherence scores of the different models. We measure the similarity between topics in different languages with Jaccard index. Our experimental outcomes demonstrate the practicality of our proposed model in finding topics by processing multi-lingual documents.

The remainder of this work is organized as follows: In Sect. 2, we explain in detail how to construct DP-based latent generalized Dirichlet allocation (DP-LGDA) and DP-based latent Beta-Liouville allocation (DP-LBLA) models, respectively. This is followed by Sect. 3 which is devoted to experimental results. In Sect. 4, we conclude and discuss future works.

2   Model Description

Let us consider a set of D documents in M different languages, where, $d = \{1, 2, ..., D\}$ and $m = \{1, 2, ..., M\}$ represent the dth document and m^{th} language, respectively. Each document d in language m can be represented as a word vector $\boldsymbol{w}_{md} = (w_{md1}, w_{md2}, ..., w_{mdN_{md}})$, where, N_{md} is the number of words in that particular document. The nth word in a document can be represented by an indicator vector which is V_m dimensional, corresponding to the vocabulary size of language m following the rule, $w_{mdnv} = 1$ when the word w_{mdn} is the same as the word v_m in the vocabulary and 0, otherwise. Similarly, we also define a latent indicator variable $\mathcal{Z}_m = \{z_{md}\} = \{z_{mdn}\}$ showing which of the K topics the word belongs to based on the criteria $z_{mdnk} = 1$ if word w_{mdn} is present in topic

k and 0 if not. Each language has a separate variable β_{mk} which describes the distribution of words in each topic, given by, $\beta_{mk} = (\beta_{mk1}, \beta_{mk2}, ..., \beta_{mkV_m})$. To define the prior for the topic distribution for each document in a general manner, let's say $p(\boldsymbol{\theta} \mid \boldsymbol{\Phi})$ is the prior given the parameter of that distribution $\boldsymbol{\Phi}$. In the case of LDA this distribution is Dirichlet. It is to be noted that in our case the topic probabilities are drawn from an infinite mixture model. $\mathcal{Y} = (\boldsymbol{y}_1, \boldsymbol{y}_2, ..., \boldsymbol{y}_D)$ is the indicator matrix which stipulates which cluster the document belongs to, where $\boldsymbol{y}_d$ is L dimensional with $y_{dl} = 1$ when the document d belongs to cluster l. Here L is the truncation level set for the DP mixture. $\mathcal{Y}$ has a multinomial distribution with parameters $\boldsymbol{\pi} = (\pi_1, \pi_2, ..., \pi_L)$ corresponding to the mixing coefficient and follows the constraint $\sum_{l=1}^{L} \pi_l = 1$. The mixing coefficients here will follow a stick-breaking process to construct a DP model. The main idea here is to define a common set of topic proportion vectors $\boldsymbol{\theta}$ which is shared by all the languages. This restricts the topic proportion vectors that each document can take and forces the topic word proportions across multiple languages to have a similar structure. Doing this helps us to extract parallel topics across languages. The generative process for our multilingual model can be written as follows:

– For each language corpus m in the dataset:
  • For each word vector $\boldsymbol{w}_{md}$ in that corpus:
    * Draw component l from the mixture $y_d = l \backsim DirichletProcess(\boldsymbol{\pi})$
    * Draw topic proportions $\boldsymbol{\theta}_d \mid y_d = l$ from a mixture of L distributions
    * For each word n of the N_d words in document $\boldsymbol{w}_{md}$
      · Draw topic $z_{mdn} = k \backsim Multinomial(\boldsymbol{\theta}_d)$
      · Draw word $w_{mdn} = v_m \mid z_{mdn} = k \backsim Multinomial(\boldsymbol{\beta}_{z_{mdn}})$

The marginal likelihood of this multilingual topic model can thus be written as,

$$p(W \mid \boldsymbol{\pi}, \boldsymbol{\Phi}, \boldsymbol{\beta}) = \prod_{m=1}^{M} \prod_{d=1}^{D} \int \left[\left(\sum_{y_d} p(\boldsymbol{\theta}_d \mid y_d, \boldsymbol{\Phi}) p(y_d \mid \boldsymbol{\pi}) \right) \right.$$
$$\left. \times \prod_{n=1}^{N_d} \sum_{z_{mdn}} p(w_{mdn} \mid z_{mdn}, \boldsymbol{\beta}_m) p(z_{mdn} \mid \boldsymbol{\theta}_d) \right] d\boldsymbol{\theta}_d \qquad (1)$$

for a multilingual corpus W.

2.1 Dirichlet Process Based Latent Generalized Dirichlet Allocation

When we use a generalized Dirichlet prior for the topic proportions $\boldsymbol{\theta}_d$, the distribution for each topic k takes the form,

$$p(\theta_{dk} \mid \sigma_{lk}, \tau_{lk}) = \frac{\Gamma(\tau_{lk} + \sigma_{lk})}{\Gamma(\tau_{lk}) \Gamma(\sigma_{lk})} \theta_{dk}^{\sigma_{lk}-1} \left(1 - \sum_{j=1}^{k} \theta_{dj} \right)^{\gamma_{lk}} \qquad (2)$$

where $(\sigma_{l1}, \sigma_{l2}, ..., \sigma_{lN_d}, \tau_{l1}, \tau_{l2}, ..., \tau_{lN_d})$ are the parameters of GD distribution and $\gamma_k = \tau_k - \tau_{k+1} - \sigma_{k+1}$ for $k = 1, 2, ..., K - 1$ and $\gamma_k = \sigma_k - 1$ for $k = K$.

Since considering mixture of distributions helps us to improve flexibility, we consider a mixture of GD distributions as prior for our model. Thus we can write the prior for our topic proportions as,

$$p(\boldsymbol{\theta}_d \mid \boldsymbol{y}_d, \boldsymbol{\sigma}, \boldsymbol{\tau}) = \prod_{l=1}^{\infty} \prod_{k=1}^{K} \Big(p(\theta_{dk} \mid \sigma_{lk}, \tau_{lk}) \Big)^{y_{dl}} \tag{3}$$

Since, $\boldsymbol{y}_d$ follows a multinomial with parameter π, we can write $p(\boldsymbol{y}_d)$ as, $p(\boldsymbol{y}_d) = \prod_{l=1}^{\infty} \pi_l^{y_{dl}}$. By using a stick-breaking reconstruction of DP, replacing π_j as a function of λ_j, the equation becomes,

$$p(\mathcal{Y} \mid \boldsymbol{\lambda}) = \prod_{d=1}^{D} \prod_{l=1}^{\infty} \left[\lambda_l \prod_{o=1}^{l-1} (1 - \lambda_o) \right]^{y_{dl}} \tag{4}$$

The first part of Eq. 1 can hence be written as,

$$p(\boldsymbol{\theta}_d \mid \boldsymbol{y}_d, \boldsymbol{\sigma}, \boldsymbol{\tau}) p(\boldsymbol{y}_d \mid \boldsymbol{\pi}) \Big) = \prod_{l=1}^{\infty} \prod_{k=1}^{K} \left[\Big(\lambda_l \prod_{o=1}^{l-1} (1 - \lambda_o) \Big) \Big(p(\theta_{dk} \mid \sigma_{lk}, \tau_{lk}) \Big) \right]^{y_{dl}} \tag{5}$$

$p(w_{mdn} \mid z_{mdn}\boldsymbol{\beta}_m)$ and $p(z_{mdn} \mid \boldsymbol{\theta}_d)$ are multinomials given by,

$$p(w_{mdn} \mid z_{mdn}, \boldsymbol{\beta}_m) = \prod_{k=1}^{K} \Big(\prod_{v=1}^{V} \beta_{mkv}^{w_{mdnv}} \Big)^{z_{mdnk}} \tag{6}$$

$$p(z_{mdn} \mid \boldsymbol{\theta}_d) = \prod_{k=1}^{K} \theta_{dk}^{z_{mdnk}} \tag{7}$$

We use Gamma priors which have proven to be adequate [4]. Hence the priors for the parameters of GD are given by,

$$p(\sigma_{lk}) = \mathcal{G}(\sigma_{lk} \mid \upsilon_{lk}, \nu_{lk}) = \frac{\nu_{lk}^{\upsilon_{lk}}}{\Gamma(\upsilon_{lk})} \sigma_{lk}^{\upsilon_{lk}-1} e^{-\nu_{lk}\sigma_{lk}} \tag{8}$$

$$p(\tau_{lk}) = \mathcal{G}(\tau_{lk} \mid s_{lk}, t_{lk}) = \frac{t_{lk}^{s_{lk}}}{\Gamma(s_{lk})} \tau_{lk}^{s_{lk}-1} e^{-t_{lk}\tau_{lk}} \tag{9}$$

where $\mathcal{G}(\cdot)$ represents a Gamma distribution. Authors in [2] introduced a concept called variational smoothing to handle data which are sparse in nature. Following this idea we establish a Dirichlet prior for $\boldsymbol{\beta}_m$ as,

$$p(\boldsymbol{\beta}_{mk} \mid \boldsymbol{\kappa}_{mk}) = \frac{\Gamma(\sum_{v=1}^{V_m} \kappa_{mkv})}{\prod_{v=1}^{V_m} \Gamma(\kappa_{mkv})} \prod_{v=1}^{V_m} \beta_{mkv}^{\kappa_{mkv}-1} \tag{10}$$

Assuming a variational prior for $\boldsymbol{\theta}_d$ helps us to simplify the inference process. Hence we define the equation,

$$p(\boldsymbol{\theta}_d \mid \boldsymbol{g}_d, \boldsymbol{h}_d) = \prod_{k=1}^{K} \frac{\Gamma(g_{dk} + h_{dk})}{\Gamma(g_{dk})\Gamma(h_{lk})} \theta_{dk}^{g_{dk}-1} \Big(1 - \sum_{j=1}^{k} \theta_{dj} \Big)^{\zeta_{dk}} \tag{11}$$

where, $\zeta_{dk} = h_{dk} - g_{d(k-1)} - h_{d(k-1)}$ while $k \leq K - 1$ and $\zeta_{dk} = h_{dk} - 1$ when $k = K$. Similarly, we also place a Beta distribution to define λ with hyperparameters ω which gives, $p(\lambda \mid \omega) = \prod_{l=1}^{\infty} Beta(1, \omega_l) = \prod_{l=1}^{\infty} \omega_l(1 - \lambda_l)^{\omega_l - 1}$. We introduce Gamma priors to the stick lengths as [5], $p(\omega) = \mathcal{G}(\omega \mid a, b) = \prod_{l=1}^{\infty} \frac{b_l^{a_l}}{\Gamma(a_l)} \omega_j^{a_l - 1} e^{-b_l \omega_l}$. Based on these equations, we can write the joint distribution of the posterior as,

$$
\begin{aligned}
p(W, \Theta) &= p(W \mid \mathcal{Z}, \beta, \theta, \sigma, \tau, \mathcal{Y}) \\
&= p(W \mid \mathcal{Z}, \beta)p(z \mid \theta)p(\theta \mid \sigma, \tau, \mathcal{Y})p(\mathcal{Y} \mid \lambda)p(\lambda \mid \omega)p(\omega) \\
&\quad p(\theta \mid g, h)p(\beta \mid \kappa)p(\sigma \mid \upsilon, \nu)p(\tau \mid s, t)
\end{aligned}
\tag{12}
$$

Given $\Theta = \{\mathcal{Z}, \beta, \theta, \sigma, \tau, \mathcal{Y}\}$ which represents all the parameters in our model. We can represent our model as a plate diagram shown in Fig. 1.(a).

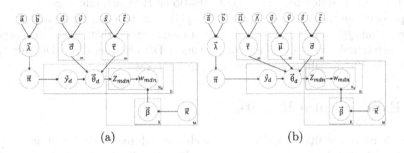

(a)   (b)

Fig. 1. Plate model of (a) DP-LGDA, (b) DP-LBLA

2.2 Dirichlet Process Based Latent Beta-Liouville Mixture Allocation

We can construct the DP-LBLA with the same definitions considered for DP-LGDA just by replacing the GD prior used in Eq. 2 with the BL distribution. The prior in this case can be written as,

$$
\begin{aligned}
p(\theta_d \mid y_d, \mu, \sigma, \tau) &= \prod_{l=1}^{L} \prod_{k=1}^{K} \left[\frac{\Gamma(\sum_{k=1}^{K} \mu_{lk})}{\prod_{k=1}^{K} \Gamma(\mu_{lk})} \frac{\Gamma(\sigma_l + \tau_l)}{\Gamma(\sigma_l)\Gamma(\tau_l)} \theta_{dk}^{\mu_{lk} - 1} \right. \\
&\quad \left. \times \left[\sum_{k=1}^{K} \theta_{dk} \right]^{\sigma_l - \sum_{k=1}^{K} \mu_{lk}} \left[1 - \sum_{k=1}^{K} \theta_{dk} \right]^{\tau_l - 1}
\end{aligned}
\tag{13}
$$

where $(\mu_{l1}, \mu_{l2}, ..., \mu_{lN_d}, \sigma_l, \tau_l)$ are the parameters of Beta-Liouville distribution. The Gamma priors for DP-LBLA are $p(\mu_{lk}) = \mathcal{G}(\mu_{lk} \mid \upsilon_{lk}, \nu_{lk})$, $p(\sigma_l) = \mathcal{G}(\sigma_l \mid s_l, t_l)$, and $p(\tau_l) = \mathcal{G}(\sigma_l \mid \Omega_l, \Lambda_l)$. Changing the prior to BL distribution also changes the variational prior in Eq. 11 to,

$$p(\boldsymbol{\theta}_d \mid \boldsymbol{f}_d, g_d, h_d) = \prod_{k=1}^{K} \frac{\Gamma(\sum_{k=1}^{K} f_{dk})}{\prod_{k=1}^{K} \Gamma(f_{dk})} \frac{\Gamma(g_d + h_d)}{\Gamma(g_d)\Gamma(h_d)} \theta_{dk}^{f_{dk}-1}$$

$$\times \left[\sum_{k=1}^{K} \theta_{dk} \right]^{g_d - \sum_{k=1}^{K} f_{dk}} \left[1 - \sum_{k=1}^{K} \theta_{dk} \right]^{h_d - 1} \tag{14}$$

The joint likelihood can now be written with respect to the parameters Θ as,

$$p(W, \Theta) = p(W \mid \mathcal{Z}, \boldsymbol{\beta}, \boldsymbol{\theta}, \boldsymbol{\mu}, \boldsymbol{\sigma}, \boldsymbol{\tau}, \mathcal{Y}) \tag{15}$$

$$= p(W \mid \mathcal{Z}, \boldsymbol{\beta}) p(z \mid \boldsymbol{\theta}) p(\boldsymbol{\theta} \mid \boldsymbol{\mu}, \boldsymbol{\sigma}, \boldsymbol{\tau}, \mathcal{Y}) p(\mathcal{Y} \mid \boldsymbol{\lambda}) p(\boldsymbol{\lambda} \mid \boldsymbol{\omega}) p(\boldsymbol{\omega})$$

$$p(\boldsymbol{\theta} \mid \boldsymbol{f}, \boldsymbol{g}, \boldsymbol{h}) p(\boldsymbol{\beta} \mid \boldsymbol{\kappa}) p(\boldsymbol{\mu} \mid \boldsymbol{v}, \boldsymbol{\nu}) p(\boldsymbol{\sigma} \mid \boldsymbol{s}, \boldsymbol{t}) p(\boldsymbol{\tau} \mid \boldsymbol{\Omega}, \boldsymbol{\Lambda})$$

The plate model of DP-LBLA is shown in Fig. 1.(b). In this work we use the variational approach used by [4] as opposed to the regular approach used by [2] which is slightly different. The idea of variational Bayesian inference in general is to approximate the posterior distribution $p(W \mid \Theta)$ by considering a variational distribution $Q(\Theta)$ and then reduce the distance between them until convergence. The variational solutions when considering a DP of generalized Dirichlet and Beta-Liouville distributions are detailed in [3,5], respectively.

3   Experimental Results

To evaluate our multi-lingual model, we choose a dataset which comprises transcripts of TED talks on varied topics from Kaggle[1]. The parallel dataset consists of talks from various disciplines like physics, environment, politics, relationships, pollution, space, etc. In order to perform a deep analysis pertaining to the quality of the extracted topics we keep things simple by choosing 99 talks which comprises of around 30 transcripts of talks closely related to three different topics namely, astrophysics, relationships and climate change. These talks do not exactly belong to the same class and might be slightly different in many cases. For example, a talk from astrophysics might be about space travel and aliens or experiments on dark matter. The variance in these topics with a small dataset helps us to see how our model is able to perform in situations where only limited data is available for learning. When it comes to evaluating topic extraction models, it can be done in two ways: extrinsic methods and intrinsic methods. Extrinsic evaluation involves validating the learned topics with respect to the task at hand. For example, in classification tasks we can check how our topics help classifying a test set. Since, our model is unsupervised, using an intrinsic method like topic coherence score makes more sense. In our experiments we use the UMass coherence score [11] which calculates a score based on the probability of two words occurring together in the text. It is given by the equation,

---

[1] https://www.kaggle.com/datasets/miguelcorraljr/ted-ultimate-dataset.

$$score_{UMass}(k) = \sum_{i=2}^{U_k} \sum_{j=1}^{U_k-1} \log \frac{p(w_i, w_j) + 1}{p(w_i)} \tag{16}$$

Here, U_k is the number of top words in the topic k. As seen from the formula, we are basically calculating the ratio between two words w_i and w_j occurring together and the probability of the word under consideration w_i occurring alone. This helps us to measure the relevancy of extracted topics.

The first experiment we conducted is to test the quality of topics extracted by our models compared to other standard models. We use transcripts from only French and English to simplify analysis, however, the models will perform equally if compared with more languages as well. LDA being the basic and widely used topic extraction model will be our benchmark to compare, followed by Poly-LDA [10] which is a multilingual model based on LDA. We compare the coherence scores of the extracted topics with respect to each language separately. We also wanted to test how the model performs if a Dirichlet process mixture is not used for the mixture. In this case the parameter π_l will act as mixing coefficients of the model and the equations will transform accordingly. These models will be represented as 'Mix-LGDA' and 'Mix-LBLA' respectively. To study the effect of modifying the prior distributions, we also compare with 'DP-LDA' and 'mix-LDA' which are the LDA counterparts of our models. Tables 1 and 2 show the coherence score for the different models for English and French languages, respectively, while varying the number of topics K. We can see that the coherence score is the highest when the number of topics is set as 5. Even though our data consisted of documents from three main categories, most of them had multidisciplinary concepts which were clearly captured by almost all of the models. LDA falters here when extracting topics in English in addition to mix-LBLA.

Table 1. Average coherence scores of topics for TED talks transcripts in English.

Model	K = 3	K = 5	K = 7	K = 9
LDA	−0.44	−0.45	−0.53	−0.61
Poly-LDA	−0.44	−0.43	−0.52	−0.59
Mix-LDA	−0.45	−0.44	−0.52	−0.52
DP-LDA	−0.42	−0.41	−0.48	−0.54
Mix-LGDA	−0.47	−0.43	−0.48	−0.53
DP-LGDA	−0.46	−0.43	−0.45	−0.51
Mix-LBLA	−0.39	−0.38	−0.44	−0.47
DP-LBLA	−0.39	−0.35	−0.41	−0.44

In general, we can observe that the coherence score is higher for mixture models without DP assumption compared to LDA and poly-LDA. Similarly, with the DP assumption, the models perform to the best compared to the rest. This

Table 2. Average coherence scores of topics for TED talks transcripts in French.

Model	K = 3	K = 5	K = 7	K = 9
LDA	-5.97	-5.29	-6.29	-6.84
Poly-LDA	-5.67	-5.67	-6.59	-7.02
Mix-LDA	-5.45	-5.25	-5.73	-6.88
DP-LDA	-5.33	-5.16	-5.75	-6.57
Mix-LGDA	-5.59	-5.05	-5.56	-6.88
DP-LGDA	-5.36	-4.90	-5.33	-5.88
Mix-LBLA	-5.27	-4.79	-5.43	-5.99
DP-LBLA	-5.10	-4.50	-5.25	-5.25

pattern is clearly observed in Figs. 2 and 3, respectively. Though the coherence scores are different in scale for French and English, both languages follow a similar pattern.   To analyze deeper, let us consider the topics extracted by

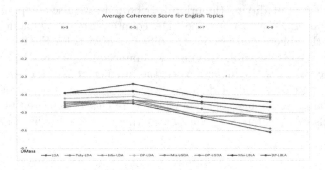

Fig. 2. Coherence scores by varying number of topics in English.

our baseline poly-LDA and the best performing model in terms of coherence, DP-LBLA. Tables 3 and 4 show the French and English topics extracted by poly-LDA and DP-LBLA, respectively. Looking at the topics we can see how good DP-LBLA is able to extract parallel topics. Topic 1 represents astrophysics, topic 2 is a set of common words in all topics, topic 3 is an intersection between science and energy, topic 4 shows words corresponding to relationships and topic 5 is related to climate change. Interestingly, the word 'like' is present in almost all the languages as a major word. This is because it was used in almost all the talks by different people when they pause and give examples. To check how similar the extracted topics are to each other, we calculate the Jaccard index $(Jac(F, E) = \frac{F \cap E}{F \cup E})$ between the set of words in each topic in both languages. F and E represent the set of words in the French topic and English topic, respectively.

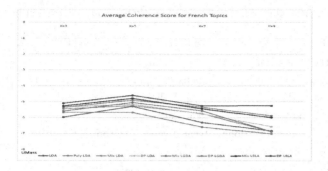

Fig. 3. Coherence scores by varying number of topics in French.

Table 3. Jaccard Index between English and French topics extracted by DP-LBLA.

French	English	F∩E	F∪E	Jac
'lumi', 'donc', 'cette', 'toiles', 'espace', 'mati', 'galaxie', 'particules', 'galaxies', 'univers'	'like', 'dark', 'matter', 'black', 'light', 'universe', 'galaxy', 'galaxies', 'space', 'stars',	7	13	0.54
'quand', 'chose', 'cette', 'cela', 'bien', 'parce', 'comme', 'donc', 'alors', 'tout'	'find', 'mars', 'well', 'time', 'like', 'actually', 'think', 'life', 'going', 'know',	2	18	0.11
'soleil', 'surface','cette', 'solaire', 'comme', 'milliards', 'atmosph', 'syst', 'terre', 'plan'	'solar', 'planet', 'water', 'ocean', 'surface', 'energy', 'atmosphere', 'years', 'system', 'earth',	6	14	0.43
'tout', 'gens', 'cette', 'autre', 'cela', 'amour', 'quelqu', 'personne', 'rires', 'quand'	'want', 'brain', 'feel', 'person', 'really', 'think', 'laughter', 'love', 'like', 'people'	4	16	0.25
'donc', 'gens', 'pays', 'probl', 'climatique', 'cette', 'changement', 'tout', 'monde', 'cela',	'really', 'much', 'think', 'going', 'need', 'climate', 'global', 'change', 'world', 'people'	5	16	0.31

Table 4. Jaccard Index between English and French topics extracted by Poly-LDA.

French	English	F∩E	F∪E	Jac
'mati', 'lumi', 'plastique', 'espace', 'particules', 'toiles', 'galaxies', 'donc', 'cette', 'univers'	'dark', 'galaxy', 'galaxies', 'plastic', 'black', 'light', 'universe' 'like', 'stars', 'space'	6	14	0.43
'devons', 'monde', 'donc', 'cette', 'changement', 'climatique', 'missions', 'tout', 'probl', 'cela'	'year', 'world', 'carbon', 'people', 'going', 'climate', 'global', 'need', 'change', 'energy'	5	16	0.31
'trois', 'fois', 'moins', 'deux', 'comme', 'gens', 'monde', 'cette', 'jour', 'bien'	'much', 'many', 'make', 'first', 'percent', 'system', 'people', 'years', 'world', 'water'	2	18	0.11
'alors', 'comme', 'gens', 'autre', 'amour', 'cette', 'rires', 'quand' 'cela', 'tout'	'want', 'life', 'know', 'really', 'like', 'going', 'laughter', 'people', 'think', 'love'	4	16	0.25
'soleil', 'cela', 'donc', 'tout', 'mars', 'cette', 'surface', 'plan' 'comme', 'terre'	'years', 'look', 'mars', 'life', 'earth' 'going', 'actually', 'like', 'planet', 'know'	4	16	0.25

Words which mean the same in both languages are considered to be intersection between the two sets in our case. For example, 'lumi' which is the shortened word for 'lumiere' in French, means 'light' in English which would be counted as intersection. There are some cases where a word might have multiple equivalents in the other language. For example, in topic 5 for DP-LBLA, the French word 'monde' can mean both 'global and 'world' in English. In these cases we consider both the English words as intersection. Based on our analysis, we found that DP-LBLA had 24 similar words overall in the 5 topics averaging a Jaccard index of 0.33 whereas poly-LDA had only 21 words in common with an average Jaccard index of 0.27. In addition to these metrics, eye-balling the topics would clearly indicate the quality of topics derived by DP-LBLA. Furthermore, there are ways to improve the quality of the topics which would help more in language understanding tasks. For example, authors in [9] use an interactive approach to modify the probabilities for each word within topics. Incorporating this method we aim to improve the quality of the discovered topics. This can be done by simply splitting the topic word proportionals β into two terms, β_o and β_u, respectively, corresponding to the objective probability learned by the DP-LBLA or DP-LGDA model and the subjective probability defined by the user. The relationship between the two variables is given by:

$$\beta = \eta_1\beta_o + \eta_2\beta_u \tag{17}$$

η_1 and η_2 in the above equation denote the weights given to the objective and subjective properties, respectively, with the condition $\eta_1 + \eta_2 = 1$. This helps us to control the impact of the probabilities modified by the user since it is not necessary that the user is well versed in the content of the documents. Keeping a low value of η_2 will ensure that these probabilities do not modify the extracted topics to a greater extent. Since DP-LBLMA is the model which gave the best result, we try to improve the result further by using this interactive model after convergence. We ask the users for input on probabilities concerning the T_k words within each of the kth topic. Let $\{p_{o1}^k, p_{o2}^k, ..., p_{oT}^k\}$ be the probabilities that are obtained by running our model on the documents and T_k' be the set of words among T_k words selected to be modified by the user. Consequently, let's consider $\{p_{u1}^k, p_{u2}^k, ..., p_{uT'}^k\}$ be the modified probabilities for T_k' words. We modify the subjective topic word probabilities based on user input for the tth word using the formula $\beta_{ut}^K = p_{ot}^k * p_{ut}^k$. The remaining probabilities are spread out to the rest of the words apropos to normalizing β based on the equation,

$$\beta_{ut}^k = p_{ot} * \left(1 + \frac{p_r}{\sum_t p_{ot}}\right); \forall t \notin T_k' \tag{18}$$

given, $p_r = \sum_t^{T_k'} p_{ot}^k * \left(1 - p_{ut}^k\right)$. β_u can now be substituted in Eq. 17 to obtain the new set of values. We use the modified values of β in the next iteration to improve the topics. Table 5 and Figs. 4 and 5 show the improvement in results based on this idea.

Table 5. Improvement in coherence score for DP-LBLA with interactive learning.

Model (Language)	$\eta_1 = 0.2$	$\eta_1 = 0.4$	$\eta_1 = 0.6$	$\eta_1 = 0.8$	$\eta_1 = 1$
DP-LBLA (En)	−0.31	−0.32	−0.32	−0.35	−0.35
DP-LBLA (Fr)	−3.96	−3.96	−4.15	−4.39	−4.50

The experiment was conducted by varying the weights for the objective and subjective probabilities. We can see that as we keep increasing the value of η_1, the coherence decreases. It is noteworthy that, at $\eta_1 = 1$, the model acts as a regular DP-LBLA model. The observations show the effect of varying the impact of user defined probabilities. The pattern is visible in Fig. 4 and Fig. 5. In Fig. 4 and Fig. 5, η denotes η_1 in general for simplicity. In case, the user modifying the probabilities is new to the topics involved in the documents, keeping a higher value for η_1 will help maintaining the performance of our model.

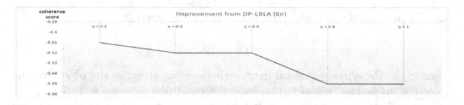

Fig. 4. Improvement in coherence score from DP-LBLA with interactive learning for English topics.

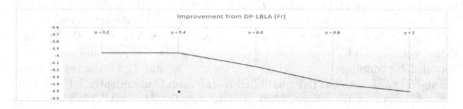

Fig. 5. Improvement in coherence score from DP-LBLA with interactive learning for French topics.

4   Conclusion

We have proposed two novel models for parallel topic extraction in multiple languages. We provide an overview about these models and discuss them from high-level assumptions down to their foundations. We apply DP as an elegant nonparametric Bayesian method to provide flexibility to our model to find various topics and use variational inference to learn our model. The performance of our methods validates the effectiveness of our model design. Considering the results for English and French language documents, both the models perform better than the baseline models LDA and poly-LDA. Though DP-LBLA performs a little better than DP-LGDA their metrics are still closely on par with each other. Our experiments also reveal the advantages of using GD and BL distributions in place of Dirichlet. The performance boost achieved by interactive learning proves to be promising. Future works could be devoted to testing alternative priors and considering other languages.

References

1. Bakhtiari, A.S., Bouguila, N.: A variational bayes model for count data learning and classification. Eng. Appl. Artif. Intell. **35**, 176–186 (2014)
2. Blei, D.M., Ng, A.Y., Jordan, M.I.: Latent dirichlet allocation. J. Mach. Learn. Res. **3**, 993–1022 (2003)
3. Fan, W., Bouguila, N.: Model-based clustering based on variational learning of hierarchical infinite beta-liouville mixture models. Neural Process. Lett. **44**(2), 431–449 (2016)

4. Fan, W., Bouguila, N., Ziou, D.: Variational learning for finite dirichlet mixture models and applications. IEEE Trans. Neural Netw. Learn. Syst. **23**(5), 762–774 (2012)
5. Fan, W., Sallay, H., Bouguila, N., Bourouis, S.: A hierarchical dirichlet process mixture of generalized dirichlet distributions for feature selection. Comput. Electr. Eng. **43**, 48–65 (2015)
6. Gutiérrez, E.D., Shutova, E., Lichtenstein, P., de Melo, G., Gilardi, L.: Detecting cross-cultural differences using a multilingual topic model. Trans. Assoc. Comput. Linguist. **4**, 47–60 (2016)
7. Ihou, K.E., Bouguila, N.: Stochastic topic models for large scale and nonstationary data. Eng. Appl. Artif. Intell. **88**, 103364 (2020)
8. Liu, X., Duh, K., Matsumoto, Y.: Multilingual topic models for bilingual dictionary extraction. ACM Trans. Asian Low-Resour. Lang. Inf. Process. (TALLIP) **14**(3), 1–22 (2015)
9. Liu, Y., Du, F., Sun, J., Jiang, Y.: ilda: An interactive latent dirichlet allocation model to improve topic quality. J. Inf. Sci. **46**(1), 23–40 (2020)
10. Mimno, D., Wallach, H.M., Naradowsky, J., Smith, D.A., McCallum, A.: Polylingual topic models. In: Proceedings of the 2009 Conference on Empirical Methods in Natural Language Processing, vol. 2, pp. 880–889 (2009)
11. Mimno, D., Wallach, H.M., Talley, E., Leenders, M., McCallum, A.: Optimizing semantic coherence in topic models. In: Proceedings of the Conference on Empirical Methods in Natural Language Processing, pp. 262–272. USA (2011)
12. Reber, U.: Overcoming language barriers: assessing the potential of machine translation and topic modeling for the comparative analysis of multilingual text corpora. Commun. Methods Measures **13**(2), 102–125 (2019)
13. Yang, W., Boyd-Graber, J., Resnik, P.: A multilingual topic model for learning weighted topic links across corpora with low comparability. In: Proceedings of EMNLP-IJCNLP, pp. 1243–1248 (2019)
14. Yuan, M., Durme, B.V., Ying, J.L.: Multilingual anchoring: interactive topic modeling and alignment across languages. In: Annual Conference on Neural Information Processing Systems 2018, pp. 8667–8677 (2018)

Adapting Code-Switching Language Models with Statistical-Based Text Augmentation

Chaiyasait Prachaseree[1(✉)], Kshitij Gupta[2(✉)], Thi Nga Ho[1(✉)],
Yizhou Peng[3], Kyaw Zin Tun[1], Eng Siong Chng[1], and G. S. S. Chalapthi[2]

[1] Nanyang Technological University, Singapore, Singapore
prac0003@e.ntu.edu.sg, {ngaht,ztkyaw,asieschng}@ntu.edu.sg
[2] BITS Pilani, Pilani, India
{f20190212,gssc}@pilani.bits-pilani.ac.in
[3] National University of Singapore, Singapore, Singapore
yizhou.p@nus.edu.sg

Abstract. This paper introduces a statistical augmentation approach to generate code-switched sentences for code-switched language modeling. The proposed technique converts monolingual sentences from a particular domain into their corresponding code-switched versions using pretrained monolingual Part-of-Speech tagging models. The work also showed that adding 150 handcrafted formal to informal word replacements can further improve the naturalness of augmented sentences. When tested on an English-Malay code-switching corpus, a relative decrease of 9.7% in perplexity for ngram language model interpolated with the language model trained with augmented texts and other monolingual texts was observed, and 5.9% perplexity reduction for RNNLMs.

Keywords: Code-switching · Language Modeling · Data Augmentation

1 Introduction

Language modeling estimates the probability of word occurrences in a sentence given previous word contexts. It is critical in many language processing applications such as text generation or Automatic Speech Recognition(ASR) [11]. However, LMs are susceptible to domain mismatches between training and testing. For example, [34] has shown that the change in speaking style for an LM trained on read and evaluated on conversational texts decreases the accuracy of ASR system by 9%. Therefore, it is important to have in-domain texts for training LMs.

In the modern world, there is an increasing amount of multilingual speakers, i.e. people with the ability to speak more than one language. Multilingual speakers tend to mix up between languages in their conversations. This occurrence is called code-switching (CS) speech. Specifically, the act of alternating

N. T. Nguyen et al. (Eds.): ACIIDS 2023, LNAI 13996, pp. 310–322, 2023.
https://doi.org/10.1007/978-981-99-5837-5_26

two or more different languages or dialects within a single conversation is usually referred to as inter-sentential code-switch, while switching within a single utterance is referred to as intra-sentential code-switch. This phenomenon is becoming prevalent among many multilingual societies, which motivates the need for CS speech technologies. However, as CS occurs primarily in spoken form and not in written texts, it is difficult to curate a large amount of CS texts. Thus, CS texts are scarce relative to monolingual languages. Consequentially, language modeling for CS text remains a challenging task.

In Natural Language Processing (NLP) field, data scarcity has been addressed by augmentation. For example, text augmentation aims to diversify training texts for the Named Entity Recognition [5,22,38]. Previously, text augmentation techniques have also been used improve to LMs for ASR tasks [3,16,37]. Similarly, augmentation may alleviate scarcity issues for CS language modeling by generating synthetic conversational CS text. However, such methods do not work to generate realistic CS texts, as linguists believe that natural conversations do not code-switch randomly, but follow particular rules or constraints [9,25,26]. This work utilizes POS information to produce artificial English-Malay code-switched sentences. These generated sentences are then used to train CS LMs to reduce the perplexity of the models on CS test sets.

The rest of this paper is organized as follows: Sect. 2 explores the background, Sect. 3 introduces the proposed methodology, Sect. 4 establishes the experimental setup, Sect. 5 analyses the experimental results and shows the improvements from using the proposed method, and Sect. 6 concludes the overall paper and indicates some potential directions for future works.

2 Background

2.1 Code-Switching Data Generation

Realistic code-switching transcripts are difficult to emulate as random or incorrect generation does not help in improving the performance of LMs [18,27]. Previous works in CS text augmentation can be categorised into rule-based, statistical-based, and neural methods. Rule-based techniques use linguistic constraints such as the Equivalence Constraint (EC), Functional Head Constraint (FHC), and Matrix Language Framework (MLF) to generate CS texts. In [27], parallel English and Spanish sentences were parsed so that the CS data used for training follows the EC. Similar methods have been used for English-Chinese in [10,20,36]. Works by [19,21] showed that employing the FHC on English-Chinese improved performance on WFST-based language models and speech recognition. MLF has also been shown to produce more correct synthetic CS sentences [15] when evaluated on English-Chinese corpus. On the other hand, statistical-based methods utilize ngram counts as well as other information such as language and Part of Speech (POS) [30].

Neural methods in text generation require additional model training to produce more representing CS data. Generative adversarial network (GAN) methods consist of a generator that produces CS sentences and a discriminator that differentiates real and synthetic sentences. The generator generates CS sentences from scratch [7], or with monolingual sentences as inputs [4,6], or with cyclic adversarial networks [17]. Other neural methods without GANs have also been studied. Pointer-generator networks, [32] by copying parts of concatenated parallel monolingual sentences, was able to improve LMs [36]. Employing generative methods as data augmentation for CS using variational autoencoders has also been investigated in [29]. Overall, neural methods implicitly learn the linguistics patterns for realistic CS text generation.

2.2   Part-of-Speech in Code-Switching Language Modelling

Part-of-Speech (POS) are grammar categories for each word in a sentence that performs a specific grammatical function. Even though distinct languages have their own grammatical structures, most languages and dialects have comparable part-of-speech tags, including nouns, verbs, adjectives, and adverbs. Still, while there are similar tags, its syntactic role and positions in the sentence may vary across languages.

Previous works have shown improvements of code-switching language models when utilising POS information. Coupling POS information and other CS point triggers, such as language identification, trigger words, etc., in factored language models have shown to reduce perplexity [2,8,31]. Indirect methods of incorporating POS tags in LMs by multi-task learning also outperform perplexity baseline [35]. These results also provide support for linguists' notion that code-switching has particular syntactic structures and is not random.

Our paper proposes using this POS on English-Malay corpus as supplementary information to generate realistic English-Malay CS sentences for language modelling task. A similar method [30] has shown a naive bayes method in predicting code-switching points for English-Spanish using POS were able to generate convincing CS texts, but was not tested for language modelling and assumed independence between language and POS tag as in our case.

3   Methodology

In this section, we describe our proposed augmentation methodology for converting monolingual sentences to their corresponding code-switched peers. Figure 1 shows the overall pipeline of the proposed approach that incorporates 3 modules. Firstly, POS frequencies are collected from training data with assistance from a pre-trained POS tagger. These frequencies are then used as probabilities to generate the corresponding code-switched versions of monolingual sentences. A CS corpus is used to train the Bayesian classifier in predicting code-switching points. Then, using the Bayesian classifier and input monolingual texts, several potential CS texts from monolingual sentences were generated. The CS augmentations generated from this method will be used as additional training texts for CS LM. Details of each step in the pipeline will be described in the following sub-sections.

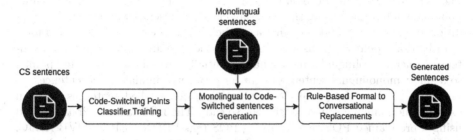

Fig. 1. Overall Methodology

3.1 Code-Switching Points Classifier Training

A POS tagger is employed to detect POS tags for each input sentence. However, as pretrained POS taggers work only on monolingual sentences, the input sentence is first separated into phrases respective to their languages. These monolingual phrases are then fed through a POS tagger to obtain POS tags for each word and then those tags are concatenated together to form a POS tag sequence that corresponds to the original input sentence. This sequence of a CS sentence will be used as input to train a classifier that predicts the code-switching points.

The POS tags are then used to train a Bayesian classifier to predict the language of a word. Input features to the classifier include the current POS tag, the previous POS tag, and the language label of the previous word. A sentence s composed of n words is denoted as $s = w_1, w_2, ..., w_n$. Then, the probability a word w_i has its language label l_i is estimated using the current POS pos_i, previous POS pos_{i-1}, and previous word language label l_{i-1} as shown in Eq. 1. Using the Bayes theorem in Eq. 2, the probability of the language of a given word is proportional to the joint probability of the three features. These joint probability parameters are estimated by the Maximum Likelihood Estimation (MLE) on the corpus through frequency counts, as shown in Eq. 3. Note that we do not use the Naive Bayes assumption since the independence assumption between input features language and POS is unlikely.

$$P(l_i \mid w_i) = P(l_i \mid pos_i, pos_{i-1}, l_{i-1}) \tag{1}$$

$$\propto P(pos_i, pos_{i-1}, l_{i-1} \mid l_i) \tag{2}$$

$$\approx \frac{Counts(pos_i, pos_{i-1}, l_{i-1}, l_i)}{Counts(pos_i, pos_{i-1}, l_{i-1})} \tag{3}$$

3.2 Generating Code-Switched Sentences from Monolingual Sentences

This work proposes an approach to generate a codeswitch sentence from a monolingual sentence. Specifically, we first employed a POS tagger to generate POS tags for each word in the monolingual sentence. Subsequently, code-switching points are determined using the Bayesian classifier described in Sect. 3.1, as conditioned on the POS and the language label. Words between these points form phrases that could be translated into the other language to simulate CS sentences. Each monolingual sentence has potentially several code-switching points, so a single monolingual sentence might have various simulated CS sentences.

Figure 2 shows the process of generating code-switched sentences from an example monolingual sentence "she got so sad". The POS tags are first acquired using a pre-trained POS tagger, giving POS tags PRON VERB ADVB ADJC. MLE values are obtained using the method described in Sect. 3.1 are indicated as "Code-Switching Ratios" table. These MLE values indicate the probabilities of a word being the code-switching point with respect to current and previous POS tags and the previous language label. Selected phrases based on code-switching points are determined by these switching points. These phrases replace the original English phrases to form the CS versions. This process continues until the end of the sentence for k times, using random sampling.

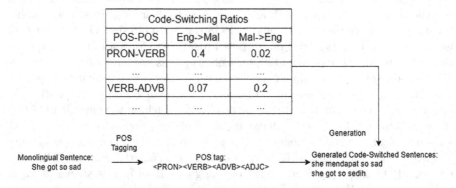

Fig. 2. Example conversion of monolingual sentence "she got so sad" into code-switching sentences

4 Experimental Setup

This section discusses the corpus used in the experiments and the configurations of ngrams and RNNLMs training and evaluation.

4.1 Corpus

The National Speech Corpus (NSC) of Singapore is an extensive Singaporean speech corpus assembled by the Info-communications and Media Development Authority of Singapore (IMDA) [14]. Totalling up to 2000 h hours of speech, it consists of a variety of Singaporean demographics and conversational topics. The transcripts' language change boundaries are fully annotated. This work employs a subset of 200 h from the entire, named NSC-ENGMAL, which contains English-Malay inter- and intra-sentential code-switching conversations. The selected corpus contains 206,712 utterances in total, in which 132,556 utterances are monolingual English, 7,967 utterances are monolingual Malay, and 66,189 are code-switched utterances.

Table 1 shows the text partitions used in the experiments. NSC-ENGMAL texts were further partitioned into utterances with English-Malay code-switching, monolingual English only, and monolingual Malay only to train the LM. For each English sentence, up to 5 CS sentences are generated, with duplicated sentences removed. Thus, from 92,789 monolingual English, 316,750 synthetic CS sentences produced are used for LM training.

Table 1. Number of utterances in different text partitions by language and training, validation, and testing

Partition	CS	English	Malay
Train	58,645	92,789	5,576
Validation	5,000	19,883	1,195
Test	5,000	19,88	1,196
Total	**66,189**	**132,556**	**7,967**

4.2 Normalizing and Generating CS Sentences

A monolingual sentence goes through the following to generate a conversational CS sentence.

- **Text Normalization:** We first normalized all texts by replacing numbers with their written form, e.g. from 1 to "one", and lowercased all capitalized letters. Furthermore, all contractions are replaced with their full form before getting tagged.

- **POS taggers:** We adopt the POS tagger built by NLTK [23] for predicting POS tags for English and Malay built by [12] for predicting POS tags for Malay. For the training set of NSC-ENGMAL corpus, the top POS tags that are most likely to indicate a change in languages from English to Malay and from Malay to English are depicted in Fig. 3.

- **Generating codeswitched sentences:** We employed Deep-Translate library[1] to translate determined codeswitched phrases from either English to Malay or Malay to English. For each monolingual sentence, we chose to generate 5 different versions of its CS peers.

- **Handcrafted word-replacement rules:** Sometimes translations may be too formal relative to a conversational corpus. This issue possibly occurs when the translation model is trained on more formal texts as compared to natural-sounding conversational ones. Thus, we used a set of handcrafted rules to convert 150 formal words into their corresponding informal words in Malay.

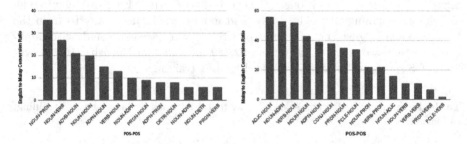

Fig. 3. (Left) Top Pairwise English to Malay POS ratios, (Right) Top Pairwise Malay to English POS ratios

4.3 Language Model Setup

Language Models are evaluated using perplexity (Eqn. 4) on the English-Malay CS testing partition.

$$PPL_{LM}(T) = \sqrt[N]{\frac{1}{P(T)}} \tag{4}$$

Ngrams were trained with a vocabulary size of 24,836 tokens. LMs are trained at an order of 4 with Kneser-Key Smoothing with SRILM toolkit [1]. Multiple LMs of the different data sources are joined with linear interpolation (Eqn. 5).

$$P_{LM}(w \mid h) = \sum_i \lambda_i \cdot P_{LM_i}(w \mid h) \tag{5}$$

Recurrent Neural Networks (RNN) LMs are trained through ESPNet [33] toolkit which uses Pytorch as the backend. All experiments use a two-layered

---

[1] https://github.com/nidhaloff/deep-translator.git.

LSTM with hidden units of 2,024. 10,000 subwords were obtained by Byte-Pair Encoding training over a large audio transcript dataset. A stochastic gradient descent optimizer is used for training, with a maximum epoch of 100, stopped early when there are no improvements in perplexity over 3 epochs.

5   Results

To evaluate the effectiveness of augmented CS text in reducing perplexity in the n-gram language model, we have tried multiple combinations of the original text and augmented text to train different n-gram language models. These abbreviations are summarised in Table 2.

Table 2. LM abbreviations trained on specific subcorpus for ngrams

LM abbreviation	Training Texts
LM-CS	CS texts train partition
LM-ENG	mono English texts train partition
LM-MAL	mono Malay texts train partition
LM-AUG	Augmented Texts from mono English
LM-RAND	Random Switching from mono English
LM-ECB	EC Augmented Texts with benepar parser
LM-ECS	EC Augmented Texts with Stanford parser

Table 3 presents the perplexity results of n-grams for each of the single models mentioned above, and the different interpolations of them together. Overall, a 9.7% decrease in the n-grams language model when the augmented texts, monolingual English, and monolingual Malay are interpolated with the baseline CS only LM is observed, from 206.4 to 186.3. The improvement from the augmentation is quite significant, 9.3% are decreased by simply interpolating augmentation to CS only baseline, which almost performs as well as simply adding more monolingual English and Malay texts. Adding only more monolingual English texts does not have the same effect as augmentation.

Results with Equivalence Constraint (EC) generation method adapted for English-Malay, previously used on English-Hindi [28], performed worse than the proposed generation method. This can be attributed to the rigid constraints of linguistic theories. In fact, relying on language-specific parsers results in only a small subset of usable code-switching sentences, only 1,550 additional English-Malay CS texts generated when using Berkeley Neural Parser [13] and 423 additional texts using Stanford Parser [24], which has a negligible effect on reducing perplexity. We also showed that random language switches from monolingual English sentences only very slightly improve the perplexity of n-gram models denoted as LM-RAND. Thus, interpolating all real CS, monolingual English, monolingual Malay, and augmented texts results in the best-performing LM.

Table 3. Ngram Perplexity Results

LM	Perplexity
LM-ENG	1510.0
LM-MAL	967.7
LM-CS	206.4
LM-ENG + LM-MAL	364.0
LM-ENG + LM-MAL + LM-AUG	356.9
LM-CS + LM-ENG	201.1
LM-CS + LM-MAL	193.0
LM-CS + LM-RAND	191.1
LM-CS + LM-AUG	187.1
LM-CS + LM-ENG + LM-MAL	186.9
LM-CS + LM-ENG + LM-MAL + LM-RAND	186.8
LM-CS + LM-ENG + LM-MAL + LM-AUG	**186.3***
LM-CS + LM-ENG + LM-MAL + LM-ECS	186.9
LM-CS + LM-ENG + LM-MAL + LM-ECB	186.8

5.9% perplexity reduction in RNNLMs when pretraining the LM with augmented texts before finetuning with the real CS sentences are observed, from 155.3 to 146.1 (Table 4). In contrast, when the RNNLM is pre-trained using English only, the improvement from the baseline was negligible only down to 154.7, further supporting the hypothesis that the generated texts work in improving LMs. Using the EC method also has a very negligible effect on RNNLM perplexity, pretraining with EC reduces perplexity by less than one-tenths.

Overall, improvements are observed in both n-grams and RNNLMs. While the improvement is more substantial in the case of RNNLMs, this also shows the benefits of data augmentation as it is model agnostic.

Table 4. RNNLM Perplexity Results

Text	Perplexity
CS Only	155.3
Aug Only	827.2
Mono Pretrained, CS finetune	154.7
ECB Aug Pretrained, CS finetune	155.3
Aug Pretrained, CS finetune	**146.1***

6 Conclusion

This paper proposed a data augmentation method for scarce CS texts for LM. This technique utilizes monolingual part-of-speech taggers on real code-switching data to find language switch boundaries that occur naturally in a sentence. Then using these language switch classifiers, monolingual sentences are converted into their potential code-switched sentences. These texts are used to augment code-switching texts to reduce 9.7% in perplexity for n-grams and 5.9% for RNNLMs.

Some limitations of the proposed method can be further explored. The accuracy of code-switched sentences is bounded by the performance of POS taggers on conversational texts. However, these taggers were trained on more formal sentences. Additionally, as POS tags do not work on contractions that are used in daily conversations, the speaking style similarity between train and test sets is slightly diminished. This leads to incomplete n-gram counts when real CS texts are not used, such as "Im gonna", which frequently appears otherwise. In such a case, it may be beneficial to combine different augmentation techniques that aid on these disadvantages, similar to [10].

Acknowledgment. This research is supported by the Ministry of Education, Singapore, under its Academic Research Fund Tier 2(MOE2019-T2-1-084). Any opinions, findings and conclusions or recommendations expressed in this material are those of the author(s) and do not reflect the views of the Ministry of Education, Singapore. We would like to acknowledge the High Performance Computing Centre of Nanyang Technological University Singapore, for providing the computing resources, facilities, and services that have contributed significantly to this work. This research is supported by ST Engineering Mission Software & Services Pte. Ltd under a collaboration programme (Research Collaboration No: REQ0149132)

References

1. Srilm - an extensible language modeling toolkit. In: Hansen, J.H.L., Pellom, B.L. (eds.) INTERSPEECH. ISCA (2002). http://dblp.uni-trier.de/db/conf/interspeech/interspeech2002.html#Stolcke02
2. Adel, H., Vu, T., Kirchhoff, K., Telaar, D., Schultz, T.: Syntactic and semantic features for code-switching factored language models. IEEE/ACM Trans. Audio Speech Lang. Process. **23**, 1 (2015). https://doi.org/10.1109/TASLP.2015.2389622
3. Beneš, K., Burget, L.: Text augmentation for language models in high error recognition scenario. In: Proceedings of Interspeech 2021, pp. 1872–1876 (2021). https://doi.org/10.21437/Interspeech.2021-627
4. Chang, C.T., Chuang, S.P., VI Lee, H.: Code-switching sentence generation by generative adversarial networks and its application to data augmentation. In: Interspeech (2018)
5. Ding, B., et al.: DAGA: data augmentation with a generation approach for low-resource tagging tasks. In: Proceedings of the 2020 Conference on Empirical Methods in Natural Language Processing (EMNLP), pp. 6045–6057. Association for Computational Linguistics, November 2020. https://doi.org/10.18653/v1/2020.emnlp-main.488, http://aclanthology.org/2020.emnlp-main.488

6. Gao, Y., Feng, J., Liu, Y., Hou, L., Pan, X., Ma, Y.: Code-switching sentence generation by bert and generative adversarial networks, pp. 3525–3529, September 2019. https://doi.org/10.21437/Interspeech.2019-2501

7. Garg, S., Parekh, T., Jyothi, P.: Code-switched language models using dual RNNs and same-source pretraining. In: Conference on Empirical Methods in Natural Language Processing (2018)

8. Gonen, H., Goldberg, Y.: Language modeling for code-switching: evaluation, integration of monolingual data, and discriminative training. In: Proceedings of the 2019 Conference on Empirical Methods in Natural Language Processing and the 9th International Joint Conference on Natural Language Processing (EMNLP-IJCNLP), pp. 4175–4185. Association for Computational Linguistics, Hong Kong, China, November 2019. https://doi.org/10.18653/v1/D19-1427, http://aclanthology.org/D19-1427

9. Gumperz, J.J.: Discourse Strategies. Studies in Interactional Sociolinguistics, Cambridge University Press, Cambridge (1982). https://doi.org/10.1017/CBO9780511611834

10. Hu, X., Zhang, Q., Yang, L., Gu, B., Xu, X.: Data augmentation for code-switch language modeling by fusing multiple text generation methods. In: Proceedings of Interspeech 2020, pp. 1062–1066 (2020). https://doi.org/10.21437/Interspeech.2020-2219

11. Huang, W.R., Peyser, C., Sainath, T.N., Pang, R., Strohman, T.D., Kumar, S.: Sentence-select: large-scale language model data selection for rare-word speech recognition (2022). http://arxiv.org/abs/2203.05008

12. Husein, Z.: Malaya-speech (2020), speech-Toolkit library for bahasa Malaysia, powered by Deep Learning Tensorflow https://github.com/huseinzol05/malaya-speech

13. Kitaev, N., Klein, D.: Constituency parsing with a self-attentive encoder. In: Proceedings of the 56th Annual Meeting of the Association for Computational Linguistics (Volume 1: Long Papers), pp. 2676–2686. Association for Computational Linguistics, Melbourne, Australia, July 2018. https://doi.org/10.18653/v1/P18-1249, http://aclanthology.org/P18-1249

14. Koh, J.X., et al.: Building the Singapore English national speech corpus. In: Interspeech (2019)

15. Lee, G., Yue, X., Li, H.: linguistically motivated parallel data augmentation for code-switch language modeling. In: Proceedings of Interspeech 2019, pp. 3730–3734 (2019). https://doi.org/10.21437/Interspeech.2019-1382

16. Li, C., Vu, N.T.: Improving code-switching language modeling with artificially generated texts using cycle-consistent adversarial networks. In: Meng, H., Xu, B., Zheng, T.F. (eds.) Interspeech 2020, 21st Annual Conference of the International Speech Communication Association, Virtual Event, Shanghai, China, 25–29 October 2020, pp. 1057–1061. ISCA (2020). https://doi.org/10.21437/Interspeech.2020-2177

17. Li, C.Y., Vu, N.T.: Improving code-switching language modeling with artificially generated texts using cycle-consistent adversarial networks. In: Proceedings of Interspeech 2020, pp. 1057–1061 (2020). https://doi.org/10.21437/Interspeech.2020-2177

18. Li, S.S., Murray, K.: Language agnostic code-mixing data augmentation by predicting linguistic patterns. ArXiv abs/2211.07628 (2022)

19. Li, Y., Fung, P.: Language modeling with functional head constraint for code switching speech recognition. In: EMNLP 2014–2014 Conference on Empirical Methods in Natural Language Processing, Proceedings of the Conference, pp. 907–916, January 2014. https://doi.org/10.3115/v1/D14-1098

20. Li, Y., Fung, P.: Code-switch language model with inversion constraints for mixed language speech recognition. In: Proceedings of COLING 2012, pp. 1671–1680. The COLING 2012 Organizing Committee, Mumbai, India, December 2012).https://aclanthology.org/C12-1102

21. Li, Y., Fung, P.: Code switch language modeling with functional head constraint. In: 2014 IEEE International Conference on Acoustics, Speech and Signal Processing (ICASSP), pp. 4913–4917 (2014). https://doi.org/10.1109/ICASSP.2014.6854536

22. Liu, L., Ding, B., Bing, L., Joty, S., Si, L., Miao, C.: MulDA: a multilingual data augmentation framework for low-resource cross-lingual NER. In: Proceedings of the 59th Annual Meeting of the Association for Computational Linguistics and the 11th International Joint Conference on Natural Language Processing (Volume 1: Long Papers), pp. 5834–5846. Association for Computational Linguistics, August 2021. https://doi.org/10.18653/v1/2021.acl-long.453, https://aclanthology.org/2021.acl-long.453

23. Loper, E., Bird, S.: Nltk: The natural language toolkit. arXiv preprint cs/0205028 (2002)

24. Manning, C.D., Surdeanu, M., Bauer, J., Finkel, J.R., Bethard, S., McClosky, D.: In: ACL (System Demonstrations) (2014)

25. Myers-Scotton, C.: Duelling languages: Grammatical structure in codeswitching (1993)

26. Poplack, S.: Sometimes i'll start a sentence in spanish y termino en espaNol: toward a typology of code-switching 1. Linguistics **18**, 581–618 (1980). https://doi.org/10.1515/ling.1980.18.7-8.581

27. Pratapa, A., Bhat, G., Choudhury, M., Sitaram, S., Dandapat, S., Bali, K.: Language modeling for code-mixing: the role of linguistic theory based synthetic data. In: Proceedings of the 56th Annual Meeting of the Association for Computational Linguistics (Volume 1: Long Papers), pp. 1543–1553. Association for Computational Linguistics, Melbourne, Australia, July 2018. https://doi.org/10.18653/v1/P18-1143, https://aclanthology.org/P18-1143

28. Rizvi, M.S.Z., Srinivasan, A., Ganu, T., Choudhury, M., Sitaram, S.: GCM: a toolkit for generating synthetic code-mixed text. In: Proceedings of the 16th Conference of the European Chapter of the Association for Computational Linguistics: System Demonstrations, pp. 205–211. Association for Computational Linguistics, April 2021. https://aclanthology.org/2021.eacl-demos.24

29. Samanta, B., Nangi, S., Jagirdar, H., Ganguly, N., Chakrabarti, S.: A deep generative model for code switched text, pp. 5175–5181, August 2019. https://doi.org/10.24963/ijcai.2019/719

30. Solorio, T., Liu, Y.: Learning to predict code-switching points. In: Proceedings of the 2008 Conference on Empirical Methods in Natural Language Processing, pp. 973–981. Association for Computational Linguistics, Honolulu, Hawaii, October 2008. https://aclanthology.org/D08-1102

31. Soto, V., Hirschberg, J.: Improving code-switched language modeling performance using cognate features, pp. 3725–3729, September 2019. https://doi.org/10.21437/Interspeech.2019-2681

32. Vinyals, O., Fortunato, M., Jaitly, N.: Pointer networks. In: Cortes, C., Lawrence, N., Lee, D., Sugiyama, M., Garnett, R. (eds.) Advances in Neural Information Processing Systems, vol. 28. Curran Associates, Inc. (2015). https://proceedings.neurips.cc/paper/2015/file/29921001f2f04bd3baee84a12e98098f-Paper.pdf

33. Watanabe, S., et al.: ESPnet: end-to-end speech processing toolkit. In: Proceedings of Interspeech, pp. 2207–2211 (2018). https://doi.org/10.21437/Interspeech.2018-1456, https://dx.doi.org/10.21437/Interspeech.2018-1456

34. Weintraub, M., Taussig, K., Hunicke-Smith, K., Snodgrass, A.: Effect of speaking style on lvcsr performance. In: Proceedings of ICSLP, vol. 96, pp. 16–19. Citeseer (1996)

35. Winata, G.I., Madotto, A., Wu, C.S., Fung, P.: Code-switching language modeling using syntax-aware multi-task learning. In: Proceedings of the Third Workshop on Computational Approaches to Linguistic Code-Switching, pp. 62–67. Association for Computational Linguistics, Melbourne, Australia, July 2018. https://doi.org/10.18653/v1/W18-3207, https://aclanthology.org/W18-3207

36. Winata, G.I., Madotto, A., Wu, C.S., Fung, P.: Code-switched language models using neural based synthetic data from parallel sentences. In: Proceedings of the 23rd Conference on Computational Natural Language Learning (CoNLL), pp. 271–280. Association for Computational Linguistics, Hong Kong, China, November 2019. https://doi.org/10.18653/v1/K19-1026, https://aclanthology.org/K19-1026

37. Yılmaz, E., van den Heuvel, H., Van Leeuwen, D.: Acoustic and textual data augmentation for improved ASR of code-switching speech, September 2018. https://doi.org/10.21437/Interspeech.2018-52

38. Zhou, R., et al.: MELM: data augmentation with masked entity language modeling for low-resource NER. In: Proceedings of the 60th Annual Meeting of the Association for Computational Linguistics (Volume 1: Long Papers), pp. 2251–2262. Association for Computational Linguistics, Dublin, Ireland, May 2022. https://doi.org/10.18653/v1/2022.acl-long.160, https://aclanthology.org/2022.acl-long.160

Improving Hotel Customer Sentiment Prediction by Fusing Review Titles and Contents

Xuan Thang Tran[1], Dai Tho Dang[2(✉)], and Ngoc Thanh Nguyen[3(✉)]

[1] Information Technology Department, Tay Nguyen University, Buon Ma Thuot, Vietnam
txthang@ttn.edu.vn
[2] Vietnam - Korea University of Information and Communication Technology, The University of Danang, Danang, Vietnam
ddtho@vku.udn.vn
[3] Department of Applied Informatics, Wroclaw University of Science and Technology, Wroclaw, Poland
Ngoc-Thanh.Nguyen@pwr.edu.pl

Abstract. The large volume of online customer reviews is a valuable source of information for potential customers when making decisions and for companies seeking to improve their products and services. While many researchers have focused on the content of reviews and their impact on customers' opinions using deep learning approaches, the mechanism by which review titles and contents influence sentiment analysis (SA) has received inadequate attention. This study proposes a deep learning-based fusion method that reveals the importance of reviewing titles and contents in predicting customer opinions. Our experiments on a crawled TripAdvisor dataset showed that the performance of the document-level SA task could be improved by 2.68% to 12.36% compared to baseline methods by effectively fusing informative review titles and contents.

Keywords: sentiment analysis · document-level · deep learning · hotel review · review fusion

1 Introduction

The advent of the Internet in the contemporary information age has facilitated the dissemination of user opinions, experiences, and complaints regarding products and services through social media platforms. According to Capoccia [1], Online Consumer Reviews (OCRs) have emerged as a crucial source of information for pre-purchase decision-making, with over 90% of customers relying on them to inform their purchasing decisions. A study in the United States in 2022 by Statista[1] revealed that 34% of customers used online reviews to gather

---

[1] https://www.statista.com/forecasts/997051/sources-of-information-about-products-in-the-us.

N. T. Nguyen et al. (Eds.): ACIIDS 2023, LNAI 13996, pp. 323–335, 2023.
https://doi.org/10.1007/978-981-99-5837-5_27

product information, placing it second only to search engines such as Google with 55%.

Sentiment Analysis (SA), as defined by Bing Liu [8], is a discipline that examines individuals' opinions, sentiments, evaluations, attitudes, and emotions expressed through written language. SA has become a wide research field within artificial intelligence, specifically in the context of social media. There are three levels at which SA can be conducted: document, sentence, and aspect. While document-level SA is a well-studied problem, recent advancements in deep learning-based approaches have garnered significant attention among researchers. These approaches aim to automatically classify the sentiment expressed in an entire document as positive, negative, or neutral.

Table 1. The contribution of title and content to star-rating of reviews

No.	Review Title	Review Content	Rating
1	Very friendly excellent hotel	We stayed here 3 nights, split by a night at Halong Bay, so we experienced 2 rooms. Rooms are a good size with King sized beds...very comfortable beds and lovely pillows. Bedroom has all you need. [...]	5
2	Beds like laying on carpeted floor!	I have loved to give this place 5 star, the staff were very helpful and spoke reasonable English. I was very surprised with how HARD the bed was! I have never seen or payed on a bed so solid. [...]	3
3	Poor service. Awful hotel	This hotel was like going back 100 years. The stuff was extremely unhelpful & spoke no English. The room was horrid. No shower nor bath but a shower head next to the loo [...]	1
4	Mixed Feelings		3
5.	Great hostel!		4

Previous studies in the SA domain have primarily centered on processing customer reviews, mainly focusing on the content of the reviews [10] and neglecting a vital aspect namely the review titles. According to examples in Table 1, both the title and content of reviews can contribute to the final rating given by customers in the hospitality dataset. For example, reviews No. 4 and No. 5 with no content can be predicted as 3-star and 4-star based solely on their titles. Hence, integrating review titles into the SA field has become a topic of interest, leading us to develop a new deep learning-based approach that considers both review titles and contents for a more systematic prediction of customer's polarity.

By analyzing review titles and review contents simultaneously, our research demonstrates remarkable performance improvements compared to baseline models using the same TripAdvisor dataset. Our empirical findings prove that incorporating review titles and contents as input to the deep learning models significantly enhances sentiment prediction accuracy.

2 Related Works

The use of deep learning approaches for understanding customer opinions expressed in online feedback has recently garnered significant attention among scholars. While online reviews typically consist of a title and content, previous studies utilizing deep learning-based models for SA have mainly focused on the review contents [10]. Only a limited number of recent studies have explored the role of review titles, such as the positive effect of title length on review helpfulness [2], the relationship between positive sentiment in the title and reader attraction [13], and the positive effect of title-content similarity on review helpfulness [15]. These studies have demonstrated the important role of review titles in understanding customer feedback. However, the contribution of both the review titles and contents in comprehending user opinions requires further investigation. In a recent study, Tran et al. [14] proposed a technique combining rule-based and deep learning-based approaches utilizing both review summaries and contents, which achieved first place in an SA competition. Although the results from the rule-based approach demonstrated a significant impact on solution's performance, results from the model-based approach require further clarification.

In this study, we considered both review titles and contents as valuable sources for predicting the customer's sentiment expressed in reviews based on deep learning approach. This section discusses some related SA studies using deep learning algorithms.

Since first proposed by Hochreiter [4], the long short-term memory (LSTM) model has been widely applied in various research fields. In the natural language processing, particularly in SA tasks, LSTM and Bidirectional LSTM (BiLSTM) are often used to capture the contextual information extracted from the surrounding sentences [11,12]. Another notable model, the Convolutional Neural Network (CNN), was proposed by Collobert [3] for various tasks, including semantic role labeling, chunking, and named entity recognition. A simple CNN model proposed by Kim [5] is frequently used as a robust baseline model for SA tasks. Moreover, CNN and BiLSTM can be combined to analyze public opinion effectively [9].

Our study draws motivation from those deep learning-based studies on understanding customers' opinions. Specifically, we consider whether fusing both review titles and review contents as input for the deep learning models could contribute to improving the performance of the SA task, and discover the customer sentiment expressed in hospitality reviews.

3   Research Problem

This research aims to examine whether the accuracy performance of SA tasks using a deep learning approach can be improved by combining both the titles and contents of the online review. Consider a set of n customer reviews $Rv = \{r_1, r_2, ..., r_n\}$. For each review $r_i \in Rv$, it combines a pair of title and content $r_i = [T_i, C_i]$. Let WT_i be contextualized word embedding of review title T_i and WC_i be contextualized word embedding of review content C_i for review r_i, transformed by a vector representation model.

Then, a deep learning model will be constructed for document-level SA to predict the customers' opinions. We will test our hypothesis on this model by comparing the empirical results of the model; when fusing the review titles and contents simultaneously and using them separately. We formalize the objective by finding a mapping function $F : (r_i) \rightarrow \{positive, neutral, negatives\}$ such that:

$$F(r_i) = \begin{cases} positive, & \text{if review expressed positive opinion} \\ neutral, & \text{if review expressed neutral opinion} \\ negative, & \text{if review expressed negative opinion} \end{cases} \quad (1)$$

where $F(r_i)$ would be $F([WT_i, WC_i])$, $F([WT_i])$ and $F([WC_i])$ respectively. Therefore, we aim to answer the questions as follows:

- Could the performance of the SA task using deep learning model be improved by fusing both review titles and contents?
- How much does the deep learning-based fusion model utilizing review titles and contents improve document-level SA performance compared to the pruning model, and baseline models?
- How does the combination of different embedding models, deep learning architectures, and fusion techniques affect the performance of our approach?

4   Proposed Method

4.1   Model Architecture

Our study constructs a deep learning-based model to perform document-level SA. Our hypothesis is that fusing the review titles and contents could improve the SA in customer reviews. To test this hypothesis, our model employs two parallel branches with identical architectures to extract relevant patterns from both the titles and contents of the reviews.

The model architecture, as shown in Fig. 1, involves the following steps:

1. Mapping the input sentences into word representation vectors using the pre-trained word embedding.
2. Feeding the embedded vectors into a deep learning-based structure to capture contextualized information and extract informative features.

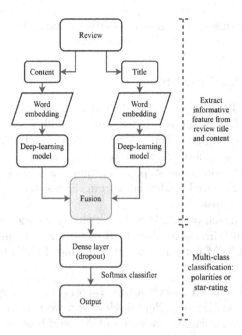

Fig. 1. The proposed architecture

3. Fusing the aggregated vectors from the two parallel branches into a single layer through fusion formula and feeding it into a fully connected layer (dense layer) and a dropout layer.
4. Applying a softmax classifier for document-level SA.

We also provide a pruning model with a single branch optimized for either review titles or contents, and then make a comparison between them to verify our hypothesis.

4.2   Word Embedding Models

In this study, we use word embedding based on the pre-trained Google word2vec[2], GloVe[3] to present the words from both the review title and content. The word embedding maps each word in text sentence into output vectors representing the entire document as follows:

$$WT = Embedding(T) \in \mathbb{R}^{m \times d_e}$$
$$WC = Embedding(C) \in \mathbb{R}^{m' \times d_e} \tag{2}$$

where WT and WC are the output vectors of the review title and review content extracted from the pre-trained word embedding (word2vec, GloVe), m and m' is

---

[2] https://code.google.com/archive/p/word2vec/.
[3] https://nlp.stanford.edu/projects/glove/.

the length of review title and content respectively, and d_e is the word embedding vector dimension.

4.3 Deep Learning-Based Structures

LSTM and BiLSTM Model. The LSTM architecture, a type of recurrent neural network (RNN), is commonly used for modelling sequential data and is optimized to capture long-term dependencies better than traditional RNNs. Like other RNNs, the LSTM network takes inputs from the current time step and outputs from the previous time step, generating a new output that is fed into the next time step. The final hidden layer from the last time step is then utilized for classification purposes.

BiLSTM is a variant of LSTM network that processes sequences in both forward and backward directions allowing it to capture surrounding information. The LSTM and BiLSTM networks closely resembles Phan's model [11].

CNN Model and Global MaxPooling Layer. The conceptual CNN model was the same as Yoon Kim's model [5]. However, we perform the Global Max-Pooling operation over each feature map of CNN and take the most informative feature with the highest value.

Ensemble of LSTM/BiLSTM and CNN Models. Although both LSTM, BiLSTM and CNN models have separately archived undoubtable performance for SA research, combining them could further boost the performance [9,14]. Generally, LSTM and BiLSTM have a mechanism to capture useful features by processing the information surrounding sentences, and CNN could look at the holistic feature of local information.

In our work, we stack the CNN to LSTM (or BiLSTM) to create a higher-level feature extraction model. The contextualized representation matrix $H \in \mathbb{R}^{m \times d_h}$ (or $H' \in \mathbb{R}^{m \times 2d_h}$), generated from the above LSTM/BiLSTM model, will be utilized sequentially as input for the CNN model. The two-dimension convolution layer (Conv2D) would be performed to yield essential patterns, known as output vector $\widetilde{C} \in \mathbb{R}^{1 \times d_k}$, where d_h and d_k are the output dimension of LSTM and CNN, respectively.

4.4 Review Title-Content Fusion and Sentiment Classifier

At the end of the parallel branches, two output vectors $\widetilde{CT} \in \mathbb{R}^{1 \times d}$ for review title and $\widetilde{CC} \in \mathbb{R}^{1 \times d}$ for review content have been calculated. We assume that both the review title and content contribute to predicting the sentiment polarity of the review. So, we investigate three different methods to deal with the outputs from parallel branches.

Method 1: Average Fusion. We perform average formula on two given vectors $\widetilde{CT}$ and $\widetilde{CC}$ as shown in Fig. 2 to yield an averaged vector (denoted as $\widetilde{C}_{avg}$) to examine whether review title and review content contribute equally on predicting customer's opinions.

$$\widetilde{C}_{avg} = Average[\widetilde{CT}, \widetilde{CC}] \in \mathbb{R}^{1 \times d} \tag{3}$$

Method 2: Maximum Fusion. As shown on Fig. 3, we perform maximum formula on two vectors $\widetilde{CT}$ and $\widetilde{CC}$ to capture the most informative features on review title and content matrices, then generate one single vector (denoted as $\widetilde{C}_{max}$).

$$\widetilde{C}_{max} = Maximum[\widetilde{CT}, \widetilde{CC}] \in \mathbb{R}^{1 \times d} \tag{4}$$

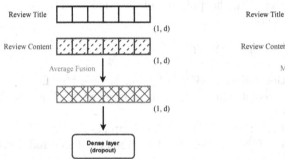

Fig. 2. The review title-content average fusion

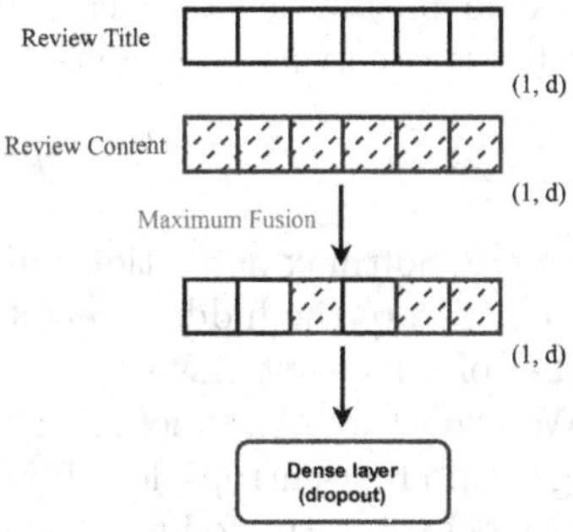

Fig. 3. The review title-content maximum fusion

Method 3: Concatenation Fusion. To evaluate the effectiveness of the fusion formula, we also utilize a concatenation fusion over review title and content matrices to keep whole informative features from them, as in Fig. 4. Unlike Method 1 and Method 2, we concatenate two vectors $\widetilde{CT}$ and $\widetilde{CC}$ and get a combined vector for reviews (denoted as $\widetilde{C}_{concat}$).

$$\widetilde{C}_{concat} = Concatenation[\widetilde{CT}, \widetilde{CC}] \in \mathbb{R}^{1 \times 2d} \tag{5}$$

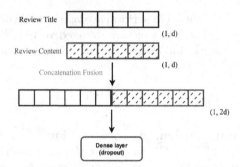

Fig. 4. The review title-content concatenation fusion

The fused vector $\widetilde{C}$, such as $\widetilde{C}_{avg}, \widetilde{C}_{max}, \widetilde{C}_{concat}$ will be fed into a fully-connected output layer to classify the polarity of the review. The sentiment classifier is constructed as follows:

$$\widehat{y} = \sigma(W \cdot \widetilde{C} + b) \tag{6}$$

where σ is Softmax activation function. $W \in \mathbb{R}^{cl \times d}$ (or $W \in \mathbb{R}^{cl \times 2d}$ for $\widetilde{C}_{concat}$) and $b \in \mathbb{R}^{cl}$ are the hidden weight matrix and bias of layer while cl denotes the number of sentiment classes.

We train the SA model on the review titles and contents by minimizing the categorical cross-entropy loss function between the predicted label and ground truth label using the following equation:

$$CE = \sum_{i=1}^{cl} y_i \cdot \log(\widehat{y_i}) \tag{7}$$

where $\widehat{y_i}$ is the $i - th$ predicted label distribution, and y_i is the $i - th$ ground truth label distribution.

5   Experiments

Our experiments aim to test the hypothesis that fusing both review titles and contents could enhance the performance of SA tasks. We will conduct experiments utilizing robust techniques, including Google word2vec, GloVe for word embedding phrase; LSTM, CNN, LSTM+CNN, and BLSTM+CNN for deep learning-based phrase; and three fusion methods. Finally, the accuracy performance of the three models including the proposed, the pruning and the baseline models will be comprehensively evaluated and compared.

5.1   Experimental Settings

Dataset. In this study, we utilized a real-world dataset from TripAdvisor [7]. The dataset comprises 527,770 English reviews from over 290,000 travelers, covering 8,297 hotels in Vietnam. The dataset includes three main criteria: *review title* (a brief summary), *review content* (detailed description of hotel services), and *star rating* (ranging from 1 to 5 stars, with 1 representing "Terrible" and 5 representing "Excellent"). For document-level sentiment analysis tasks, star rating serves as a commonly used method to explicitly express customers' opinions. Ratings of *1-star* to *2-star* indicate a *negative* sentiment, a *3-star* rating reflects a *neutral* opinion, while ratings of *4-star* to *5-star* convey a *positive* sentiment.

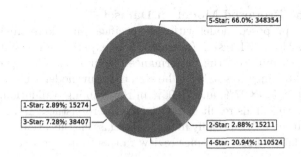

Fig. 5. The star rating distribution in dataset

The dataset seems imbalanced towards the 4-star and 5-star ratings, as shown in Fig. 5. Specifically, 5-star rating reviews occupied 66.0% of the dataset with nearly 350,000 records, followed by 4-star reviews with 110,524 records, accounting for over 20%. The remaining percentages are occupied by 3-star, 2-star, and 1-star rating reviews, with only 7.28% (38,407 reviews), 2.88% (15,211 reviews), and 2.89% (15,274 reviews), respectively. To minimize the effects of data discrepancies and estimate the model's performance, the dataset was randomly split into a training set (80%) and a testing set (20%).

Experiment Setup. We employ Google word2vec and GloVe to represent textual information. The embedding size, which is 300 and 100, respectively, is the output dimension of the word2vec and GloVe models. For deep learning-based phrase, the number of hidden states in the LSTM layer is set to 128, and the 2-dimensional convolution layer employs filter windows (l) of size 3, 4, 5 with 100 feature maps each. The output layer consists of 5 neurons, representing the 5 classes, and utilizes a softmax activation function to perform multi-class classification.

To evaluate the model's performance, we conduct 5 runs with 5 epochs each, and average the results. The batch size is set to 128. The loss function used is the sparse categorical cross-entropy loss, and the optimizer is the Adam algorithm [6] with a learning rate of 0.001 and a dropout rate of 0.5.

Baselines. We also conduct experiments on three different methods to compare the proposed model's performance with others on the TripAdvisor dataset.

- BiLSTM + static Glove and CNN + static GloVe [9], which are the modifications of CNN and LSTM, with 100-dimension GloVe embedding as the input layer.
- The ensemble of CNN and BiLSTM + static GloVe [9] computes the average probability scores of CNN and BiLSTM models to classify customer polarity.
- CNN + word2vec [5], which is a well-known model for sentence classification.

5.2   Experimental Results

Results of the Proposed Model on Dataset

We employ the proposed model with the specification: GloVe embedding, BiL-STM + CNN, Method 1 fusion (average). The experimental results are shown in Table 2, which indicates the performance of our model during the training, validation, and testing phases. It can be seen that our model achieved an average accuracy of 79.78%, 78.97%, and 79.05% in the training, validation, and testing phases, respectively. This result demonstrates the robustness and generalization capability of our model, especially in the multi-class classification scenario, where the sentiment label of the customer review ranges from 1-star to 5-star.

Table 2. The experimental results of our model using Glove, BiLSTM+CNN, Average fusion

Iteration	Train loss	Train accuracy	Val loss	Val accuracy	Test loss	Test accuracy
1	0.4855	0.7975	0.5048	0.7883	0.5076	0.7882
2	0.4842	0.7976	0.5011	0.7895	0.5029	0.7910
3	0.4846	0.7976	0.5018	0.7896	0.5036	0.7904
4	0.4840	0.7981	0.5016	0.7906	0.5047	0.7911
5	0.4841	0.7982	0.5005	0.7907	0.5051	0.7917

Comparison Between Different Word Embedding and Fusion Methods

Table 3 shows our experimental results of the proposed architecture (using title-content average fusion) compared to the pruning models (using either title or content) utilizing different deep learning-based approaches over the Google word2vec and GloVe word embedding. The model employs GloVe embedding, BiLSTM+CNN, Average fusion (Method 1) archived accuracy of 79.05%, compared to 78.42% of model used word2vec embedding. Moreover, even simple deep learning model, such as LSTM, CNN (without fine-tuning) performed on the proposed model, could gain remarkable results. It clearly proved that the fusing review titles and contents could significantly improve the performance of document-level SA tasks.

Table 3. Results of proposed model compared to pruning models using word2vec and GloVe embedding

	GloVe embedding			Word2vec embedding		
	Title	Content	Average Fusion	Title	Content	Average Fusion
LSTM	0.7404	0.7499	**0.7850**	0.7379	0.7415	**0.7754**
CNN	0.7334	0.7424	**0.7769**	0.7346	0.7428	**0.7779**
LSTM+CNN	0.7426	0.7550	**0.7892**	0.7412	0.7508	**0.7839**
BiLSTM+CNN	0.7431	0.7566	**0.7905**	0.7412	0.7518	**0.7842**

The effectiveness of GloVe embedding and Google word2vec embedding could be observed on Table 3. While the word2vec model performed efficiently on both the proposed model and pruning models utlizing CNN on deep learning-based phrase, the GloVe embedding model dominated the rest of the test cases. Thus, the GloVe word embedding could be superior in our model.

A comparison of the proposed model utilizing different fusion methods over GloVe and word2vec embedding is summarized in Table 4. Obviously, the results of Method 1 (title-content average fusion) give the best performance on almost experiments except for CNN (GloVe embedding) and LSTM (word2vec embedding) test cases. This can be explained probably because the review titles and review contents contribute equally to customers' opinions prediction, which leads to archiving better prediction results.

Table 4. Results for different fusion methods and word embeddings

	GloVe embedding			Word2vec Embedding		
	Method 1 (average)	Method 2 (maximum)	Method 3 (concat)	Method 1 (average)	Method 2 (maximum)	Method 3 (concat)
LSTM	**0.7850**	0.7847	0.7832	0.7754	**0.7774**	0.7767
CNN	0.7769	0.7771	**0.7773**	**0.7779**	0.7762	0.7768
LSTM+CNN	**0.7892**	0.7886	0.7887	**0.7839**	0.7829	0.7833
BiLSTM+CNN	**0.7905**	0.7893	0.7900	**0.7842**	0.7806	0.7838

Comparison with Baseline Methods

A comparison of the proposed method with several baseline models is shown in Table 5. The comparison models include the BiLSTM + static GloVe [9], CNN + static GloVe [9], the ensemble of BiLSTM and CNN + static GloVe [9], and CNN + word2vec [5]. It can be observed that the proposed model outperforms these baseline models with accuracy improvements of 2.68%, 3.62%, 3.06%, and 12.36% on average, respectively. This result further verifies the effectiveness of deep learning-based fusion model in accurately predicting customer sentiment based on the review titles and contents.

Table 5. Results of comparison experiments

	Model	Accuracy(%)
Baseline models	BiLSTM + static GloVe	75.68
	CNN + static GloVe	74.74
	Ensemble of BiLSTM and CNN + static GloVe	75.30
	CNN + word2vec	66.00
The proposed model	LSTM + word2vec (maximum fusion)	77.74
	CNN + GloVe (concatenation fusion)	77.73
	LSTM + CNN + GloVe (average fusion)	78.92
	BiLSTM + CNN + GloVe (average fusion)	**79.05**

6   Conclusion and Future Works

In this study, we conducted a study based on deep learning-based fusion model, which processed the review titles and review contents concurrently, demonstrated superior performances in customer sentiment prediction compared to the baseline models on the same TripAdvisor dataset. Empirical results indicated that the combination of review titles and contents as input to the model had been proved to significantly improve the accuracy of sentiment prediction, highlighting the importance of considering both review titles and contents in predicting customers' opinions expressed in textual reviews. However, there is still room for improvement in terms of considering text similarity and semantic relationships between the titles and contents. Future work, therefore, could focus on the directions: (i) Considering the text similarity between review titles and review contents, (ii) Applying graph convolutional networks (GCNs) to represent the semantic relations which we believe could archive remarkable results.

References

1. Capoccia, C.: Online reviews are the best thing that ever happened to small businesses. Forbes. Accessed 2 Feb 2019 (2018)
2. Chua, A.Y., Banerjee, S.: Analyzing review efficacy on amazon. com: does the rich grow richer? Comput. Hum. Behav. **75**, 501–509 (2017)
3. Collobert, R., Weston, J., Bottou, L., Karlen, M., Kavukcuoglu, K., Kuksa, P.: Natural language processing (almost) from scratch. J. Mach. Learn. Res. **12**(ARTICLE), 2493–2537 (2011)
4. Hochreiter, S., Schmidhuber, J.: Long short-term memory. Neural Comput. **9**(8), 1735–1780 (1997)
5. Kim, Y.: Convolutional neural networks for sentence classification. CoRR abs/1408.5882 (2014). https://arxiv.org/abs/1408.5882
6. Kingma, D.P., Ba, J.: Adam: a method for stochastic optimization. arXiv preprint arXiv:1412.6980 (2014)
7. Le, Q.H., Mau, T.N., Tansuchat, R., Huynh, V.N.: A multi-criteria collaborative filtering approach using deep learning and dempster-shafer theory for hotel recommendations. IEEE Access **10**, 37281–37293 (2022)

8. Liu, B.: Sentiment analysis and opinion mining. Synth. Lect. Hum. Lang. Technol. **5**(1), 1–167 (2012)

9. Minaee, S., Azimi, E., Abdolrashidi, A.: Deep-sentiment: sentiment analysis using ensemble of CNN and bi-lstm models. arXiv preprint arXiv:1904.04206 (2019)

10. Paredes-Valverde, M.A., Colomo-Palacios, R., Salas-Zárate, M.D.P., Valencia-García, R.: Sentiment analysis in Spanish for improvement of products and services: a deep learning approach. Sci. Program. **2017** (2017)

11. Phan, H.T., Nguyen, N.T., Mazur, Z., Hwang, D.: Sentence-level sentiment analysis using gcn on contextualized word representations. In: Groen, D., de Mulatier, C., Paszynski, M., Krzhizhanovskaya, V.V., Dongarra, J.J., Sloot, P.M.A. (eds.) International Conference on Computational Science, vol. 13351, pp. 690–702. Springer, Cham (2022). https://doi.org/10.1007/978-3-031-08754-7_71

12. Phan, H.T., Nguyen, N.T., Van Cuong, T., Hwang, D.: A method for detecting and analyzing the sentiment of tweets containing fuzzy sentiment phrases. In: 2019 IEEE International Symposium on Innovations in Intelligent Systems and Applications (INISTA), pp. 1–6. IEEE (2019)

13. Salehan, M., Kim, D.J.: Predicting the performance of online consumer reviews: a sentiment mining approach to big data analytics. Decis. Support Syst. **81**, 30–40 (2016)

14. Tran, L.Q., Van Duong, B., Nguyen, B.T.: Sentiment classification for beauty-fashion reviews. In: 2022 14th International Conference on Knowledge and Systems Engineering (KSE), pp. 1–6. IEEE (2022)

15. Yang, S., Yao, J., Qazi, A., et al.: Does the review deserve more helpfulness when its title resembles the content? locating helpful reviews by text mining. Inf. Process. Manage. **57**(2), 102179 (2020)

Resource Management
and Optimization

In Support of Push-Based Streaming for the Computing Continuum

Ovidiu-Cristian Marcu$^{(\boxtimes)}$ and Pascal Bouvry

University of Luxembourg, Esch-sur-Alzette, Luxembourg
{ovidiu-cristian.marcu,pascal.bouvry}@uni.lu

Abstract. Real-time data architectures are core tools for implementing the edge-to-cloud computing continuum since streams are a natural abstraction for representing and predicting the needs of such applications. Over the past decade, Big Data architectures evolved into specialized layers for handling real-time storage and stream processing. Open-source streaming architectures efficiently decouple fast storage and processing engines by implementing stream reads through a pull-based interface exposed by storage. However, how much data the stream source operators have to pull from storage continuously and how often to issue pull-based requests are configurations left to the application and can result in increased system resources and overall reduced application performance. To tackle these issues, this paper proposes a unified streaming architecture that integrates co-located fast storage and streaming engines through push-based source integrations, making the data available for processing as soon as storage has them. We empirically evaluate pull-based versus push-based design alternatives of the streaming source reader and discuss the advantages of both approaches.

Keywords: Streaming · Real-time storage · Push-based · Pull-based · Locality

1 Introduction

The edge-to-cloud computing continuum [10] implements fast data storage and streaming *layered architectures* that are deployed intensively in both Cloud [9] and Fog architectures [22]. Fast data processing enables high-throughput data access to streams of logs, e.g., daily processing terabytes of logs from tens of billions of events at CERN accelerator logging service [1]. Moreover, implementing sensitive information detection with the NVIDIA Morpheus AI framework enables cybersecurity developers to create optimized AI pipelines for filtering and processing large volumes of real-time data [2].

Over the past decade, Big Data architectures evolved into specialized layers for handling real-time storage and stream processing. To efficiently decouple fast storage and streaming engines, architects design a streaming source interface that implements data stream reads through *pull-based* remote procedure calls (RPC)

N. T. Nguyen et al. (Eds.): ACIIDS 2023, LNAI 13996, pp. 339–350, 2023.
https://doi.org/10.1007/978-981-99-5837-5_28

APIs exposed by storage. The streaming source operator continuously pulls data from storage while its configuration (i.e., how much data to pull and how often to issue pull requests) is left to the application, being a source of bottlenecks and overall reduced application performance. In turn, decoupled streaming sources help manage backpressure [17] and simplifies fault-tolerance implementation for system crash management.

The pull-based integration approach is opposed to monolithic architectures [28] that have the opportunity to more efficiently and safely optimize data-related tasks. Another architectural choice is to closely integrate fast storage and streaming engines through *push-based* data sources, making the data available to the processing engine as soon as it is available. The push-based source integration approach should keep control of the data flow to ensure backpressure and should promote an easy integration through storage and processing non-intrusive extensions to further promote open-source real-time storage and streaming development.

Our challenge is then **how to design and implement a push-based streaming source strategy to efficiently and functionally integrate real-time storage and streaming engines** while keeping the advantages of a pull-based approach. Towards this goal, this paper introduces a push-based streaming design to integrate co-located real-time storage and processing engines. We implement pull-based and push-based stream sources as integration between open-source KerA[1] a real-time storage system, and Apache Flink [3], a stream processing engine. We empirically evaluate the KerA-Flink push-based and pull-based approaches and we show that the push-based approach can be competitive with a pull-based design while requiring reduced system resources. Furthermore, when storage resources are constrained, the push-based approach can be up to 2x more performant compared to a pull-based design.

2　Background and Related Work

Decoupling producers and consumers through message brokers (e.g., Apache Kafka [16]) can help applications through simplified real-time architectures. This locality-poor design is preferred over monolithic architectures by state-of-the-art open-source streaming architectures. Big Data frameworks that implement MapReduce [13] are known to implement data locality (pull-based) optimizations. General Big Data architectures can thus efficiently co-locate map and reduce tasks with input data, effectively reducing the network overhead and thus increasing application throughput. However, they are not optimized for low-latency streaming scenarios. User-level thread implementations such as Arachne [24] and core-aware scheduling techniques like Shenango, Caladan [14,23] can further optimize co-located latency-sensitive stream storage and analytics systems, but are difficult to implement in practice.

Finally, it is well known that message brokers, e.g., Apache Kafka [4,16], Apache Pulsar [5], Distributedlog [25], Pravega [8], or KerA [19], can contribute

---

[1] KerA-Flink integration source code: https://gitlab.uni.lu/omarcu/zettastreams.

to higher latencies in streaming pipelines [15]. Indeed, none of these open-source storage systems implement locality and thus force streaming engines to rely on a pull-based implementation approach for consuming data streams. Consistent state management in stream processing engines is difficult [12] and depends on real-time storage brokers to provide indexed, durable dataflow sources. Therefore, stream source design is critical to the fault-tolerant streaming pipeline and potentially a performance issue.

A pull-based source reader works as follows: it waits no more than a specific timeout before issuing RPCs to pull (up to a particular batch size) more messages from stream storage. One crucial question is how much data these sources have to pull from storage brokers and how often these pull-based RPCs should be issued to respond to various application requirements. Consequently, a push-based approach can better solve these issues by pushing the following available messages to the streaming source as soon as more stream messages are available. However, a push-based source reader is more difficult to design since coupling storage brokers and processing engines can bring back issues solved by the pull-based approach (e.g., backpressure, scalability). Thus, we want to explore a non-monolithic design that integrates real-time storage and streaming engines through a push-based stream source approach and understand the performance advantages (e.g., throughput) of both approaches. Towards this goal let us introduce next a push-based streaming design that unifies real-time storage and processing engines and describe our implementation.

3   Unified Real-time Storage and Processing Architecture: Our Push-based Design and Implementation

Background. Fast storage (e.g., [19,20]) implements a layer of *brokers* to serve producers and consumers of data streams. As illustrated in Fig. 1, a multi-threaded broker is configured with one dispatcher thread polling the network and responsible for serving read/write RPC requests and multiple working threads that do the actual writes and reads. Streaming engines [3,21,27] implement a layer of workers. E.g., in Apache Flink, each worker implements a JVM process that can host multiple slots (a slot can have one core). Sources and other operators are deployed on worker slots ([3,6]) and are configured to use in-memory buffers. Backpressure is ensured as follows: sources continuously issue pull-based read requests as long as buffers are not filled.

Design Principles. To ensure backpressure, our push-based design takes source buffers outside the processing engine and shares buffer control with storage and processing through pointers and notifications. To avoid network overhead, we propose to co-locate real-time storage and streaming engines. To remove storage interference of read and write RPC requests, we separate reads and writes through dedicated push worker threads. The push-based mechanism completely removes the RPC and networking overheads of the pull-based approach.

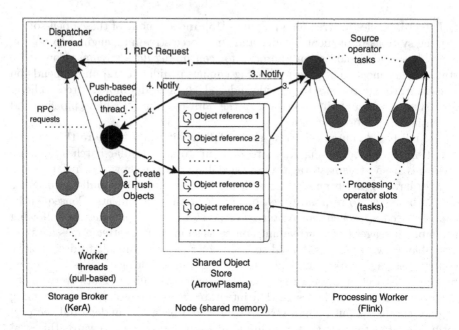

Fig. 1. Unified real-time storage and stream processing architecture. Storage broker and processing worker are co-located on a multi-core node while stream source operators implement push-based consumers through shared memory partitioned object store. At runtime, source tasks consume data from shared reusable memory objects filled by storage as data arrives.

Push-Based Architectural Design. As illustrated in Fig. 1, we co-locate a storage broker and one or multiple processing workers on a multi-core node. We propose to leverage a shared partitioned in-memory buffer between (push-based) streaming sources and storage brokers to provide backpressure support to streaming engines and to allow for transparent integration with various streaming storage and processing engines. Multiple sources coordinate to launch only one push-based RPC request (step 1). Storage creates (once) a dedicated worker thread (in black) that is responsible to continuously fill shared reusable objects with next stream data (step 2). Source tasks are notified when objects have new updates (step 3) and then process these data. The worker thread is notified (step 4) when an object was consumed, so it can reuse it an refill with new data. This flow (steps 2–4) executes continuously. Objects have a fixed-size and are shared through pointers by worker thread and source tasks. The source creates tuples and pushes them further to the stream processing operator tasks. Backpressure is ensured through shared store notifications.

Implementation. On the same node live three processes: the streaming broker, the processing worker and the shared partitioned object store. As illustrated in Fig. 1, two push-based streaming source tasks are scheduled on one processing worker. At execution time, only one of the two sources will issue (once!) the push-based RPC (e.g., based on the smallest of the source tasks' identifiers). This special RPC request (implemented by storage) contains initial partition off-sets (partitioned streams) used by sources to consume the next stream records. The storage handles the push-based RPC request by assigning a worker thread responsible for creating and pushing the next chunks of data associated with con-sumers' partition offsets. We implement the shared-memory object store based on Apache Arrow Plasma, a framework that allows the creation of in-memory buffers (named objects) and their manipulation through shared pointers. Our push-based RPC is implemented on the KerA storage engine while we transpar-ently integrate our KerA connector with Apache Flink for stream processing.

4 Evaluation

Our goal is to understand the performance advantages of push-based and pull-based streaming source integrations between real-time storage and streaming engines. While the pull-based approach simplifies implementation, the push-based approach requires a more tight integration. What performance benefits characterize each approach?

Experimental Setup and Parameter Configuration. We run our experiments on the Aion cluster[2] by deploying Singularity containers (for reproducibility) over Aion regular nodes through Slurm jobs. One Aion node has two AMD Epyc ROME 7H12 CPUs (128 cores), each with 256 GB of RAM, interconnected through Infiniband 100Gb/s network. Producers are deployed separately from the streaming architecture. The KerA broker is configured with up to 16 worker cores (for pull-based, while the push-based approach uses only one dedicated worker core) while the partition's segment size is fixed to 8 MiB. We use Apache Flink version 1.13.2. We configure several producers Np (respectively consumers Nc, values $= 1,2,4,8$), similarly to [20], that send data chunks to a partitioned stream having Ns partitions. Each partition is consumed exclusively by its asso-ciated consumer. Producers concurrently push synchronous RPCs having one chunk of CS size (values $= 1,2,4,8,16,32,64,128$ KiB) for each partition of a broker, having in total $ReqS$ size. Each chunk can contain multiple records of configurable $RecS$ size for the synthetic workloads. We configure producers to read and ingest Wikipedia files in chunks having records of 2 KiB. Flink workers correspond to the number of Flink slots NFs (values $= 8,16$) and are installed on the same Singularity instance where the broker lives.

To understand previous parameters' impact on performance, we run each experiment for 60 to 180 s while we collect producer and consumer throughput metrics (records every second). We plot 50-percentile cluster throughput per second for each experiment (i.e., by aggregating the write/read throughput of each producer/consumer every second), and we compare various configurations.

---

[2] more details at https://hpc.uni.lu/infrastructure/supercomputers the Aion section.

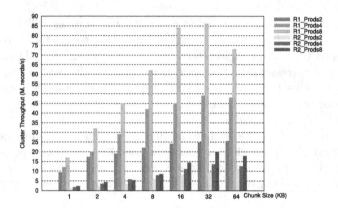

Fig. 2. Impact of replication and chunk size on ingestion cluster throughput (only producers). Parameters: Np = 2,4,8, RecS = 100 Bytes, one stream with Ns = 8 partitions. R1_Prods2 represents replication factor one and two concurrent producers writing to one stream. R2_Prods8 represents replication factor two and eight producers. Results are similar for a stream with Ns = 16.

Our second goal is to understand *who of the push-based and respectively pull-based streaming strategies is more performant, in what conditions, and what the trade-offs are in terms of configurations.*

Benchmarks. The first benchmark (relevant to use cases that transfer or duplicate partitioned datasets) implements a simple pass-over data, iterating over each record of partitions' chunks while counting the number of records. The second benchmark implements a filter function over each record, being a representative workload used in several real-life applications (*e.g.* indexing the monitoring data at the LHC [7]). The next benchmarks (CPU intensive) implement word count over Wikipedia datasets. For reproducibility, our benchmark code is open-source and a technical report [18] further describes these applications.

4.1 Results and Discussion

Synthetic Benchmarks: the Count Operator. In our first evaluation, we want to understand how our chosen parameters can impact the aggregated throughput while ingesting through several concurrent producers. As illustrated in Fig. 2, we experiment with two, four, and eight concurrent producers. Increasing the chunk size *CS*, the request size *ReqS* increases proportionally, for a fixed record size *RecS* of fixed value of 100 Bytes. While increasing the chunk size, we observe (as expected) that the cluster throughput increases; having more producers helps, although they compete at append time. We also observe that replication considerably impacts cluster throughput (as expected) since each producer has to wait for an additional replication RPC done at the broker side. Producers wait up to one millisecond before sealing chunks ready to be pushed to the broker (or the

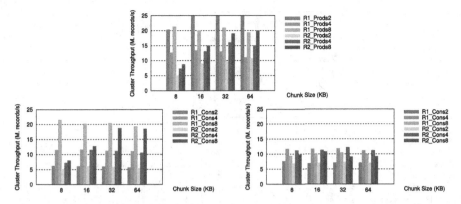

Fig. 3. Only producers (top). Pull-based (left) versus push-based (right) consumers for iterate and count benchmark for a stream with Ns = 8. Parameters: Np = Nc = 2,4,8, replication R = 1,2, consumer CS = 128 KiB, we plot producer CS = 8,16,32,64.

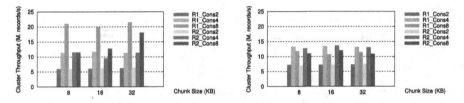

Fig. 4. Pull-based (left) versus push-based (right) consumers for iterate, count and filter benchmark. Parameters: Np = Nc = 2,4,8, replication R = 1,2, consumer CS = 128 KiB, we plot producer CS = 8,16,32,64, stream Ns = 8.

chunk gets filled and sealed) - this configuration can help trade-off throughput with latency. With only two producers, we can obtain a cluster throughput of ten million records per second, while we need eight producers to double this throughput. Next experiments introduce concurrent consumers in parallel with concurrent producers.

The subsequent evaluation looks at concurrently running producers and consumers and compares pull-based versus push-based consumers. The broker is configured with 16 working cores to accommodate up to eight producers and eight consumers concurrently writing and reading chunks of data. Since consumers compete with producers, we expect the producers' cluster throughput to drop compared to the previous evaluation that runs only concurrent producers. This is shown in Fig. 3: due to higher competition to broker resources by consumers, producers obtain a reduced cluster throughput compared to the previous experiment. We observe that consumers fail to keep up with the producers' rate.

These experiments illustrates the impact on performance of the interference between reads and writes. Increasing the chunk size, we observe that pull-based consumers obtain better performance than push-based. Increasing the number of consumers, we observe that the dedicated thread does not keep up with more than

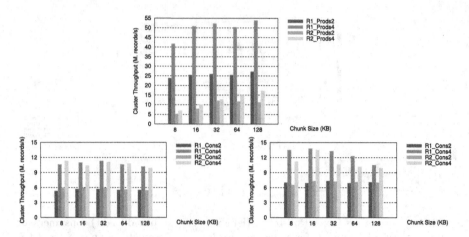

Fig. 5. Only producers (top). Pull-based (left) versus push-based (right) consumers for iterate, count and filter benchmark for a stream with 4 partitions.

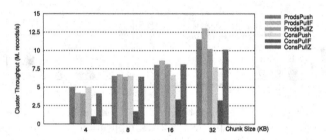

Fig. 6. Iterate, count and filter benchmark constrained broker resources. Comparing C++ pull-based consumers with Flink pull-based and push-based consumers. Prods-Push corresponds to producers running concurrently with push-based Flink consumers i.e. ConsPush. ProdsPullF corresponds to producers running concurrently with pull-based Flink consumers i.e. ConsPullF. ProdsPullZ corresponds to producers running concurrently with C++ pull-based consumers. Four producers and four consumers ingest and process a replicated stream (factor two) with eight partitions over one broker storage with four working cores. Consumer chunk size equals the producer chunk size.

four consumers; otherwise, push-based is competitive with pull-based or slightly better. However, (although with eight consumers the pull-based strategy can obtain a better cluster throughput,) for up to four consumers the push-based strategy not only can obtain slightly better cluster throughput but the number of resources dedicated to consumers reduces considerably (two threads versus eight threads for the configuration with four consumers). While pull-based consumers double the cluster throughput when using 16 threads for the source operators, push-based consumers only use two threads for the source operator.

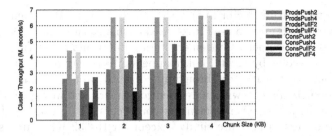

Fig. 7. Iterate and count benchmark stream with 8 partitions broker with 8 cores. Evaluation of pull-based and push-based Flink consumers. ProdsPush corresponds to producers running concurrently with push-based Flink consumers i.e. ConsPush. ProdsPullF corresponds to producers running concurrently with pull-based Flink consumers i.e. ConsPullF. Consumer chunk size equals the producer chunk size multiplied by 8. We plot producer chunk size.

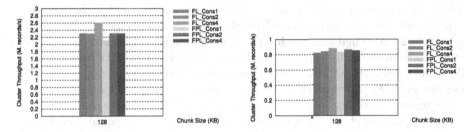

Fig. 8. Pull-based consumers versus push-based consumers for the word count benchmarks with 4 partitions. The left figure presents the word count benchmarks, the right figure corresponds to the windowed word count benchmark. FLCons2 represents two push-based consumers while FPLCons4 represents four pull-based consumers.

Synthetic Benchmarks: The Filter Operator. We further compare pull-based versus push-based consumers when implementing the filter operator, in addition, to counting for a stream with eight partitions. Similar to previous experiments, the push-based consumers are slower when scaled to eight for larger chunks, as illustrated in Fig. 4. As illustrated in Fig. 5, when experimenting with up to four producers and four consumers over a stream with four partitions, *the push-based strategy provides a cluster throughput slightly higher with smaller chunks, being able to process two million tuples per second additionally over the pull-based approach. With larger chunks, the throughput reduces: architects have to carefully tune the chunk size in order to get the best performance.* When experimenting with smaller chunks, more work needs to be done by pull-based consumers since they have to issue more frequently RPCs (see Fig. 7). Moreover, the push-based strategy provides higher or similar cluster throughput than the pull-based strategy while using fewer resources.

Next, we design an experiment with constrained resources for the storage and backup brokers configured with four cores. We ingest data from four producers into a replicated stream (factor two) with eight partitions. We concurrently run four consumers configured to use Flink-based push and pull strategies and native C++ pull-based consumers. Consumers iterate, filter and count tuples that are reported every second by eight Flink mappers. We report our results in Fig. 6 where we compare the cluster throughput of both producers and consumers. Producers compete directly with pull-based consumers, and we expect the cluster throughput to be higher when concurrent consumers use a push-based strategy. However, producers' results are similar except for the 32 KB chunk size when producers manage to push more data since pull-based consumers are slower. We observe that the C++ pull-based consumers can better keep up with producers while push-based consumers can keep up with producers when configured to use smaller chunks. *The push-based strategy for Flink is up to 2x better than the pull-based strategy of Flink consumers. Consequently, the push-based approach can be more performant for resource-constrained scenarios.*

Wikipedia Benchmarks: (Windowed) Word Count Streaming. For the following experiments, the producers are configured to read Wikipedia files in chunks with records of 2 KiB. Therefore, producers can push about 2 GiB of text in a few seconds. Consumers run for tens of seconds and do not compete with producers. As illustrated in Fig. 8, pull-based and push-based consumers demonstrate similar performance. We plot word count tuples per second aggregated for eight mappers while scaling consumers from one to four. Results are similar when we experiment with smaller chunks or streams with more partitions since this benchmark is CPU-bound. *To avoid network bottlenecks when processing large datasets like this one (e.g., tens of GBs) on commodity clusters, the push-based approach can be more competitive when pushing pre-processing and local aggregations at the storage.*

Discussion and Future Implementation Optimizations. Regarding our prototype implementation, we believe there is room for further improvements. One future step is integrating the shared object store and notifications mechanism inside the broker storage implementation. This choice will bring up two potential optimizations. Firstly, it would avoid another copy of data by leveraging existing in-memory segments that store partition data (necessary for high-throughput use cases). Secondly, we could optimize latency by implementing the notification mechanism through the asynchronous RPCs available in KerA. Furthermore, implementing pre-processing functions in-storage (e.g., as done in [11]) can further improve performance by reducing data movement.

5   Conclusion

We have proposed a unified real-time storage and processing architecture that leverages a push-based strategy for streaming consumers. Experimental evaluations show that when storage resources are enough for concurrent producers and consumers, the push-based approach is performance competitive with the pull-based one (as currently implemented in state-of-the-art real-time architectures) while consuming fewer resources. However, when the competition of concurrent producers and consumers intensifies and the storage resources (i.e., number of cores) are more constrained, the push-based strategy can enable a better throughput by a factor of up to 2x while reducing processing latency.

Acknowledgment. The experiments presented in this paper were carried out using the HPC facilities of the University of Luxembourg [26] – see hpc.uni.lu. This work is partially funded by the SnT-LuxProvide partnership on bridging clouds and supercomputers and by the Fonds National de la Recherche Luxembourg (FNR) POLLUX program under the SERENITY Project (ref. C22/IS/17395419).

References

1. Next CERN accelerator logging service architecture. https://www.slideshare.net/SparkSummit/next-cern-accelerator-logging-service-with-jakub-wozniak
2. Sensitive information detection using the NVIDIA Morpheus AI framework (2021). https://developers.redhat.com/articles/2021/10/18/sensitive-information-detection-using-nvidia-morpheus-ai-framework
3. Apache Flink (2022). https://flink.apache.org/
4. Apache Kafka (2022). https://kafka.apache.org/
5. Apache Pulsar (2022). https://pulsar.apache.org/
6. Apache Spark (2022). https://spark.apache.org/
7. Large Hadron Holider. (2022). http://home.cern/topics/large-hadron-collider
8. Pravega (2022). http://pravega.io/
9. Akidau, T., et al.: The dataflow model: a practical approach to balancing correctness, latency, and cost in massive-scale, unbounded, out-of-order data processing. Proc. VLDB Endow. **8**(12), 1792–1803 (2015). https://doi.org/10.14778/2824032.2824076
10. Antoniu, G., Valduriez, P., Hoppe, H.C., Krüger, J.: Towards integrated hardware/software ecosystems for the edge-cloud-HPC. Continuum (2021). https://doi.org/10.5281/zenodo.5534464
11. Bhardwaj, A., Kulkarni, C., Stutsman, R.: Adaptive placement for in-memory storage functions. In: 2020 USENIX Annual Technical Conference (USENIX ATC 20), pp. 127–141. USENIX Association, July 2020
12. Carbone, P., Ewen, S., Fóra, G., Haridi, S., Richter, S., Tzoumas, K.: State management in apache flink®: Consistent stateful distributed stream processing. Proc. VLDB Endow. **10**(12), 1718–1729 (2017). https://doi.org/10.14778/3137765.3137777
13. Dean, J., Ghemawat, S.: Mapreduce: simplified data processing on large clusters. Commun. ACM **51**(1), 107–113 (2008). https://doi.org/10.1145/1327452.1327492
14. Fried, J., Ruan, Z., Ousterhout, A., Belay, A.: Caladan: Mitigating Interference at Microsecond Timescales. USENIX Association, USA (2020)
15. Javed, M.H., Lu, X., Panda, D.K.D.: Characterization of big data stream processing pipeline: a case study using flink and kafka. In: Proceedings of the Fourth IEEE/ACM International Conference on Big Data Computing, Applications and Technologies, BDCAT 2017, pp. 1–10. Association for Computing Machinery, New York, NY, USA (2017). https://doi.org/10.1145/3148055.3148068
16. Jay, K., Neha, N., Jun, R.: Kafka: a distributed messaging system for log processing. In: Proceedings of 6th International Workshop on Networking Meets Databases, NetDB 2011 (2011)

17. Kalavri, V., Liagouris, J., Hoffmann, M., Dimitrova, D., Forshaw, M., Roscoe, T.: Three steps is all you need: fast, accurate, automatic scaling decisions for distributed streaming dataflows. In: Proceedings of the 13th USENIX Conference on Operating Systems Design and Implementation, OSDI 2018, pp. 783–798. USENIX Association, USA (2018)

18. Marcu, O.C., Bouvry, P.: Colocating real-time storage and processing: an analysis of pull-based versus push-based streaming (2022)

19. Marcu, O.C., et al.: Kera: scalable data ingestion for stream processing. In: 2018 IEEE 38th International Conference on Distributed Computing Systems (ICDCS), pp. 1480–1485 (2018). https://doi.org/10.1109/ICDCS.2018.00152

20. Marcu, O.C., Costan, A., Nicolae, B., Antonin, G.: Virtual log-structured storage for high-performance streaming. In: 2021 IEEE International Conference on Cluster Computing (CLUSTER), pp. 135–145 (2021). https://doi.org/10.1109/Cluster48925.2021.00046

21. Miao, H., Park, H., Jeon, M., Pekhimenko, G., McKinley, K.S., Lin, F.X.: Streambox: modern stream processing on a multicore machine. In: USENIX ATC, pp. 617–629. USENIX Association (2017)

22. Nguyen, S., Salcic, Z., Zhang, X., Bisht, A.: A low-cost two-tier fog computing testbed for streaming IoT-based applications. IEEE Internet Things J. 8(8), 6928–6939 (2021). https://doi.org/10.1109/JIOT.2020.3036352

23. Ousterhout, A., Fried, J., Behrens, J., Belay, A., Balakrishnan, H.: Shenango: achieving high CPU efficiency for latency-sensitive datacenter workloads. In: Proceedings of the 16th USENIX Conference on Networked Systems Design and Implementation, NSDI 2019, pp. 361–377. USENIX Association, USA (2019)

24. Qin, H., Li, Q., Speiser, J., Kraft, P., Ousterhout, J.: Arachne: Core-aware thread management. In: 13th USENIX Symposium on Operating Systems Design and Implementation (OSDI 18), USENIX Association, Carlsbad, CA (2018)

25. Sijie, G., Robin, D., Leigh, S.: Distributedlog: a high performance replicated log service. In: IEEE 33rd International Conference on Data Engineering, ICDE 2017 (2017)

26. Varrette, S., Bouvry, P., Cartiaux, H., Georgatos, F.: Management of an academic HpC cluster: The UL experience. In: 2014 International Conference on High Performance Computing Simulation (HPCS), pp. 959–967 (2014). https://doi.org/10.1109/HPCSim.2014.6903792

27. Venkataraman, S., et al.: Drizzle: fast and adaptable stream processing at scale. In: 26th SOSP, pp. 374–389. ACM (2017). https://doi.org/10.1145/3132747.3132750

28. Zou, J., Iyengar, A., Jermaine, C.: Pangea: monolithic distributed storage for data analytics. Proc. VLDB Endow. 12(6), 681–694 (2019). https://doi.org/10.14778/3311880.3311885

Effective Method of Implementing Constrained Optimization Problems Using Data Instances

Jarosław Wikarek and Paweł Sitek$^{(\boxtimes)}$

Kielce University of Technology, Kielce, Poland
{j.wikarek,sitek}@tu.kielce.pl

Abstract. Many practical optimization problems in the area of production, distribution, logistics, IT, etc. can be classified as constrained optimization problems. Most often these are also discrete problems of a combinatorial nature. All this makes the optimization process very time-consuming and requires extensive hardware and software resources. Therefore, for large-size problems, optimization is often unprofitable. Typically, such problems are modeled and solved in mathematical programming or constraint programming environments. The paper proposes an effective method of implementing this type of problems, which enables a significant reduction in the size of the modeled problems, resulting in reduction in computation time and the computational resources involved. The presented method uses data instances and model representation properties in mathematical programming environments.

Keywords: Constrained optimization · mathematical programming · data instances · relational model

1 Introduction

Problems of planning and scheduling, allocation of resources, routing, loading [1] occurring e.g. in production, distribution, transport, supply chains or urban logistics are characterized by numerous constraints. These usually include sequence, time, resource, cost constraints, etc., and typically take the form of discrete constrained optimization problems (Sect. 2), often of a combinatorial nature [2]. All this makes solving them time-consuming and requires large hardware and software expenditures. For problems of real/practical size, the time to solution is often unacceptable. Therefore, in practice, approximate methods are used to solve them, such as dedicated heuristics or metaheuristics, including evolutionary algorithms, ant and swarm algorithms, etc. [3, 4], which make it possible to find an approximate solution within an acceptable timeframe. However, in situations where it is necessary to find an exact solution for data instances whose sizes make it impossible to use exact methods, we propose an original method of their modeling and implementation. This method is based on data instances and the structure of the modeled problem and allows to reduce its size, i.e. the number of decision variables and the number of constraints. Consequently, it reduces the time to obtain a solution and in many cases enables the use of exact methods. The rest of the

N. T. Nguyen et al. (Eds.): ACIIDS 2023, LNAI 13996, pp. 351–362, 2023.
https://doi.org/10.1007/978-981-99-5837-5_29

paper is as follows. The second section presents the characteristics of the considered constraint problems. The third section presents the assumptions and algorithm of the proposed method of transforming the modeled problem. The fourth section shows how the method works through an illustrative example. The fifth section contains calculation examples. The last section is a summary and conclusions.

2   Constrained Optimization Problems

Constrained optimization problems (COPs) [5] are the ones for which a function $f(x)$ is to be minimized or maximized subject to a set of constraints $C(x)$. There $f(x)$: $Rn \rightarrow R$ is called the objective (goal) function and $C(x)$ is a Boolean-valued formula. The set of constraints $C(x)$ can be an arbitrary Boolean combination of equations $ce(x) = 0$, weak inequalities $cwi(x) \geq 0$, strict inequalities $csi(x) > 0$, and $x \in X$ statements. COPs are usually modeled in the form of MILP (Mixed Integer Linear Programming), BILP (Binary Integer Linear Programming), IP (Integer Programming) [6], etc. The above mentioned types of models differ in the nature of the decision variables (integer, binary, mixed). A characteristic feature of the MILP, BILP and IP models is their matrix representation. This means that both model parameters and their decision variables are presented in the form of matrices/vectors. On the other hand, the actual data of the modeled problems were most often placed in relational databases, which resulted in a different representation. The considered problems are characterized by high computational complexity (NP-hard problems). For an exact solution, the methods of mathematical programming, constraint programming, dynamic programming and hybrid methods are most often used. In practice, optimization solvers are often used to model and solve this class of problems. The most popular solvers available today include: LINGO, CPLEX, GUROBI, SCIP, BARON [7], etc. In practice, if we are unable to obtain a solution for specific data instances in an acceptable time, COPs are solved using approximate methods. Most often, these are dedicated heuristics and AI methods, which can be classified as Nature Inspired Intelligence Methods, Multi-agent Systems, Machine Learning Methods and Artificial Neural Network. On the one hand, the use of heuristics and/or AI methods to solve COPs makes it often possible to overcome the barrier of computational complexity, but this is at the expense of obtaining only approximate solutions.

3   Proposed Method of Effective Implementation

An important challenge is to find exact solutions to COPs class problems in the case of data instances for which the use of exact methods is inefficient in terms of time and resources. A certain solution is to use a proprietary hybrid approach [8], which combines methods of mathematical programming and constraint programming. The method was applied to the optimization problems of SCM (Supply Chain Management), CVRP (Capacitated Vehicle Routing), 2ECVRP (Two-Echelon Capacitated Vehicle Routing Problem), etc. [8–10].

However, the use of a hybrid approach requires a good knowledge of both mathematical and constraint programming methods. Therefore, the current research focused on developing a method that can be used as a supplement to any mathematical programming solver. The basis of this method was the observation of two different ways of representing the modeled problems when saving real data (relational representation) and models of these problems (matrix representation).

So far, the problem of different forms of representation has been dealt with as follows (Fig. 1). In order to build MILP, BILP, IP models that require matrix representation, the data retrieved from the relational database were placed in matrices describing the parameters of the models. Missing data, i.e. those for which no tuples (rows) were found in the database, were filled with zeros. Such a procedure caused the resulting matrices to reach large sizes and to be usually sparse matrices. After such transformation, an implementation model (Model M) was built in the mathematical programming solver environment which was then solved by the solver.

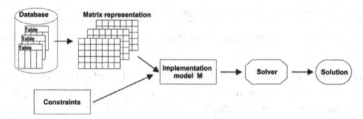

Fig. 1. The classic way to model and solve COP using a mathematical programming solver

We propose a different approach (Fig. 2). Based on the relational representation of the data instance, a transformed implementation model (Model T) is built using the proposed transformation procedure and the mathematical programming solver environment.

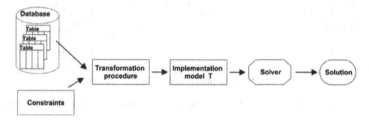

Fig. 2. Proposed way of effective implementation of modeled COP.

For this purpose, a procedure of model transformation was developed based on transformation algorithm (Fig. 3). A major advantage of the developed algorithm is its linear computational complexity.

The proposed procedure (*Proc_T*) is based on several assumptions:

- applies to constraint issues that are classified as COPs;
- data instances are stored in relational databases or as facts;
- the formalization of problems is in the form of models MILP, BILP, IP, etc., which are characterized by matrix notation;

- all or some of the decision variables are discrete (binary, integer).

Step 1. For each constraint
    Step 2. For each relation (table) that describes the problem
        Step 3. If the relation key attributes are a subset of constraint attributes.
        Step 4. If there is no tuple (row) in relation for attribute values.
           Step 5. Remove this constraint
           Step 6. Go to the next constraint.
Step 7. Otherwise.
    Step 8. For each variable in constraint.
        Step 9. If the relation's key is a subset of the variable's key.
           Step 10. If there is no tuple (row) in the relation with variable attribute values
              Step 11. Remove this variable from the constraint.

Fig. 3. Algorithm of the proposed transformation procedure (*Proc_T*) to obtain Model T

The presentation of how both approaches work and what effects they produce for an example of one type of constraints is shown in Fig. 4 and Fig. 5, respectively. It can be seen that even for such a simple example and such a small size, a significant simplification of the structure of constraints and reduction of the number of constraints and decision variables is obtained.

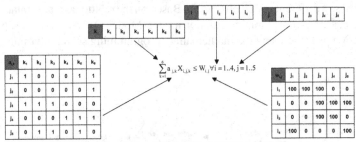

20 - constrains (4x5); 120 variables (4x5x6)

Fig. 4. The classic way to model constraints using a mathematical programming solver

The second element of the proposed approach, apart from the transformation procedure reducing the size of the modeled problem, is the introduction of additional decision variables. These are additional decision variables that do not result directly from the nature of the modeled problem and do not affect its specification. They are introduced so that the solved model always has a solution. The lack of a solution to the problem, i.e. the situation for COPs - No Solution Found (NSF), is of no importance from the decision-making point of view. However, the introduction of these additional decision variables, the values of which determine why we are unable to find a satisfactory solution, e.g. what is missing, what we will not do, will not deliver, etc., is already some information for decision-makers.

The effectiveness of the proposed approach has been tested on a simple illustrative example of COP (Sect. 4, 5).

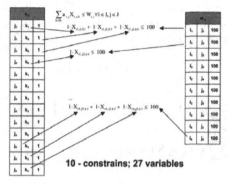

Fig. 5. Proposed way of modeling constraints using transformation procedure and a mathematical programming solver

4   Illustrative Example

The evaluation of the effectiveness and possibilities of the proposed method of implementation was made on the basis of an illustrative example of COP. This example concerns the problem of optimizing the production and delivery of a specific range of products to customers. This is a significantly simplified example that combines the issue of transport and allocation.

A general diagram showing the ideas of the illustrative example is shown in Fig. 6.

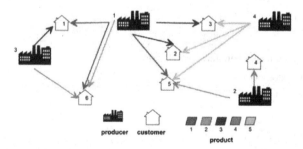

Fig. 6. Visualization of illustrative examples

Manufacturers $i \in I$ produce products $j \in J$. Products are ordered and delivered to customers $k \in K$. Our goal is to determine the optimal (in terms of profit) method of customer service, i.e. how many and what products a given manufacturer will produce and deliver to a given customer. The cost of delivery and the profit from the sale of the product to the customer are taken into account. There are many constraints in the problem, e.g. regarding the production capacity, deliveries, order fulfillment, etc.

356 J. Wikarek and P. Sitek

For illustrative example, the ILP (Integer Linear Programming) model (1)..(5) was built. Indices, parameters, decision variables and a description of the model limitations are presented in Table 1. The data instances of the modeled problem are located in a relational database, the relevant part of the structure of which in the form of an ERD (entity relationship diagram) is shown in Fig. 7. The designations on the ERD are consistent with the model designations in Table 1.

$$\sum_{k \in O} X_{i,j,k} \leq W_{i,j} \forall i \in I, j \in J \tag{1}$$

$$\sum_{i \in I} a_{i,j} \cdot X_{i,j,k} = z_{j,k} - Y_{j,k} \forall j \in J, k \in K \tag{2}$$

$$\sum_{j \in J} X_{i,j,k} \leq ST * E_{i,k} - Y_{j,k} \forall i \in i, k \in K$$

$$\sum_{j \in J} X_{i,j,k} \geq E_{i,k} - Y_{j,k} \forall i \in i, k \in K \tag{3}$$

$$X_{i,j,k} \in C \forall i \in I, j \in J, k \in K$$

$$Y_{j,k} \in C \forall j \in J, k \in K$$

$$E_{i,k} \in C \forall i \in I, k \in K \tag{4}$$

$$\max(\sum_{i \in I}\sum_{j \in J}\sum_{k \in K} G_{j,k} \cdot X_{i,j,k} - \sum_{i \in I}\sum_{j \in J}\sum_{k \in K} C_{i,j} \cdot X_{i,j,k}) - R \cdot \sum_{i \in I}\sum_{k \in K} E_{i,k} \tag{5}$$

Table 1. Indices, parameters and decision variables

Symbol	Description
Sets	
I	Set of producers (manufacturers, factories)
K	Set of customers
J	Set of products
Indexes	
i	Manufacturer's index $i \in I$
k	Customer index $k \in K$
j	Product index $j \in J$
Parameters	
$A_{i,j}$	If manufacturer i can produce product j $A_{i,j} = 1$ otherwise $A_{i,j} = 0$
$W_{i,j}$	It determines how much of the j-th product can be produced by the i-th manufacturer

(continued)

Table 1. (*continued*)

Symbol	Description
$C_{i,j}$	Specifies the cost of producing product j by manufacturer i
$Z_{j,k}$	Specifies how much of product j was ordered by customer k
$G_{j,k}$	Profit from the sale of product j to customer k
R	The cost of delivery from the factory to the customer (for simplicity, it was assumed to be constant)
ST	A very large constant
Decision variables	
$X_{i,j,k}$	It determines how much of the product j was produced by the manufacturer i for the needs of the customer's order k
$Y_{j,k}$	How much of the product j cannot be delivered to the customer k (additional decision variable introduced so that there is always a solution, e.g. in a situation where the sum of orders is greater than the production capacity)
$E_{i,k}$	If manufacturer i delivers to customer k then $E_{i,k} = 1$ otherwise $E_{i,k} = 0$
Constraints	
(1)	It ensures that production capacity is not exceeded
(2)	It ensures fulfillment of customer orders
(3)	The constraint makes it possible to relate the variables $E_{i,k}$ and $X_{i,j,k}$. This relationship of decision variables ensures that if a factory produces a product for a given customer, delivery must be made from that factory to the given customer
(4)	The constraint ensures that the decision variables are integer
(5)	Objective function (1st element - profit from sales; 2nd element - production costs)

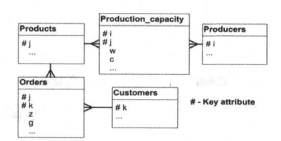

Fig. 7. Entity Relationship Diagram for illustrative example

5 Computational Examples

For an illustrative example, computational experiments were carried out using the presented method of effective implementation (Sect. 3) and for a standard implementation in the MP solver environment (Gurobi 10.0.1). The experiments were carried out using an Intel(R) Core(TM) i3-1005G1 CPU @ 1.20 GHz 1.20 GHz, RAM 16.0 GB computer running the Windows 11 operating system.

In the first stage, the experiments were carried out for data instances of the following sizes ($I = 4, J = 5, K = 6$ and $Z = 13$). Obtaining the optimal solution for both approaches took very little time, i.e. $T = 1$ s for our approach and $T = 13$ for the standard MP solver, respectively. The obtained result and the parameters of the solution are presented in Table 2 and in Fig. 8 where the optimal way of executing customer orders in terms of profit can be seen.

In the second stage of calculations, experiments were carried out for 10 different data instances with the following parameters: $I = 10, J = 20, K = 50$ and $Z = 50..500$. The main goal of the second stage of the experiments was to examine to what extent and in what area our proposed approach (Sect. 3) is better in relation to the use of MP solver environments available on the market. Both approaches were compared in several areas. Firstly, it was about computation time, which is crucial for discrete COPs. The use of our approach allowed, depending on the data instance, to reduce the optimization time from 20 to about 200 times (Table 3 and Fig. 9). This is a very good achievement, clearly pointing to the advantage of the proposed approach.

Table 2. Results for first stage experiments (data from Table 4, 5)

i	j	k	$X_{i,j,k}$	i	j	k	$X_{i,j,k}$
1	1	1	10	2	4	4	10
1	1	5	10	2	4	5	10
1	1	6	10	2	4	6	10
1	2	6	10	4	5	2	10
1	3	1	10	4	5	3	10
1	3	2	10	4	5	5	10
1	3	3	10				

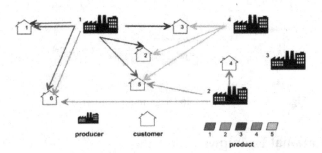

Fig. 8. Solution for data with Appendix A Fc = 92990

In the optimization of COPs, an extremely important, apart from the computation time, is the use of resources, especially hardware. Operational memory is crucial in this regard. The use of our approach meant that the resources in the form of operational memory were used only in half compared to using only MP solvers (Table 3 and Fig. 10). The last area subjected to comparative research was the size of the modeled problems for both approaches. The proposed approach allows for a 10-fold reduction in the number of decision variables and reduces the number of constraints by half. (Table 3, Fig. 11, Fig. 12).

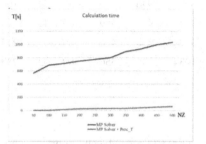

Fig. 9. Comparison of calculation times for MP Solver and MP Solver + *Proc_T*

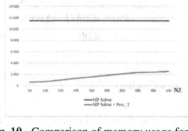

Fig. 10. Comparison of memory usage for MP Solver and MP Solver + *Proc_T*

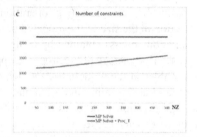

Fig. 11. Comparison of the number of decision variables for MP Solver and MP Solver + *Proc_T*

Fig. 12. Comparison of the number of constraints for MP Solver and MP Solver + *Proc_T*

Table 3. Results for second stage experiments for MP Solver and MP Solver + *Proc_T*

i	j	k	Z	Fc	MP solver				MP Solver + *Proc_T*			
					V	C	T	M	V	C	T	M
4	5	6	13	92 990	174	99	3	108	65	72	1	103
10	20	50	50	137595	11 501	2203	567	1936	630	1168	2	1047
10	20	50	100	146761	11 501	2203	685	1936	748	1180	3	1049
10	20	50	150	200222	11 501	2203	710	1936	977	1230	15	1057
10	20	50	200	284584	11 501	2203	745	1936	1297	1280	24	1095

(*continued*)

Table 3. (*continued*)

i	j	k	Z	Fc	MP solver				MP Solver + *Proc_T*			
					V	C	T	M	V	C	T	M
10	20	50	250	315112	11 501	2203	774	1936	1558	1330	27	1119
10	20	50	300	343654	11 501	2203	801	1936	1828	1380	29	1134
10	20	50	350	395010	11 501	2203	892	1936	2106	1430	34	1160
10	20	50	400	412013	11 501	2203	934	1936	2359	1480	42	1193
10	20	50	450	433532	11 501	2203	997	1936	2434	1530	53	1215
10	20	50	500	438585	11 501	2203	1034	1936	2556	1580	63	1239

6 Conclusions

The proposed method of effective implementation of COPs can be an important supplement to the application of MP solvers, especially in situations where we are obliged to obtain an exact solution for COPs. In this case, the proposed approach makes it possible to find an exact solution in a shorter time while using less resources (e.g. operational memory). All this is the result of a significant reduction in the size of the modeled problem (reducing the number of decision variables and constraints) obtained by using the transformation procedure (Fig. 3). The second important advantage of the proposed approach is the introduction of additional decision variables at the modeling stage, which do not affect the essence of the modeled problem and at the same time eliminate the occurrence of NSF (No Solution Found) during optimization. NFS does not give full feedback as to why the problem could not be resolved or what caused it.

In further works, research is planned on the application of the proposed approach to COPs in the area of logistics, distribution, routing, etc. It is also planned to extend the proposed approach to the situation when data instances are stored in NoSQL databases, e.g. in Neo4j graph [11].

Appendix a Data Instances for Numerical Experiments

Table 4. Data instances for first stage of computational experiments - part A.

i	i	$A_{i,j}$	$W_{i,j}$	$C_{i,j}$	i	i	$A_{i,j}$	$W_{i,j}$	$C_{i,j}$
1	1	1	100	20	3	1	0	0	0
1	2	1	100	30	3	2	0	0	0

(*continued*)

Table 4. (*continued*)

i	i	$A_{i,j}$	$W_{i,j}$	$C_{i,j}$	i	i	$A_{i,j}$	$W_{i,j}$	$C_{i,j}$
1	3	1	100	40	3	3	1	100	80
1	4	0	0	0	3	4	1	100	80
1	5	0	0	0	3	5	0	0	0
2	1	0	0	0	4	1	1	100	70
2	2	0	0	0	4	2	0	0	0
2	3	1	100	50	4	3	0	0	0
2	4	1	100	40	4	4	0	0	0
2	5	1	100	30	4	5	1	100	20

Table 5. Data instances for first stage of computational experiments - part B.

j	k	$Z_{j,k}$	$G_{j,k}$	i	k	$Z_{j,k}$	$G_{j,k}$
1	1	10	90	3	4	0	0
1	2	0	0	3	5	0	0
1	3	0	0	3	6	0	0
1	4	0	0	4	1	0	0
1	5	10	90	4	2	0	0
1	6	10	70	4	3	0	0
2	1	0	0	4	4	10	80
2	2	0	0	4	5	10	70
2	3	0	0	4	6	10	60
2	4	0	0	5	1	0	0
2	5	0	0	5	2	10	80
2	6	10	50	5	3	10	70
3	1	10	80	5	4	0	0
3	2	10	90	5	5	10	80
3	3	10	60	5	6	0	0

The cost of delivery from the factory to the customer $R = 10$.

References

1. Paraskevopoulos, D.C., Panagiotis, G.L., Repoussis, P., Tarantilis, C.: Resource constrained routing and scheduling: review and research prospects. Eur. J. Oper. Res. **263**(3), 737–754 (2017). https://doi.org/10.1016/j.ejor.2017.05.035
2. Kumar, K., Paulo Davim, J.: Optimization for Engineering Problems, First published 8 July 2019. ISTE Ltd. (2019). https://doi.org/10.1002/9781119644552

3. Sheth, P.D., Umbarkar, A.J.: Constrained optimization problems solving using evolutionary algorithms: a review. In: 2015 International Conference on Computational Intelligence and Communication Networks (CICN), Jabalpur, India, pp. 1251–1257 (2015). https://doi.org/10.1109/CICN.2015.241

4. Rahimi, I., Gandomi, A.H., Chen, F., et al.: A review on constraint handling techniques for population-based algorithms: from single-objective to multi-objective optimization. Arch. Comput. Methods Eng. **30**, 2181–2209 (2023). https://doi.org/10.1007/s11831-022-09859-9

5. Meisels, A.: Constraints Optimization Problems – COPs, Distributed Search by Constrained Agents. Advanced Information and Knowledge Processing. Springer, London (2018). https://doi.org/10.1007/978-1-84800-040-7_3

6. MirHassani, S.A., Hooshmand, F.: Methods and Models in Mathematical Programming. Springer, Cham (2019). https://doi.org/10.1007/978-3-030-27045-2

7. Kronqvist, J., Bernal, D.E., Lundell, A., et al.: A review and comparison of solvers for convex MINLP. Optim. Eng. **20**, 397–455 (2019). https://doi.org/10.1007/s11081-018-9411-8

8. Sitek, P., Wikarek, J., Rutczyńska-Wdowiak, K., Bocewicz, G., Banaszak, Z.: Optimization of capacitated vehicle routing problem with alternative delivery, pick-up and time windows: a modified hybrid approach. Neurocomputing **423**, 670–678 (2021). ISSN 0925-2312. https://doi.org/10.1016/j.neucom.2020.02.126

9. Sitek, P., Wikarek, J.: A multi-level approach to ubiquitous modeling and solving constraints in combinatorial optimization problems in production and distribution. Appl. Intell. **48**, 1344–1367 (2018). https://doi.org/10.1007/s10489-017-1107-9

10. Sitek, P., Wikarek, J., Grzybowska, K.: A multi-agent approach to the multi-echelon capacitated vehicle routing problem. In: Corchado, J.M., et al. (eds.) PAAMS 2014. CCIS, vol. 430, pp. 121–132. Springer, Cham (2014). https://doi.org/10.1007/978-3-319-07767-3_12

11. Sharma, M., Sharma, V.D., Bundele, M.M.: Performance analysis of RDBMS and No SQL databases: PostgreSQL, MongoDB and Neo4j. In: 2018 3rd International Conference and Workshops on Recent Advances and Innovations in Engineering (ICRAIE), Jaipur, India, pp. 1–5 (2018). https://doi.org/10.1109/ICRAIE.2018.8710439

Scheduling Deep Learning Training in GPU Cluster Using the Model-Similarity-Based Policy

Panissara Thanapol[1], Kittichai Lavangnananda[2(✉)], Franck Leprévost[1],
Julien Schleich[1], and Pascal Bouvry[3]

[1] Department of Computer Science, Faculty of Science, Technology and Medicine,
University of Luxembourg, Esch-sur-Alzette, Luxembourg
`{panissara.thanapol,franck.leprevost,julien.schleich}@uni.lu`
[2] School of Information Technology, King Mongkut's University of Technology
Thonburi, Bangkok, Thailand
`kitt@sit.kmutt.ac.th`
[3] Department of Computer Science, Faculty of Science, Technology and Medicine and
Interdisciplinary Centre for Security, Reliability and Trust,
University of Luxembourg, Esch-sur-Alzette, Luxembourg
`pascal.bouvry@uni.lu`

Abstract. Training large neural networks with huge amount of data using multiple Graphic Processing Units (GPUs) became widespread with the emergence of Deep Learning (DL) technology. It is usually operated in datacenters featuring multiple GPU clusters, which are shared amongst users. However, different GPU architectures co-exist on the market and differ in training performance. To maximise the utilisation of a GPU cluster, the scheduler plays an important role in managing the resources by dispatching the jobs to the GPUs. An efficient scheduling strategy should take into account that the training performance of each GPU architecture varies for the different DL models. In this work, an original model-similarity-based scheduling policy is introduced that takes into account the GPU architectures that match with the DL models. The results show that using the model-similarity-based scheduling policy for distributed training across multiple GPUs of a DL model with a large batch size can reduce the makespan.

Keywords: Deep learning · Distributed Training · GPU Cluster · Scheduling · Scheduling Policy · Similarity Measurement

1 Introduction

Deep Learning (DL) had numerous applications in recent years, ranging from face recognition to voice assistant to autonomous driving. In order for a DL model to acquire the ability to solve a particular problem, it needs to be trained with a given data set. The model usually learns from a huge amount of data through many layers of a deep neural network over several epochs [5]. As a result, the

ⓒ The Author(s), under exclusive license to Springer Nature Singapore Pte Ltd. 2023
N. T. Nguyen et al. (Eds.): ACIIDS 2023, LNAI 13996, pp. 363–374, 2023.
https://doi.org/10.1007/978-981-99-5837-5_30

training process is a time-consuming and resource-intensive task. To accelerate the training process, DL takes benefits from Graphic Processing Units (GPUs) to utilise distributed training over multiple GPUs. The GPUs work together in a parallel and synchronous manner in order to decrease the training time [6].

The recent upsurge in the popularity of DL has significantly impacted the requirement of GPUs to serve as accelerators for training DL models. GPU clusters become a choice as it comprises a group of GPUs connected with high-speed interconnect [2,14]. It is common for GPU clusters to contain heterogeneous GPU architectures. In considering the training performance, neither selecting the best GPU nor utilising as many GPUs as available in training can guarantee minimal training time. The training performance varies depending on several factors, such as GPU architecture, neural network architecture, and model configurations [7,26].

The scheduler plays an important role in managing a large-scale cluster among shared users. The scheduler needs to make a scheduling decision based on the knowledge of the workload, especially for DL training tasks that are distinct from other Machine Learning (ML) approaches. This raises many challenges for schedulers to deal with the workload. For example, the DL training process is an iterative process in fitting the model with training data. Another unique behaviour of DL is that it requires hyperparameter tuning, which is an adjustment to the model configuration to yield better performance. In order to adjust the model configuration, it is required to repeat several times the training process, which is computing intensive [4].

Recent works have proposed approaches for scheduling DL training tasks [1,17,18,27]. Those approaches do not, however, take into consideration the variability in terms of training performance due to the DL model. This particular impact has been highlighted in [3,16]. Based on those studies, this work proposes a scheduling policy that takes into account DL model characteristics to maximise the training performances of different DL jobs.

The organisation of this paper is as follows. Related work is first discussed in Sect. 2. Section 3 describes the DL models and GPU architectures used throughout this study. Section 4 provides the preliminary studies. The design of the scheduling policy is elaborated in Sect. 5. Section 6 is concerned with experimentation in this work. The results of the experimentation are discussed in Sect. 7. This paper is concluded in Sect. 8, where the contributions are summarised, and possible future works are suggested.

2   Related Works

There have been several studies on scheduling DL training tasks in a GPU cluster. They focus on different perspectives of DL. SchedTune [1] analyses the job characteristics and predicts the required memory to avoid the out-of-memory problem. Optimus [17] predicts a number of epochs for a model to be converged. It uses this prediction to allocate resources to a job. SLAQ [27] utilises a loss reduction while training the model to allocate resources. The job which achieves a significant loss reduction is prioritised. DL2 [18] builds a Reinforcement Learning (RL) based scheduling policy by using past job information to assist in making scheduling decisions for an incoming job. As interference occurs between models in distributed training, Gandiva [23] studies the collocation where to achieve the best training performance.

DL has several components specified when implementing the model, for example, the number of layers, batch size, and epoch. They differ from model to model. Therefore, it varies in training behaviour. Previous work has observed the training behaviour. In [1], memory usage is investigated when training the model on various GPU architectures. Prior work studies the effect of DL model characteristics on GPU utilisation [25]. To design a scheduling policy for DL training tasks, the work in [24] highlights several uniqueness of DL training behaviours that help to understand DL workload while training in GPU cluster.

To evaluate the performance of a scheduling policy, the makespan is a common metric, which refers to the finish time of the last job in the cluster. For example, previous works in [1, 25] evaluated the quality of their scheduling policies on the resulting makespan. Nevertheless, in a multi-tenant GPU cluster, a user can submit a job anytime without a pattern on job arrival. The makespan objective ought to be concerned with additional management in job density of the cluster to maximise the cluster utilisation.

3   DL Models and GPU Architectures Used in This Study

This section provides an overview of jobs that are used for experiments throughout this work. First, the details of the DL models are described. Second, the considered GPU cluster comprised of several GPU architectures is illustrated. Finally, the approach to record the training time of those DL models on each GPU architecture is described.

3.1   DL Models

This work focuses on training Convolutional Neural Networks (CNNs) [22] that perform image processing. The CNN models are implemented with the Keras framework. Their detail of model size, number of parameters, and number of layers are specified in Table 1. To have the ability to tackle the problem, the model learns from given data with multiple rounds passing the entire data or epochs. In this study, the popular CIFAR10 dataset [13] is used as the training data given to the model.

In DL, the amount of training data usually required is usually tremendous, for example, the CIFAR10 dataset contains 60,000 images. The limitation of memory of computational hardware is problematic to processing a large amount of data. To issue this challenge, distributed training utilises data parallelism in that the identical model is copied to multiple GPUs, and the training data is proportionally split into different slices. The illustration of the data parallelism approach is shown in Fig. 1. As the model is copied into multiple GPUs to

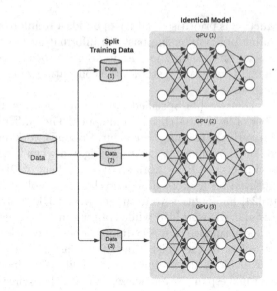

Fig. 1. Data parallelism approach

function in their training, it requires synchronisation to update its model after finishing each epoch. This study uses a distributed training function with the All-Reduce communication method of TensorFlow framework to manage the training across multiple GPUs.

The batch size is an important parameter influencing the training speed. Its value is usually a power of two. In this study, each model is configured with these batch sizes: 32, 64, 128, 256, and 512. The maximum is 512 because of the limitation of GPU memory.

3.2 GPU Architectures

The GPU cluster used in this work is provided by Grid'5000 [14]. It includes six different GPU architectures, as shown in Table 2. The number of GPUs available per machine varies, indicated in column #GPU. The memory size of each machine also varies. This work carries out the training process on a single machine that contains multiple GPUs.

3.3 Recorded Training Time

To execute the job in Grid'5000, a user has to request resources and submits a source code to run on the given resources. The models in Table 1 are submitted for a specific batch size (from 32 to 512) to six GPU architectures in Table 2. For distributed training, the number of required GPUs is part of the job submission. These models are trained on a number of GPUs that range from one to the maximum number of GPUs available on a specific machine. For instance, Nvidia

Table 1. Deep Learning models used for experiments

Model	Size (MB)	# Parameter (Millon)	# Layer
VGG16 [12]	528	138.4	16
VGG19 [12]	549	143.7	19
ResNet50 [9]	98	25.6	107
ResNet50V2 [9]	98	25.6	103
ResNet101 [9]	171	44.7	209
ResNet101V2 [9]	171	44.7	205
ResNet152 [9]	232	60.4	311
ResNet152V2 [9]	232	60.4	307
MobileNet [10]	16	4.3	55
MobileNetV2 [19]	14	3.5	105
DenseNet121 [11]	33	8.1	242
DenseNet169 [11]	57	14.3	338
DenseNet201 [11]	80	20.2	402
EfficientNetB0 [20]	29	5.3	132
EfficientNetB1 [20]	31	7.9	186
EfficientNetB2 [20]	36	9.2	186
EfficientNetB3 [20]	48	12.3	210
EfficientNetB4 [20]	75	19.5	258
EfficientNetB5 [20]	118	30.6	312
EfficientNetB6 [20]	166	43.3	360
EfficientNetB7 [20]	256	66.7	438

RTX2080Ti machines are equipped with four GPU cards, and thus the training performs on 1, 2, 3 and 4 GPUs. The training time of all jobs is recorded at job runtime for further analysis. In total, there are 1, 680 jobs in experimentation in this study.

Training the DL model involves several epochs. The training time of every epoch is approximately the same except the first one, which is common as spends more time is needed to initialise a model into the specified GPUs. Hence, the training time of the first two epochs of each job is recorded.

4   Preliminary Studies

This section presents the preliminary studies in order to understand the factors of distributed DL training on GPU clusters. The preliminary studies provide the basis for the design of scheduling policy in this study. They are described in two aspects as follows.

4.1   The Performance of Different GPU Architectures

The cluster usually contains heterogeneous GPU architectures, which perform computations with different performances. Figure 2 shows the training time of examples of DL models on different GPU architectures. This figure illustrates three factors that influence the training process to compare a decrease in training time. First, the batch size: this is set from 32 to 512. Second, GPU architectures: the training performs on different GPU architectures. Third, training method: the distribution training is across multiple GPUs.

Table 2. Hardware specification of machines used for experiments

CPU Architecture	GPU Architecture	GPU Family	# CPU	Core/CPU	# GPU	GPU Mem. (GiB)	Machine Mem. (GiB)	Storage (GB)	Added Storage (GB)	Machine Network (Gbps)
AMD EPYC 7452	Nvidia A100	Ampere	2	32	2	40	128	1920 (SSD)	960 (SSD)	25
AMD EPYC 7352	Nvidia A40	Ampere	2	24	2	48	256	1920 (SSD)	–	25
Intel Xeon Silver 4110	Nvidia RTX2080Ti	Turing	2	8	4	11	128	479 (HDD)	–	10
Intel Xeon E5-2650 v4	Nvidia RTX1080Ti	Pascal	2	12	2	11	128	299 (HDD)	299 (HDD)	10
AMD EPYC 7351	Nvidia Tesla T4	Tesla	2	16	4	16	128	479 (HDD)	–	10
Intel Xeon E5-2603 v3	Nvidia Tesla K40M	Kepler	2	6	2	11	64	1000 (HDD)	-	10

The above factors definitely affect the training time. In considering the choice of GPU architectures, Nvidia A40 can be an alternative to Nvidia A100 despite being the most powerful GPU architecture among all in our study. For training ResNet50 (Fig. 2a) and EfficientNetB7 (Fig. 2b) models, the Nvidia Tesla T4 is better than Nvidia K40M, while it is worst for VGG16 model (Figure 2c). The facts presented in Figs. 2a–2c lead to a conclusion that the choice of GPU architecture can be based on the DL model.

4.2   The Relation Between Batch Size and the Number of GPUs

The batch size has two primary effects on the DL model: the precision of updating the model parameter and the speed of the training process. It is widely known that training the DL model is a time-consuming process [15], therefore, an adjustment of batch size can help reduce the training time. With respect to model accuracy, previous work has shown that increasing the batch size of the model with distributed training achieves approximately the same accuracy compared to small batch size [21].

Referring to Fig. 2, it is apparent that increasing the batch size despite training with only one GPU can decrease the training time. With distributed training, the training process tends to accelerate when the model is configured with a batch size of more than 256. In contrast, the training time of the model with batch size from 32 to 128 gradually increases over the number of GPUs. This leads to the conclusion that a small batch-size model cannot benefit from distributed training. The batch size needs to be large, along with increasing the number of GPUs in distributed training to accelerate the training process.

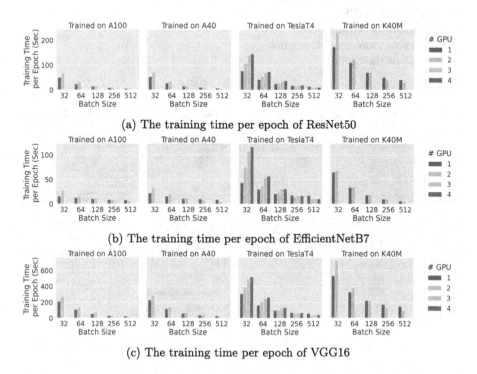

(a) The training time per epoch of ResNet50

(b) The training time per epoch of EfficientNetB7

(c) The training time per epoch of VGG16

Fig. 2. The training time of DL models for specific batch size on several GPU architectures

5   Scheduling Policy

Referring to Sect. 4.1, the performance of GPU architecture varies depending on the DL models. Therefore, the scheduling policy for DL training tasks cannot be fixed as a rule-based approach for scheduling jobs with different models. This work records the training time of the models on several GPU architectures as stated in Sect. 3.3, and uses this information to establish a scheduling policy that selects the GPU architecture based on DL models.

To make a scheduling decision properly for a specific DL model, this work adopts similarity analysis. It is applied to discover a similar model. In comparison between models, each model has its own model characteristic as represented by the model configurations. A configuration of a model comprises the following:

- Number of layers
- Total parameters
- Number of trainable parameters
- Number of non-trainable parameters
- Number of FLOPS
- GPU memory requirement
- Memory to store model weights
- File size of the model

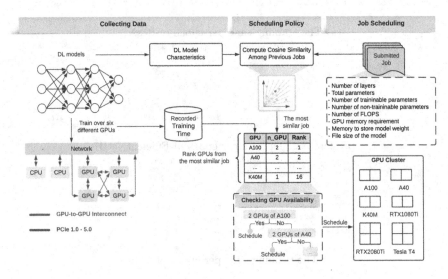

Fig. 3. An overview of model-similarity-based scheduling policy

Cosine similarity [8] is a promising strategy and can be utilised in this work. Therefore, it is applied to compare the model configurations. Cosine similarity is advantageous to work with multi-dimensional data (as it is the case in this study). Furthermore, this technique is an instance-based learning that uses all existing instances as the measurement system. Therefore, it does not require a training process.

Once a job is submitted to the GPU cluster, its model configurations are first compared to the existing models in the system to discover a similar model. The training time of this similar model on several GPU architectures is sorted in ascending order. Then, the scheduler needs to check on the sorted list of which GPU architecture is available at the time of job submission because the GPU cluster is commonly shared with many users. In case of no available resources at the job's arrival time, the scheduler chooses the GPU architecture, which can start executing a job as soon as possible. The overview diagram of the model-similarity-based scheduling policy is illustrated in Fig. 3.

6   Experimentation

In the preliminary studies in Sect. 4, it has been noticed that the batch size and the number of GPUs are the most important factors impacting training performance. In order to thoroughly evaluate the impact of the aforementioned factors, we proposed the following configurations:

1. **Configuration 1 (Baseline):** The baseline is the scheduling policy with a First Come, First Serve (FCFS) algorithm. To simulate the jobs, the DL models in Table 1 are selected, and the batch size and epoch are randomly generated. As the observations described in Sect. 4.1, the performance of GPU architectures is based on the model. It is difficult to predict without knowledge. The resource requested from a user can be arbitrary. Therefore, GPU architecture and its number used in the execution of each job in this experiment is also randomly generated.

2. **Configuration 2 (Model-similarity-based scheduling policy):** This configuration uses the model-similarity-based scheduling policy to make scheduling decisions. To compare the result with the baseline, the same DL models for specific batch size and epochs in the baseline experiment are selected. GPU architecture and its number in training are decided by model-similarity-based scheduling policy.

3. **Configuration 3 (Model-similarity-based scheduling policy with the addition of setting batch size for 512):** As the batch size is one of the factors to speed up the training process, this configuration allows the effect of batch size in making scheduling decisions to be investigated. It is the same as it is carried out in Configuration 2, except the batch size of all DL models is set to 512.

4. **Configuration 4 (Model-similarity-based scheduling policy with the addition of setting batch size for 512 and utilising a maximum number of selected GPU architecture contained in a machine):** According to the finding in Sect. 4.2, the training time can decrease by a large batch size together with utilising multiple GPUs in distributed training. This is an extension of Configuration 3 by adjusting the number of GPUs in training to the maximum of the selected GPU architecture contained in a single machine. The maximum number of GPUs in each machine is specified in the #GPU column in Table 2.

Experiments above six GPU architectures, as stated in Table 2, are based on simulation due to resource availability. The workload in simulations comprises thirty models, which are randomly and uniformly selected from Table 1. For the training time of a job simulation in our experiments, the total training time of the model equals the training time of the first epoch added to the product of the training time of the latter epoch and the number of epochs. The experimentation is carried out extensively for hundred sets of the workload in order to ensure the statistical significance of the results. The results shown in Fig. 4 are average values of these hundred sets.

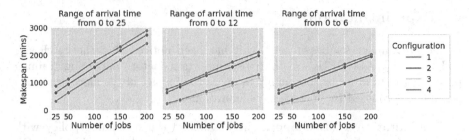

Fig. 4. Comparison among the four configurations for three ranges of arrival time

7   Results and Discussions

The main objective of this work is to minimise the makespan, and the lower values are preferable to higher ones. Figure 4 compares the makespan of all four configurations described in Sect. 6. It illustrates three scenarios of job density over a certain period of time, which refers to the range of arrival time or the duration between the previous job and the current job submitted. The $0 - 25$, $0 - 12$, and $0 - 6$ minutes are selected for the range of arrival.

Overall, referring to Fig. 4, the makespan usually increases over the number of jobs executed in the experiments. In the first scenario, the makespan of Configuration 2 is slightly better than the baseline. Configurations 3 and 4 are comparable and yield the best makespan. In the second scenario, the makespan of the baseline and Configuration 2 are comparable, while Configurations 3 and 4 are much lower. Lastly, as the range of arrival time is smaller, Configurations 1 and 2 achieve approximately the same results, while Configuration 3 yields the best performance, and there is a sharp rise in Configuration 4 over the number of jobs.

In summary, Configuration 3 (the model-similarity-based scheduling policy for training the DL model with a large batch size) significantly improves the makespan when the GPU cluster is a high job density. Configuration 4 does not perform well. Based on these findings, it can be concluded that using the maximum number of GPUs available in a machine is unnecessary, as there may be a trade-off between the shortest training time and the waiting time for all required GPUs to become available.

8   Conclusions and Future Works

DL can use GPUs for distributed training to accelerate its training process. However, there are many alternative GPU architectures in the GPU cluster, and their performance in training depends on the DL models. To make a scheduling decision for the DL training task properly, a scheduler can benefit from model characteristics, such as the number of parameters and the number of layers, to guide the scheduler in selecting the GPU architecture based on the model similarity. This work demonstrates that the model-similarity-based scheduling policy is beneficial and can improve cluster efficiency in training the DL model with a large batch size in terms of makespan. Furthermore, this scheduling policy can reduce the makespan even in the scenario where jobs in the cluster are dense over a period of time.

Future works can be extended by increasing the diversity of DL models as new and novel DL models have been proposed continuously. From the scheduling perspective, other optimisation objectives, such as minimising job completion time and minimising waiting time, can also be investigated. In considering the computational performance for training the DL model, when the new GPU architecture is released, it may even be necessary to reconsider or reevaluate the existing scheduling decision in order to optimise resources available.

Acknowledgements. The authors are grateful for Grid'5000, which provides computing resources throughout this research.

References

1. Albahar, H., Dongare, S., Du, Y., Zhao, N., Paul, A.K., Butt, A.R.: Schedtune: a heterogeneity-aware gpu scheduler for deep learning. In: 2022 22nd IEEE International Symposium on Cluster, Cloud and Internet Computing (CCGrid), pp. 695–705 (2022). https://doi.org/10.1109/CCGrid54584.2022.00079
2. Amazon web services inc.: deep learning AMI: Developer guide. Technical Report (2022)
3. Chaudhary, S., Ramjee, R., Sivathanu, M., Kwatra, N., Viswanatha, S.: Balancing efficiency and fairness in heterogeneous GPU clusters for deep learning. In: The Fifteenth European Conference on Computer Systems, pp. 1–16 (2020)
4. Chollet, F.: Deep learning with Python. Manning Publications, Shelter Island (2017)
5. Goodfellow, I., Bengio, Y., Courville, A.: Deep Learning. MIT Press, Cambridge (2016)
6. Goyal, P., et al.: Accurate, large minibatch sgd: Training imagenet in 1 hour. Technical Report (2017)
7. Gu, J., et al.: Tiresias: A GPU cluster manager for distributed deep learning. In: 16th USENIX Symposium on Networked Systems Design and Implementation (NSDI 19), pp. 485–500 (2019)
8. Han, J., Kamber, M., Pei, J.: Data Mining, pp. 39–82. The Morgan Kaufmann Series in Data Management Systems, Morgan Kaufmann, Boston, 3 edn. (2012). https://doi.org/10.1016/B978-0-12-381479-1.00002-2
9. He, K., Zhang, X., Ren, S., Sun, J.: Deep residual learning for image recognition. In: Proceedings of the IEEE Conference on Computer Vision and Pattern Recognition (CVPR) (2016)
10. Howard, A.G., et al.: Mobilenets: efficient convolutional neural networks for mobile vision applications. arXiv preprint arXiv:1704.04861 (2017)
11. Huang, G., Liu, Z., Van Der Maaten, L., Weinberger, K.Q.: Densely connected convolutional networks. In: Proceedings of the IEEE Conference on Computer Vision and Pattern Recognition, pp. 4700–4708 (2017)
12. Karen, S., Andrew, Z.: Very deep convolutional networks for large-scale image recognition. In: International Conference on Learning Representations (2015)

13. Krizhevsky, A., Hinton, G., et al.: Learning multiple layers of features from tiny images. Technical Report (2009)
14. Margery, D., Morel, E., Nussbaum, L., Richard, O., Rohr, C.: Resources description, selection, reservation and verification on a large-scale testbed. In: TRIDENT-COM - 9th International Conference on Testbeds and Research Infrastructures for the Development of Networks & Communities (2014)
15. Mattson, P., et al.: Mlperf training benchmark. Proc. Mach. Learn. Syst. **2**, 336–349 (2020)
16. Narayanan, D., Santhanam, K., Kazhamiaka, F., Phanishayee, A., Zaharia, M.: Heterogeneity-aware cluster scheduling policies for deep learning workloads. In: 14th USENIX Symposium on Operating Systems Design and Implementation (OSDI 20), pp. 481–498. USENIX Association (2020)
17. Peng, Y., Bao, Y., Chen, Y., Wu, C., Guo, C.: Optimus: an efficient dynamic resource scheduler for deep learning clusters. In: The Thirteenth EuroSys Conference, pp. 1–14 (2018). https://doi.org/10.1145/3190508.3190517
18. Peng, Y., Bao, Y., Chen, Y., Wu, C., Meng, C., Lin, W.: Dl2: A deep learning-driven scheduler for deep learning clusters. IEEE Trans. Parallel Distrib. Syst. **32**(8), 1947–1960 (2021)
19. Sandler, M., Howard, A., Zhu, M., Zhmoginov, A., Chen, L.C.: Mobilenetv 2: Inverted residuals and linear bottlenecks. In: 2018 IEEE/CVF Conference on Computer Vision and Pattern Recognition, pp. 4510–4520 (2018). https://doi.org/10.1109/CVPR.2018.00474
20. Tan, M., Le, Q.: Efficientnet: rethinking model scaling for convolutional neural networks. In: International Conference on Machine Learning, pp. 6105–6114. PMLR (2019)
21. Torres, J.: Train a neural network on multi-gpu with tensorflow - supercomputing for artificial intelligence, March 2023. https://towardsdatascience.com/train-a-neural-network-on-multi-gpu-with-tensorflow-42fa5f51b8af
22. Wu, J.: Introduction to convolutional neural networks. Natl Key Lab Novel Softw. Technol. **5**(23), 495 (2017)
23. Xiao, W., et al.: Gandiva: introspective cluster scheduling for deep learning. In: 13th USENIX Symposium on Operating Systems Design and Implementation (OSDI 18), pp. 595–610 (2018)
24. Xiao, W., et al.: Antman: Dynamic scaling on GPU clusters for deep learning. In: 14th USENIX Symposium on Operating Systems Design and Implementation (OSDI 20), pp. 533–548. USENIX Association, November 2020. https://www.usenix.org/conference/osdi20/presentation/xiao
25. Yeung, G., Borowiec, D., Yang, R., Friday, A., Harper, R., Garraghan, P.: Horus: Interference-aware and prediction-based scheduling in deep learning systems. IEEE Trans. Parallel Distrib. Syst. **33**(1), 88–100 (2022). https://doi.org/10.1109/TPDS.2021.3079202
26. Yu, G.X., Gao, Y., Golikov, P., Pekhimenko, G.: Habitat: a runtime-based computational performance predictor for deep neural network training. In: 2021 USENIX Annual Technical Conference (USENIX ATC 21), pp. 503–521. USENIX Association (2021)
27. Zhang, H., Stafman, L., Or, A., Freedman, M.J.: Slaq: quality-driven scheduling for distributed machine learning. In: The 2017 Symposium on Cloud Computing, pp. 390–404 (2017)

Author Index

Printed in the United States
by Baker & Taylor Publisher Services